Some Physical Constants

Speed of light	c	3.00×10^{8} m/s
Gravitational constant	G	6.67×10^{-11} N·m²/kg²
Coulomb's constant	$1/4\pi\varepsilon_0$	8.99×10^{9} N·m²/C²
Permittivity constant	ε_0	8.85×10^{-12} C²/(N·m²)
Permeability constant	μ_0	$4\pi \times 10^{-7}$ N/A²
Planck's constant	h	6.63×10^{-34} J·s
Boltzmann's constant	k_B	1.38×10^{-23} J/K
Elementary charge	e	1.602×10^{-19} C
Electron mass	m_e	9.11×10^{-31} kg
Proton mass	m_p	1.673×10^{-27} kg
Neutron mass	m_n	1.675×10^{-27} kg
Avogadro's number	N_A	6.02×10^{23}

Standard Metric Prefixes (for powers of 10)

Power	Prefix	Symbol
10^{18}	exa	E
10^{15}	peta	P
10^{12}	tera	T
10^{9}	giga	G
10^{6}	mega	M
10^{3}	kilo	k
10^{-2}	centi	c
10^{-3}	milli	m
10^{-6}	micro	μ
10^{-9}	nano	n
10^{-12}	pico	p
10^{-15}	femto	f
10^{-18}	atto	a

Commonly Used Physical Data

Gravitational field strength $g = \lvert\vec{g}\rvert$ (near the earth's surface)	9.80 N/kg = 9.80 m/s²
Mass of the earth M_e	5.98×10^{24} kg
Radius of the earth R_e	6380 km (equatorial)
Mass of the sun $M_\odot$	1.99×10^{30} kg
Radius of the sun $R_\odot$	696,000 km
Mass of the moon	7.36×10^{22} kg
Radius of the moon	1740 km
Distance to the moon	3.84×10^{8} m
Distance to the sun	1.50×10^{11} m
Density of water†	1000 kg/m³ = 1 g/cm³
Density of air†	1.2 kg/m³
Absolute zero	0 K = −273.15°C = −459.67°F
Freezing point of water‡	273.15 K = 0°C = 32°F
Boiling point of water‡	373.15 K = 100°C = 212°F
Normal atmospheric pressure	101.3 kPa

†At normal atmospheric pressure and 20°C.
‡At normal atmospheric pressure.

Useful Conversion Factors

1 meter = 1 m = 100 cm = 39.4 in = 3.28 ft
1 mile = 1 mi = 1609 m = 1.609 km = 5280 ft
1 inch = 1 in = 2.54 cm
1 light-year = 1 ly = 9.46 Pm = 0.946×10^{16} m
1 minute = 1 min = 60 s
1 hour = 1 h = 60 min = 3600 s
1 day = 1 d = 24 h = 86.4 ks = 86,400 s
1 year = 1 y = 365.25 d = 31.6 Ms = 3.16×10^{7} s
1 newton = 1 N = 1 kg·m/s² = 0.225 lb
1 joule = 1 J = 1 N·m = 1 kg·m²/s² = 0.239 cal
1 watt = 1 W = 1 J/s
1 pascal = 1 Pa = 1 N/m² = 1.45×10^{-4} psi
1 kelvin (temperature difference) = 1 K = 1°C = 1.8°F
1 radian = 1 rad = 57.3° = 0.1592 rev
1 revolution = 1 rev = 2π rad = 360°
1 cycle = 2π rad
1 hertz = 1 Hz = 1 cycle/s

1 m/s = 2.24 mi/h = 3.28 ft/s
1 mi/h = 1.61 km/h = 0.447 m/s = 1.47 ft/s
1 liter = 1 l = (10 cm)³ = 10^{-3} m³ = 0.0353 ft³
1 ft³ = 1728 in³ = 0.0283 m³
1 gallon = 1 gal = 0.00379 m³ = 3.79 l ≈ 3.8 kg H_2O
Weight of 1-kg object near the earth = 9.8 N = 2.2 lb

1 pound = 1 lb = 4.45 N
1 calorie = energy needed to raise the temperature of 1 g of H_2O by 1 K = 4.186 J
1 horsepower = 1 hp = 746 W
1 pound per square inch = 6895 Pa
1 food calorie = 1 Cal = 1 kcal = 1000 cal = 4186 J
1 electron volt = 1 eV = 1.602×10^{-19} J

$$T = \left(\frac{1\text{K}}{1^\circ\text{C}}\right)(T_{[C]} + 273.15^\circ\text{C}) \qquad T_{[C]} = \left(\frac{5^\circ\text{C}}{9^\circ\text{F}}\right)(T_{[F]} - 32^\circ\text{F})$$

$$T = \left(\frac{5\text{K}}{9^\circ\text{F}}\right)(T_{[F]} + 459.67^\circ\text{F}) \qquad T_{[F]} = 32^\circ\text{F} + \left(\frac{9^\circ\text{F}}{5^\circ\text{C}}\right)T_{[C]}$$

Relativistic Units, Conversion Factors, and Benchmarks

SR unit system: distance is measured in seconds (so that $c = 1$)
1 s of distance = 299,792,458 m
1 ns of distance = 0.300 m = 0.984 ft
1 min of distance = 18×10^6 km = 18 Gm
1 h of distance = 1.08×10^9 km = 1.08 Tm
1 day of distance = 2.59×10^{10} km = 25.9 Tm
1 month of distance ≈ 780 Tm
1 year of distance = 1 light-year = 1 ly = 0.946×10^{16} m = 9.46 Pm
Average distance between the earth and the moon: 1.28 s.
Average distance between the earth and the sun: 8.33 min.
Average distance between Mars and the sun: 12.69 min.
Average distance between Neptune and the sun: 4.17 h
Average distance between Pluto and the sun: 5.53 h.
Distance to the nearest star: 4.3 y
Distance to the galactic center: ~30,000 y
Diameter of the Milky Way galaxy ~100,000 y
Light travel time from the edge of the visible universe: ~13.75 Gy

$|\vec{g}| = 3.27 \times 10^{-8}\text{s}^{-1} = (0.969\ \text{y})^{-1}$
1 kg (energy) = 9.0×10^{16} J
Energy released by a 100-kt atom bomb: 4.184×10^{14} J = 4.66 g.
1 J = 1.1×10^{-17} kg (of energy)

Table RA.1 Some important equations and their equivalents in SI units

Equation	SR Version	SI Equivalent
Metric	$\Delta s^2 = \Delta t^2 - \Delta x^2 - \Delta y^2 - \Delta z^2$	$\Delta s^2 = \Delta t^2 - \dfrac{\Delta x^2 + \Delta y^2 + \Delta z^2}{c^2}$
Proper time	$d\tau = dt\sqrt{1-\lvert\vec{v}\rvert^2}$	$d\tau = dt\sqrt{1-\lvert\vec{v}/c\rvert^2}$
Lorentz transformations (t and x)	$\gamma = 1/\sqrt{1-\beta^2}$ $t' = \gamma(t - \beta x)$ $x' = \gamma(-\beta t + x)$	$\gamma = 1/\sqrt{1-(\beta/c)^2}$ $t' = \gamma(t - \beta x/c^2)$ $x' = \gamma(-\beta t + x)$
Lorentz contraction	$L = L_R\sqrt{1-\lvert\vec{v}\rvert^2}$	$L = L_R\sqrt{1-\lvert\vec{v}/c\rvert^2}$
Transformation for x-velocity	$v'_x = \dfrac{v_x - \beta}{1 - \beta v_x}$	$v'_x = \dfrac{v_x - \beta}{1 - \beta v_x/c^2}$
Energy in terms of speed	$E = \dfrac{m}{\sqrt{1-\lvert\vec{v}\rvert^2}}$	$E = \dfrac{m}{\sqrt{1-\lvert\vec{v}/c\rvert^2}}$
Relativistic momentum magnitude	$\lvert\vec{p}\rvert = \dfrac{m\lvert\vec{v}\rvert}{\sqrt{1-\lvert\vec{v}\rvert^2}}$	$\lvert\vec{p}\rvert = \dfrac{m\lvert\vec{v}\rvert}{\sqrt{1-\lvert\vec{v}/c\rvert^2}}$
Mass in terms of E and $\lvert\vec{p}\rvert$	$m^2 = E^2 - \lvert\vec{p}\rvert^2$	$(mc^2)^2 = E^2 - \lvert\vec{p}c\rvert^2$
Speed in terms of E and $\lvert\vec{p}\rvert$	$\lvert\vec{v}\rvert = \dfrac{\lvert\vec{p}\rvert}{E}$	$\dfrac{\lvert\vec{v}\rvert}{c} = \dfrac{\lvert\vec{p}c\rvert}{E}$
Photon energy in terms of f and λ	$E = hf = \dfrac{h}{\lambda}$	$E = hf = \dfrac{hc}{\lambda}$

Six Ideas That Shaped Physics

Unit R: The Laws of Physics Are Frame-Independent

Third Edition

Thomas A. Moore

SIX IDEAS THAT SHAPED PHYSICS, UNIT R:
THE LAWS OF PHYSICS ARE FRAME-INDEPENDENT, THIRD EDITION

Published by McGraw-Hill Education, 2 Penn Plaza, New York, NY 10121.

This book is printed on acid-free paper.
5 6 7 8 9 LMN 21 20 19

ISBN 978-0-07-760095-2
MHID 0-07-760095-9

Senior Vice President, Products & Markets: *Kurt L. Strand*
Vice President, General Manager, Products & Markets: *Marty Lange*
Vice President, Content Design & Delivery: *Kimberly Meriwether David*
Managing Director: *Thomas Timp*
Brand Manager: *Thomas M. Scaife, Ph.D.*
Product Developer: *Jolynn Kilburg*
Marketing Manager: *Nick McFadden*
Director of Development: *Rose Koos*
Digital Product Developer: *Dan Wallace*
Director, Content Design & Delivery: *Linda Avenarius*
Program Manager: *Faye M. Herrig*
Content Project Managers: *Melissa M. Leick, Tammy Juran, Sandy Schnee*
Design: *Studio Montage, Inc.*
Content Licensing Specialists: *Deanna Dausener*
Cover Image: *NASA*
Compositor: *SPi Global*

Dedication

For My Parents, Stanley and Elizabeth,
who taught me the joy of wondering

Library of Congress Cataloging-in-Publication Data
Names: Moore, Thomas A. (Thomas Andrew), author.
Title: Six ideas that shaped physics. Unit R, The laws of physics are frame-independent/Thomas A. Moore.
Other titles: Laws of physics are frame-independent
Description: Third edition. | New York, NY : McGraw-Hill Education, [2016] | 2017 | Includes index.
Identifiers: LCCN 2015043354 | ISBN 9780077600952 (alk. paper) | ISBN 0077600959 (alk. paper)
Subjects: LCSH: Special relativity (Physics)—Textbooks.
Classification: LCC QC173.65 .M657 2016 | DDC 530.11—dc23 LC record available at http://lccn.loc.gov/2015043354

www.mhhe.com

Contents: Unit R
The Laws of Physics Are Frame-Independent

About the Author

Thomas A. Moore graduated from Carleton College (magna cum laude with Distinction in Physics) in 1976. He won a Danforth Fellowship that year that supported his graduate education at Yale University, where he earned a Ph.D. in 1981. He taught at Carleton College and Luther College before taking his current position at Pomona College in 1987, where he won a Wig Award for Distinguished Teaching in 1991. He served as an active member of the steering committee for the national Introductory University Physics Project (IUPP) from 1987 through 1995. This textbook grew out of a model curriculum that he developed for that project in 1989, which was one of only four selected for further development and testing by IUPP.

He has published a number of articles about astrophysical sources of gravitational waves, detection of gravitational waves, and new approaches to teaching physics, as well as a book on general relativity entitled *A General Relativity Workbook* (University Science Books, 2013). He has also served as a reviewer and as an associate editor for *American Journal of Physics*. He currently lives in Claremont, California, with his wife Joyce, a retired pastor. When he is not teaching, doing research, or writing, he enjoys reading, hiking, calling contradances, and playing Irish traditional fiddle music.

Preface

Introduction

This volume is one of six that together comprise the text materials for *Six Ideas That Shaped Physics*, a unique approach to the two- or three-semester calculus-based introductory physics course. I have designed this curriculum (for which these volumes only serve as the text component) to support an introductory course that combines two elements that rarely appear together: (1) a thoroughly 21st-century perspective on physics (including a great deal of 20th-century physics), and (2) strong support for a student-centered classroom that emphasizes active learning both in and outside of class, even in situations where large-enrollment sections are unavoidable.

This course is based on the premises that innovative metaphors for teaching basic concepts, explicitly instructing students in the processes of constructing physical models, and active learning can help students learn the subject much more effectively. In the course of executing this project, I have completely rethought (from scratch) the presentation of every topic, taking advantage of research into physics education wherever possible. I have done nothing in this text just because "that is the way it has always been done." Moreover, because physics education research has consistently underlined the importance of active learning, I have sought to provide tools for professors (both in the text and online) to make creating a coherent and self-consistent course structure based on a student-centered classroom as easy and practical as possible. All of the materials have been tested, evaluated, and rewritten multiple times. The result is the culmination of more than 25 years of continual testing and revision.

I have not sought to "dumb down" the course to make it more accessible. Rather, my goal has been to help students become *smarter*. I have intentionally set higher-than-usual standards for sophistication in physical thinking, but I have also deployed a wide range of tools and structures that help even average students reach this standard. I don't believe that the mathematical level required by these books is significantly different than that in most university physics texts, but I do ask students to step beyond rote thinking patterns to develop flexible, powerful, conceptual reasoning and model-building skills. My experience and that of other users is that normal students in a wide range of institutional settings can (with appropriate support and practice) meet these standards.

Each of six volumes in the text portion of this course is focused on a single core concept that has been crucial in making physics what it is today. The six volumes and their corresponding ideas are as follows:

Unit C: **C**onservation laws constrain interactions
Unit N: The laws of physics are universal (**N**ewtonian mechanics)
Unit R: The laws of physics are frame-independent (**R**elativity)
Unit E: **E**lectric and Magnetic Fields are Unified
Unit Q: Particles behave like waves (**Q**uantum physics)
Unit T: Some processes are irreversible (**T**hermal physics)

I have listed the units in the order that I *recommend* they be taught, but I have also constructed units R, E, Q, and T to be sufficiently independent so they can be taught in any order after units C and N. (This is why the units are lettered as opposed to numbered.) There are *six* units (as opposed to five or seven) to make it possible to easily divide the course into two semesters, three quarters, or three semesters. This unit organization therefore not only makes it possible to dole out the text in small, easily-handled pieces and provide a great deal of flexibility in fitting the course to a given schedule, but also carries its own important pedagogical message: *Physics is organized hierarchically*, structured around only a handful of core ideas and metaphors.

Another unusual feature of all of the texts is that they have been designed so that each chapter corresponds to what one might handle in a single 50-minute class session at the *maximum possible pace* (as guided by years of experience). Therefore, while one might design a syllabus that goes at a *slower* rate, one should not try to go through *more* than one chapter per 50-minute session (or three chapters in two 70-minute sessions). A few units provide more chapters than you may have time to cover. The preface to such units will tell you what might be cut.

Finally, let me emphasize again that the text materials are just one part of the comprehensive *Six Ideas* curriculum. On the *Six Ideas* website, at

`www.physics.pomona.edu/sixideas/`

you will find a wealth of supporting resources. The most important of these is a detailed instructor's manual that provides guidance (based on *Six Ideas* users' experiences over more than two decades) about how to construct a course at your institution that most effectively teaches students physics. This manual does not provide a one-size-fits-all course plan, but rather exposes the important issues and raises the questions that a professor needs to consider in creating an effective *Six Ideas* course at their particular institution. The site also provides software that allows professors to post selected problem solutions online where their students alone can see them and for a time period that they choose. A number of other computer applets provide experiences that support student learning in important ways. You will also find there example lesson plans, class videos, information about the course philosophy, evidence for its success, and many other resources.

There is a preface for students appearing just before the first chapter of each unit that explains some important features of the text and assumptions behind the course. I recommend that *everyone* read it.

Comments about the Current Edition

My general goals for the current edition have been to correct errors, improve the presentation in some key areas, make the book more flexible, and especially to improve the quality and range of the homework problems, as well as significantly increase their number. Users of previous editions will note that I have split the old "Synthetic" homework problem category into "Modeling" and "Derivations" categories. "Modeling" problems now more specifically focus on the process of building physical models, making appropriate approximations, and binding together disparate concepts. "Derivation" problems focus more on supporting or extending derivations presented in the text. I thought it valuable to more clearly separate these categories.

The "Basic Skills" category now includes a number of multipart problems specifically designed for use in the *classroom* to help students practice basic issues. The instructor's manual discusses how to use such problems.

I have also been more careful to give instructors more choice about what to cover, making it possible for instructors to omit chapters without loss of continuity. See the unit-specific part of this preface for more details.

Users of previous editions will also note that I have dropped the menu-like chapter location diagrams, as well as the glossaries and symbol lists, that appeared at the end of each volume. There was no evidence that these were actually helpful to students. Units C and N still instruct students very carefully on how to construct problem solutions that involve translating, modeling, solving, and checking, but examples and problem solutions for the remaining units have been written in a more flexible format that includes these elements implicitly but not so rigidly and explicitly. Students are rather guided in Unit N to start recognizing these elements in more generally formatted solutions, something that I think is an important skill.

The only general notation change is that now I use $|\vec{v}|$ exclusively and universally for the magnitude of a vector $\vec{v}$. I still think it is very important to have notation that clearly distinguishes vector magnitudes from other scalars, but the old mag($\vec{v}$) notation is too cumbersome to use exclusively, and mixing it with using just the simple letter has proved confusing. Unit C contains some specific instruction about the standard notation that most other texts use (as well as discussing its problems).

Finally, at the request of *many* students, I now include short answers to selected homework problems at the end of each unit. This will make students happier without (I think) significantly impinging on professors' freedom.

Specific Comments About Unit R

Unit R is a relatively short unit that focuses on developing the theory of special relativity as a logical consequence of the principle of relativity. Typically, one spends little time in a traditional introductory physics course exploring relativity, and as a result, few students understand or appreciate the beauties it has to offer. The experience of those of us who have used this text is that if two to three weeks of class time are devoted to the study of relativity using the approach outlined in this unit, students at almost any level can develop a robust and satisfying understanding of the logic and meaning of relativity, and many will become genuinely excited about really being able to *understand* such a well-known but counterintuitive topic in physics (the intensity of this excitement actually surprised some of our early users).

Special relativity is probably the best and most accessible example in all of physics for illustrating how carefully thinking through the consequences of an idea can uncover unexpected truths beyond the realm of daily experience. Therefore, studying relativity can provide students with a glimpse of both the process and the rewards of theoretical physics, as well as help them take an important first step into the world of contemporary physics, where reaching beyond the level of our daily experience requires an increasing reliance on logical reasoning and abstract models.

This text therefore emphasizes the logical structure of relativity, clearly showing how well-known and bizarre relativistic effects such as length contraction and time dilation are the inevitable consequences of the principle of relativity. If students come away from this unit feeling that the universe not only *is* consistent with relativity, but indeed almost *has* to be, then this unit has been successful. I urge instructors to tailor their efforts toward this goal.

This unit should follow a treatment of Newtonian mechanics that includes nonrelativistic kinematics, Newton's second law, conservation of momentum and energy, and some study of reference frames. This book can be used as a supplement to a traditional introductory text any time after these topics are covered. In a *Six Ideas* course, this unit should definitely follow units C and N.

On the other hand, I think it is good to go against history and schedule unit R before unit E for several reasons. First, knowing some relativity can make certain aspects of electricity and magnetism simpler, and unit E takes some advantage of the relativistic perspective in general and Lorentz contraction and the cosmic speed limit in particular (see the preface to unit E for more details). Second, I think it is good for students to get a taste of some exciting contemporary physics between the many weeks of classical physics represented by units C, N, and E. This is especially true if this is the last unit discussed in the first semester: ending the first semester with unit R means that many students will leave the course excited and intrigued about physics and (perhaps) more eager to continue with the second semester.

Unit Q uses relativity quite sparingly. The chapters on nuclear physics do draw on the idea that mass can be converted to energy and vice versa, but otherwise, references to relativity are mostly confined to problems. Unit T does not use relativity at all.

This unit has been only lightly revised for this edition compared to some of the other units. The most significant change is that I have compressed the material in the second edition's chapters R1–R3 into the third edition's chapters R1 and R2. Chapter R3 of the second edition was particularly underweight, and in recent years I have been successfully able to teach the three old chapters in two class sessions. I hope that compressing the text this way will give you more flexibility in allotting time to other topics.

In previous editions, I required that the Other Frame must move in the $+x$ direction with respect to the Home Frame (meaning that β was always positive). This was a bit artificial, and the restriction proved tricky for some students. In this edition, I have relaxed this restriction, allowing β to become negative (though I almost always take it to be positive).

I have also used hyperbola graph paper more extensively in this edition, and have spent a bit more time showing students how to use such paper. I have included some graph paper at the end of chapter R5 that students can xerox (one can also download graph paper from the *Six Ideas* website).

Otherwise, I have made the notation for vector magnitudes consistent with the other revised units, corrected errors, added some new homework problems, and cleaned up the writing.

As a result of the compression, the unit now has nine chapters. The shortest possible treatment of relativity using this book would be to omit chapters R4 and R7 through R9. This would yield a five-session introduction to basic relativistic kinematics (with no dynamics or $E = mc^2$). Adding chapter R4 and/or R7 would provide a richer introduction to pure kinematics.

The shortest introduction that includes dynamics would be to omit chapters R4, R6, and R7 and add a single class session devoted to sections R4.1 through R4.4 and section R7.4 (and possibly section R7.1). This would go over everything that is useful for units E and Q within seven class sessions.

However, students find the material in chapters R4 and R6 some of the most interesting in the book, and chapter R6 is also where they really test their understanding of relativistic kinematics in the context of tough paradoxes. Therefore, I really recommend doing the whole unit if you have time. I hope that by shortening it by one chapter, I have made this a bit easier.

If you want to assign appendix RB on the Doppler shift, you can go over it any time after section R4.2. It can displace some of the latter sections of chapter R4, displace some of the middle sections of chapter R7, or supplement chapter R6 (which involves fewer new ideas than the other chapters).

Appreciation

A project of this magnitude cannot be accomplished alone. A list including everyone who has offered important and greatly appreciated help with this project over the past 25 years would be much too long (and such lists appear in the previous editions), so here I will focus for the most part on people who have helped me with this particular edition. First, I would like to thank Tom Bernatowicz and his colleagues at Washington University (particularly Marty Israel and Mairin Hynes) who hosted me for a visit to Washingtion University where we discussed this edition in detail. Many of my decisions about what was most important in this edition grew out of that visit. Bruce Sherwood and Ruth Chabay always have good ideas to share, and I appreciate their generosity and wisdom. Benjamin Brown and his colleagues at Marquette University have offered some great suggestions as well, and have been working hard on the important task of adapting some *Six Ideas* problems for computer grading.

I'd like to thank Michael Lange at McGraw-Hill for having faith in the *Six Ideas* project and starting the push for this edition, and Thomas Scaife for continuing that push. Eve Lipton and Jolynn Kilburg have been superb at guiding the project at the detail level. Many others at McGraw-Hill, including Melissa Leick, Ramya Thirumavalavan, Kala Ramachandran, David Tietz, and Deanna Dausener were instrumental in proofreading and producing the printed text. I'd also like to thank my students in Physics 70 at Pomona College, Alma Zook, Jim Supplee, David Horner and his students at North Central College, and Frank Van Steenwijk and his students at the University of Groningen for helping me track down errors in the manuscript form and offering useful feedback. Finally a very special thanks to my wife Joyce, who sacrificed and supported me and loved me during this long and demanding project. Heartfelt thanks to all!

Thomas A. Moore
Claremont, California

SMARTBOOK®

SmartBook is the first and only adaptive reading experience designed to change the way students read and learn. It creates a personalized reading experience by highlighting the most impactful concepts a student needs to learn at that moment in time. As a student engages with SmartBook, the reading experience continuously adapts by highlighting content based on what the student knows and doesn't know. This ensures that the focus is on the content he or she needs to learn, while simultaneously promoting long-term retention of material. Use SmartBook's real-time reports to quickly identify the concepts that require more attention from individual students–or the entire class. The end result? Students are more engaged with course content, can better prioritize their time, and come to class ready to participate.

connect

Learn Without Limits

Continually evolving, McGraw-Hill Connect® has been redesigned to provide the only true adaptive learning experience delivered within a simple and easy-to-navigate environment, placing students at the very center.

- Performance Analytics – Now available for both instructors and students, easy-to-decipher data illuminates course performance. Students always know how they're doing in class, while instructors can view student and section performance at-a-glance.
- Mobile – Available on tablets, students can now access assignments, quizzes, and results on-the-go, while instructors can assess student and section performance anytime, anywhere.
- Personalized Learning – Squeezing the most out of study time, the adaptive engine in Connect creates a highly personalized learning path for each student by identifying areas of weakness, and surfacing learning resources to assist in the moment of need. This seamless integration of reading, practice, and assessment, ensures that the focus is on the most important content for that individual student at that specific time, while promoting long-term retention of the material.

Introduction for Students

Introduction

Welcome to *Six Ideas That Shaped Physics!* This text has a number of features that may be different from science texts you may have encountered previously. This section describes those features and how to use them effectively.

Why Is This Text Different?

Why *active learning* is crucial

Research into physics education consistently shows that people learn physics most effectively through *activities* where they practice applying physical reasoning and model-building skills in realistic situations. This is because physics is not a body of facts to absorb, but rather a set of thinking skills acquired through practice. You cannot learn such skills by listening to factual lectures any more than you can learn to play the piano by listening to concerts!

This text, therefore, has been designed to support *active learning* both inside and outside the classroom. It does this by providing (1) resources for various kinds of learning activities, (2) features that encourage active reading, and (3) features that make it as easy as possible to use the text (as opposed to lectures) as the primary source of information, so that you can spend class time doing activities that will actually help you learn.

The Text as Primary Source

Features that help you use the text as the primary source of information

To serve the last goal, I have adopted a conversational style that I hope you will find easy to read, and have tried to be concise without being too terse.

Certain text features help you keep track of the big picture. One of the key aspects of physics is that the concepts are organized *hierarchically*: some are more fundamental than others. This text is organized into six units, each of which explores the implications of a single deep idea that has shaped physics. Each unit's front cover states this **core idea** as part of the unit's title.

A two-page **chapter overview** provides a compact summary of that chapter's contents to give you the big picture before you get into the details and later when you review. **Sidebars** in the margins help clarify the purpose of sections of the main text at the subpage level and can help you quickly locate items later. I have highlighted technical terms in bold type (like **this**) when they first appear: their definitions usually appear nearby.

A physics **formula** consists of both a mathematical equation and a *conceptual frame* that gives the equation physical meaning. The most important formulas in this book (typically, those that might be relevant outside the current chapter) appear in **formula boxes**, which state the equation, its ***purpose*** (which describes the formula's meaning), a description of any ***limitations*** on the formula's applicability, and (optionally) some other useful ***notes***. Treat everything in a box as a unit to be remembered and used together.

Active Reading

What is *active reading*?

Just as passively listening to a lecture does not help you really learn what you need to know about physics, you will not learn what you need by simply

scanning your eyes over the page. **Active reading** is a crucial study skill for all kinds of technical literature. An active reader stops to pose internal questions such as these: Does this make sense? Is this consistent with my experience? Do I see how I might be able to use this idea? This text provides two important tools to make this process easier.

Features that support developing the habit of active reading

Use the **wide margins** to (1) record *questions* that arise as you read (so you can be sure to get them answered) and the *answers* you eventually receive, (2) flag important passages, (3) fill in missing mathematical steps, and (4) record insights. Writing in the margins will help keep you actively engaged as you read and supplement the sidebars when you review.

Each chapter contains three or four **in-text exercises**, which prompt you to develop the habit of *thinking* as you read (and also give you a break!). These exercises sometimes prompt you to fill in a crucial mathematical detail but often test whether you can *apply* what you are reading to realistic situations. When you encounter such an exercise, stop and try to work it out. When you are done (or after about 5 minutes or so), look at the answers at the end of the chapter for some immediate feedback. Doing these exercises is one of the more important things you can do to become an active reader.

SmartBook (TM) further supports active reading by continuously measuring what a student knows and presenting questions to help keep students engaged while acquiring new knowledge and reinforcing prior learning.

Class Activities and Homework

Read the text BEFORE class!

This book's *entire purpose* is to give you the background you need to do the kinds of *practice* activities (both in class and as homework) that you need to genuinely learn the material. *It is therefore ESSENTIAL that you read every assignment BEFORE you come to class.* This is *crucial* in a course based on this text (and probably more so than in previous science classes you have taken).

Types of practice activities provided in the text

The homework problems at the end of each chapter provide for different kinds of practice experiences. **Two-minute problems** are short conceptual problems that provide practice in extracting the implications of what you have read. **Basic Skills** problems offer practice in straightforward applications of important formulas. Both can serve as the basis for classroom activities: the letters on the book's back cover help you communicate the answer to a two-minute problem to your professor (simply point to the letter!). **Modeling** problems give you practice in constructing coherent mental models of physical situations, and usually require combining several formulas to get an answer. **Derivation** problems give you practice in mathematically extracting useful consequences of formulas. **Rich-context** problems are like modeling problems, but with elements that make them more like realistic questions that you might actually encounter in life or work. They are especially suitable for collaborative work. **Advanced** problems challenge advanced students with questions that involve more subtle reasoning and/or difficult math.

Note that this text contains perhaps fewer examples than you would like. This is because the goal is to teach you to *flexibly reason from basic principles*, not slavishly copy examples. You may find this hard at first, but real life does not present its puzzles neatly wrapped up as textbook examples. With practice, you will find your power to deal successfully with realistic, practical problems will grow until you yourself are astonished at how what had seemed impossible is now easy. *But it does take practice*, so work hard and be hopeful!

R1

The Principle of Relativity

Chapter Overview

Introduction

In units C and N, we have explored the Newtonian model of mechanics. In this unit, we will explore a *different* model, called the *special theory of relativity*, that better explains the behavior of objects, especially objects moving at close to the speed of light. This chapter lays the foundations for that exploration by describing the core idea of the theory and linking it to Newtonian mechanics.

Section R1.1: Introduction to the Principle

We can informally state this unit's great idea, the **principle of relativity**, as follows:

> The laws of physics are the same inside a laboratory moving at a constant velocity as they are inside a laboratory at rest.

The theory of special relativity simply spells out the logical consequences of this idea.

This unit is divided into four subsections. The first (this chapter) discusses the principle itself. The second (chapters R2 through R4) explores the relativistic concept of time. The third (chapters R5 through R7) discusses how observers in different reference frames view a sequence of events. Finally, the fourth (chapters R8 and R9) examines the consequences for the laws of mechanics.

Section R1.2: Events, Coordinates, and Reference Frames

What exactly does the principle mean by a "laboratory"? The first step to understanding this better is to describe *operationally* how we measure a particle's motion. An **event** is something that happens at a well-defined place and time. An event's **spacetime coordinates** are a set of four numbers that locate the event in space and time. A particle's motion is a series of events.

A **reference frame** is a tool for assigning spacetime coordinates to events. We can visualize a reference frame as being a cubical lattice with a clock at every intersection. This ensures that there is a clock present at every event, but it also implies that we must synchronize the clocks somehow. An **observer** is a person who *interprets* results obtained in a reference frame to reconstruct the motions of particles. A real reference frame does not actually consist of a cubical lattice of clocks, but must be functionally equivalent.

Section R1.3: Inertial Reference Frames

An **inertial reference frame** is a frame in which an isolated object is *always* and *everywhere* observed to move at a constant velocity. We can check whether a frame is inertial by distributing **first-law detectors** around the frame to test for violations of Newton's first law.

A consequence of this definition is that two inertial frames in the same region of space must move at a constant velocity relative to each other. Conversely, if a given frame moves at a constant velocity relative to another inertial frame in the same region of space, the first must be inertial also.

Section R1.4: The Final Principle of Relativity

Note that in our original statement of the principle of relativity, the "laboratory moving at a constant velocity" and the "laboratory at rest" are *both* inertial frames. Moreover, the principle itself implies that there is no physical way to distinguish a frame in motion from one at rest: only the relative velocity between two inertial reference frames is measurable. Our final, polished statement of the principle therefore expresses the core issue without referring to "moving" or being at "rest":

The laws of physics are the same in all inertial reference frames.

Section R1.5: Newtonian Relativity

What does the phrase "the laws of physics are the same" mean? We can examine this issue in the context of Newtonian physics if we temporarily embrace Newton's hypothesis about time, which is that *time is universal and absolute* and thus independent of reference frame. Consider two inertial frames that have constant relative velocity $\vec{\beta}$ and which are in **standard orientation** relative to each other; that is, the axes of both point in the same directions in space and the Other (primed) Frame moves along the x axis relative to the Home (unprimed) Frame. The concept of universal time implies that time measured in both frames is the same ($t = t'$). This, together with some simple vector addition and a little bit of calculus, implies that

$$\vec{r}\,'(t') = \vec{r}\,(t) - \vec{\beta}t \tag{R1.1}$$

$$\vec{v}\,'(t') = \vec{v}\,(t) - \vec{\beta} \tag{R1.3}$$

$$\vec{a}\,'(t') = \vec{a}\,(t) \tag{R1.4}$$

- **Purpose:** These equations describe how to compute an object's position $\vec{r}\,'$, velocity $\vec{v}\,'$ and acceleration $\vec{a}\,'$ at any given time t' in the Other Frame, given the object's position, position $\vec{r}$, velocity $\vec{v}$, and acceleration $\vec{a}$ at the same time t in the Home Frame, where $\vec{\beta}$ is the velocity of the Other Frame relative to the Home Frame.
- **Limitations:** These equations assume that $t = t'$, which is not true unless both $\vec{v} \ll c$ and $\vec{\beta} \ll c$, for reasons we are about to discover.
- **Note:** Equation R1.1 and the equation $t = t'$ together comprise the **Galilean transformation equations**; equation R1.3 represents the **Galilean velocity transformation equations.**

Equation R1.3 implies that Newton's second law is frame-independent: the vector sum of the physical forces acting on an object will be equal to the object's mass times its acceleration in all inertial frames. This is an example of the laws of physics being the same in two frames (the problems explore some others).

Section R1.6: The Problem of Electromagnetic Waves

Maxwell's equations of electromagnetism predict that light (electromagnetic waves) must move with a certain specific speed c. In the 19th century, people assumed that electromagnetic waves moved through a hypothetical medium called the **ether** and that Maxwell's equations only really applied in the ether's rest frame. In *other* frames, then, the speed of light would be more or less than c according to the Galilean velocity transformation equation. However, 19th-century experiments seeking to check this showed that the speed of light seemed to have the same numerical value in *all* reference frames and failed to find *any* evidence supporting the ether hypothesis.

Einstein asserted that the simplest way to explain the evidence was to reject the ether hypothesis and instead assume that Maxwell's equations satisfy the principle of relativity. This means that the speed c of light in a vacuum must be the same in all inertial reference frames. But this contradicts the Galilean transformation equations.

R1.1 Introduction to the Principle

Everyday experience with the principle of relativity

If you have ever traveled on a jet airplane, you know that while the plane may be flying through the air at 550 mi/h, things inside the plane cabin behave pretty much as they would if the plane were sitting at the loading dock. A cup dropped from rest in the cabin, for example, will fall straight to the floor (even though the plane moves forward many hundreds of feet with respect to the earth in the time that it takes the cup to reach the floor). A ball thrown up in the air by a child in the seat in front of you falls straight back into the child's lap (instead of being swept back toward you at hundreds of miles per hour). Your watch, the attendants' microwave oven, and the plane's instruments behave just as they would if they were at rest on the ground.

Indeed, suppose you were confined to a small, windowless, and soundproofed room in the plane during a stretch of exceptionally smooth flying. Is there any physical experiment you could perform entirely within the room (that is, an experiment that would not depend on any information coming from beyond the room's walls) that would indicate whether or how fast the plane was moving?

The answer to this question appears to be "no." No one has ever found a convincing physical experiment that yields a different result in a laboratory moving at a constant velocity than it does when the laboratory is at rest. The designers of the plane's electronic instruments do not have to use different laws of electromagnetism to predict the behavior of those instruments when the plane is in flight than they do when the plane is at rest. Scientists working to improve the performance of the *Voyager* 2 space probe tested out various techniques on an identical model of the probe at rest on earth, confident that if the techniques worked for the earth-based model, they would work for the actual probe, even though the actual probe was moving relative to the earth at nearly 72,000 km/h. Astrophysicists are able to explain and understand the behavior of distant galaxies and quasars by using physical laws developed in earth-based laboratories, even though such objects move with respect to the earth at substantial fractions of the speed of light.

An informal statement of the principle of relativity

In short, all available evidence suggests that we can make the following general statement about the way the universe is constructed:

> The laws of physics are the same inside a laboratory moving at a constant velocity as they are in a laboratory at rest.

This is an unpolished statement of what we will call the **principle of relativity**. This simple idea, based on common, everyday experience, is the foundation of Einstein's **special theory of relativity**. All of that theory's mind-bending predictions about space and time follow *as logical consequences* of this simple principle! The remainder of this book is little more than a step-by-step unfolding of this statement's rich implications.

Historical notes

The principle of relativity is both a very new and a very old idea. It was not first stated by Einstein but rather by Galileo Galilei in his book *Dialog Concerning the Two Chief World Systems* (1632). (Galileo's vivid and entertaining description of the principle of relativity is a wonderful example of a style of discourse that has, unfortunately, become archaic.) In the nearly three centuries that passed between Galileo's statement and Einstein's first paper on special relativity in 1905, the principle of relativity as it applied to the laws of *mechanics* was widely understood and used (in fact, it was generally considered to be a consequence of the particular nature of Newton's laws).

What Einstein did was to assert the applicability of the principle of relativity to *all* the laws of physics, and most particularly to the laws of electromagnetism (which had just been developed and thus were completely unknown to Galileo). Thus, Einstein did not *invent* the principle of relativity; rather, his main contribution was to reinterpret it as being *fundamental*, more fundamental than Newton's laws or even than ideas about time that up to that point had been considered obvious and inescapable, and to explore insightfully its implications regarding the nature of light, time, and space.

An overview of the unit

Our task in this text is to work out the rich and unexpected consequences of this principle. Figure R1.1 illustrates how we will proceed to do this in the remaining chapters of this unit.

This unit is divided into four subsections, as shown in figure R1.1. The first (this chapter) deals with the foundation of relativity theory, the principle of relativity itself. The second subsection (chapters R2 through R4) explores the implications of the principle of relativity regarding the nature of time, with special emphasis on the *metric equation*, an equation that links space and time into a geometric unity that we call *spacetime*. The third subsection (chapters R5 through R7) examines how the metric equation and the principle of relativity together determine how observers in different reference frames will interpret the same sequence of physical events; how *disagreements* between such observers lead to the phenomenon of length contraction; and how the fact that all observers *must* agree about cause and effect implies that nothing can go faster than light. The final subsection (chapters R8 and R9) explores why and how we must redefine energy and momentum somewhat to make conservation of energy and conservation of momentum consistent with the principle of relativity.

Foundations
- The Principle of Relativity

Time
- Coordinate Time
- The Spacetime Interval
- Proper Time

Comparing Frames
- Coordinate Transformations
- Lorentz Contraction
- The Cosmic Speed Limit

Relativistic Dynamics
- Four-Momentum
- Conservation of Four-Momentum

Figure R1.1
An outline illustrating the four subsections of unit R.

Our focus in this chapter is on the principle of relativity itself, and on developing an understanding of its meaning in the context of Newtonian physics, before we proceed to explore the changes that Einstein proposes. It is important before we proceed , however, to understand two important things about the principle of relativity: (1) it is a *postulate* and (2) we must state it more precisely before we can extract any of its logical implications.

The principle of relativity is a *postulate*

The principle of relativity is one of those core physical assumptions (like Newton's second law or the law of conservation of energy) that must be accepted on faith: it cannot be *proved* experimentally (it is not possible even in principle to test *every* physical law in every laboratory moving at a constant velocity) or logically derived from more basic ideas. The value of such a postulate rests entirely on its ability to provide the foundation for a model of physics that successfully explains and illuminates experimental results.

The principle of relativity has weathered nearly a century of intense critical examination. No contradiction of the principle or its consequences has *ever* been conclusively demonstrated. Moreover, the principle of relativity has a variety of unusual and unexpected implications that have been verified (to an extraordinary degree of accuracy) to occur exactly as predicted. Therefore, while it cannot be *proved*, it has not yet been *disproved*, and physicists find it to be something in which one can confidently believe. The principle of relativity, simple as it is, is a very rich and powerful idea, and one that the physics community has found to be not only helpful but *crucial* in the understanding of much of modern physics.

Our informal statement of the principle needs clarification

Turning to the other problem, we see that the principle of relativity, as we have just stated it, suffers from certain problems of both abstraction and ambiguity. For example, what do we *mean* by "the laws of physics are the same"? What exactly do we mean by "a laboratory at rest"? How can we tell if a laboratory is "at rest" or not? If we intend to explore the logical consequences of any idea, it is essential to state that idea in such a way that its meaning is clear and unambiguous. Our task in the remaining sections of this chapter is to address this problem.

R1.2 Events, Coordinates, and Reference Frames

Our task is to specify what we really mean by "laboratory"

The principle of relativity, as we have stated it so far, asserts that the laws of physics are the same in a laboratory moving at a constant velocity as they are in a laboratory at rest. A *laboratory* in this context is presumably a place where one performs experiments that test the laws of physics. How can we more carefully state what we mean by this term?

The most fundamental physical laws describe how particles interact with one another and how they move in response to such interactions. So perhaps what a physicist seeking to specify and test the laws of physics needs most is a means of mathematically describing the *motion* of a particle in space.

As we develop the theory of relativity, we need to be *very* careful about describing exactly *how* we will measure the motion of particles (hidden assumptions about the measurement process have plagued thinkers both before and after Einstein). In what follows, I will describe how we can measure the motion of a particle in terms of simple and well-defined concepts that are based on a minimum of supporting assumptions.

Definition of *event*

The first of these concepts is described by the technical term **event**. An *event* is any physical occurrence that we can consider to happen at a definite place in space and at a definite instant in time. The explosion of a small firecracker at a particular location in space and at a definite instant in time is a vivid example of an event. The collision of two particles or the decay of a single particle at a certain place and time also defines an event. Even the simple passage of a particle through a given mathematical point in space can be treated as an event (simply imagine that the particle sets off a firecracker at that point as it passes by).

Because an event occurs at a specific point in space and at a specific instant of time, four numbers quantify when and where the event occurs: three numbers that specify the event's location in some three-dimensional spatial coordinate system and one number that specifies what time the event happened. We call these four numbers the event's **spacetime coordinates**.

We can describe motion in terms of events

Note that the exact values of an event's spacetime coordinates depend on certain arbitrary choices, such as the origin and orientation of the spatial coordinate axes and what time we define to be $t = 0$. Once these choices are made and consistently used, however, specifying the coordinates of physical events provides a useful method of mathematically describing motion.

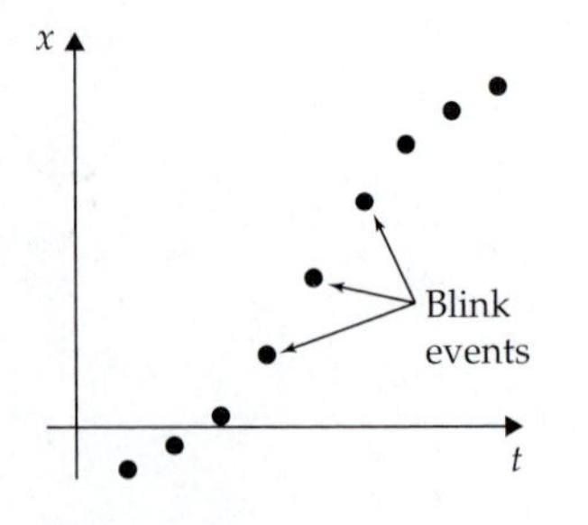

Figure R1.2
We can sketch out a graph of an object's motion (position vs. time) by plotting the "blink events" that occur along the object's path.

Specifically, we can quantify a particle's motion by treating it as a *series* of events. For example, imagine an airplane moving along the x axis of some coordinate system. The airplane carries a blinking light. Each blink of the light is an event in the sense that we are using the word here: it occurs at a definite place in space and at a definite instant of time. We can describe the plane's motion by plotting a graph of the position coordinate of each "blink event" versus the time coordinate of the same, as illustrated in figure R1.2. If we decrease the time between blink events, we get an even more detailed picture of the plane's motion. We can in fact describe the plane's motion to whatever accuracy we need by listing the spacetime coordinates of a sufficiently large number of blink events distributed along its path.

The preceding is a specific illustration of a general idea: the motion of *any* particle can be mathematically described to arbitrary accuracy by specifying the spacetime coordinates of a sufficiently large number of events suitably distributed along its path. Studying the motion of particles is the most basic way to discover and test the laws of physics. Therefore, *the most fundamental task of a "laboratory"* (as a place in which the laws of physics are to be tested) *is to provide a means of measuring the spacetime coordinates of events.*

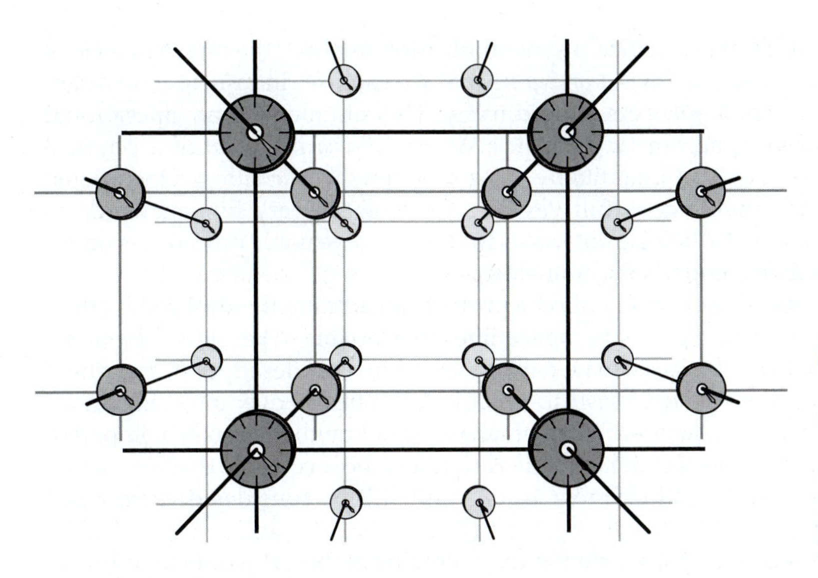

Figure R1.3
A reference frame visualized as a cubical lattice with a clock at every lattice intersection. This figure (and indeed the whole approach in this section) is adapted from E. F. Taylor and J. A. Wheeler, *Spacetime Physics*, San Francisco: Freeman, 1966, pp. 17–18.

Now, we have already discussed in chapter C3 how we can quantify the *spatial* coordinates of an object (and thus presumably an event) using a cubical lattice of measuring sticks (or something equivalent). In our past discussions of reference frames, however, we did not really face the issue of how one might measure the *time* of an event: we simply *assumed* this could be done in some simple and obvious way. To proceed with our discussion of relativity, we now need to face this issue squarely. In what follows, we will extend the cubical lattice model of a reference frame to include a mechanism for measuring time in such a way that we can clearly distinguish the approaches to time implicit in Newtonian mechanics and special relativity.

The operational definition of spacetime coordinates

The trick is to imagine that we attach a clock to every lattice intersection (see figure R1.3). We can then define the *time* coordinate of an event (such as a firecracker explosion) to be the time displayed on the lattice clock nearest the event (relative to some specified time $t = 0$) and the event's three *spatial* coordinates to be the lattice coordinates of that nearest clock, specified in the usual way by stating the distances along the lattice directions that one has to travel (from some specified spatial origin) to the clock. We can determine these four numbers to whatever precision we want by sufficiently decreasing the lattice spacing and the time between clock ticks.

Why must we have a clock at *every* lattice intersection? The point is to ensure that there is a clock essentially *at* the location of any event we wish to measure. If we attempt to read the the event's time by using a clock located a substantial distance away, we need to make assumptions about how long it took the information that the event has occurred to *reach* that clock. For example, if we read the time when the *sound* from an event reaches the distant clock, then we should correct that value by subtracting the time it takes sound to travel from the event to the clock. However, to do this, we have to know the speed of sound in our lattice. We can avoid this problem if we require that an event's time coordinate be measured by a clock *present* at that event.

Note the clocks must all be *synchronized* in some meaningful and self-consistent manner if we are to get meaningful results. If these clocks are not synchronized, adjacent clocks might differ wildly, thus giving one a totally incoherent picture of when a particle moving through the lattice passes various lattice points. What exactly we *mean* when we say that our "lattice clocks are synchronized" is precisely where Newton's and Einstein's models diverge. We will discuss this issue later: for now, it is sufficient to recognize that we must synchronize the clocks *somehow*.

Once we have specified a synchronization method, the image of a clock lattice completely defines a *procedure* that we can use (in principle) to determine an event's spacetime coordinates. This amounts to an **operational definition** of spacetime coordinates: An *operational definition* of a physical quantity defines that quantity by describing how we *measure* it. Operational definitions provide a useful way of anchoring slippery human words to physical reality by linking the words to specific, repeatable procedures rather than to vague comparisons or analogies.

The procedure just described represents an admittedly idealized method for determining an event's spacetime coordinates. The actual methods employed by physicists may well differ from this description, but these methods should be *equivalent* to what is described above: the clock lattice method defines a standard against which actual methods can be compared. It is such a simple and direct method that it is inconceivable that any actual technique could yield different results and still be considered correct and meaningful.

Technical terms involving reference frames

With this in mind, we define the following technical words to aid us in future discussions:

> A **reference frame** is defined to be a rigid cubical lattice of appropriately synchronized clocks *or its functional equivalent*.

> An event's **spacetime coordinates** in a given reference frame are an ordered set of four numbers, the first specifying the *time* of the event as registered by the nearest clock in the lattice, followed by three that specify the spatial coordinates of that clock in the lattice.

(For example, in a frame oriented in the usual way on the earth's surface, a firecracker explosion whose spacetime coordinates are [3 s, −3 m, 6 m, −1 m] thus happened 3 meters west, 6 meters north, and 1 meters below the frame's spatial origin, and 3 seconds after whatever event defines $t = 0$.)

> An **observer** is defined to be a (possibly hypothetical) person who interprets measurements made in a reference frame (for example, a person who interprets the spacetime coordinates collected by a central computer receiving information from all the lattice clocks).

Note that the act of "observing" in the last definition is an act of *interpretation* of measurements generated by the frame apparatus, and that act may have little or nothing to do with what that observer sees with his or her own eyes. When we say that "an observer in such-and-such reference frame observes such-and-such," we are actually referring to *conclusions* that the observer draws from measurements performed using the reference frame lattice. Figure R1.4 illustrates a computer's "observation" of particle tracks from a collision between two protons in the Large Hadron Collider. The computer has reconstructed the particle tracks by collecting data from the functional equivalent of a clock-lattice in the particle detector.

R1.3 Inertial Reference Frames

We can in principle attach a frame to any object

While exploring relativity theory, we will often speak of a reference frame in connection with some object. For example, one might refer to "the reference frame of the surface of the earth" or "the reference frame of the cabin of the plane" or "the reference frame of the particle." In these cases, we are being asked to imagine a clock lattice (or equivalent) fixed to the object in question. Sometimes the actual frame is referred to only obliquely, as in the phrase

Figure R1.4
The aftermath of a collision between two protons in the Large Hadron Collider. The computer reconstructs the particle tracks by collecting and interpreting data about particle detection events in the detector. This represents an act of "observation." (Credit: © CERN/Science Source)

"an observer in the plane cabin finds . . . ." Since *observer* in this text refers to someone using a reference frame to determine event coordinates, the phrase presumes the existence of a reference frame attached to the plane's cabin.

A reference frame may be moving or at rest, accelerating, or even rotating about some axis. The beauty of the definition of spacetime coordinates given earlier is that we can measure the coordinates of events (and thus measure the motion of objects) in a reference frame no matter how it is moving (provided only that the clocks in the frame can be synchronized in some meaningful manner).

Distinguishing inertial from noninertial frames

However, not all reference frames are equally useful for doing physics. We saw in chapter N8 that we can divide reference frames into two general classes: **inertial frames** and **noninertial frames**. An *inertial frame* is one in which an isolated object is always and everywhere observed to move at a constant velocity (as required by Newton's first law); in a *noninertial frame,* such an object moves with a nonconstant velocity in at least some situations.

We can operationally distinguish inertial from noninertial frames using a **first-law detector**. Figure R1.5 shows a simple first-law detector. Electrically actuated "fingers" hold an electrically uncharged and nonmagnetic ball at rest in the center of an evacuated spherical container. When the ball is released, it should remain at rest according to Newton's first law. If it does not, the frame to which the detector container is attached is noninertial.

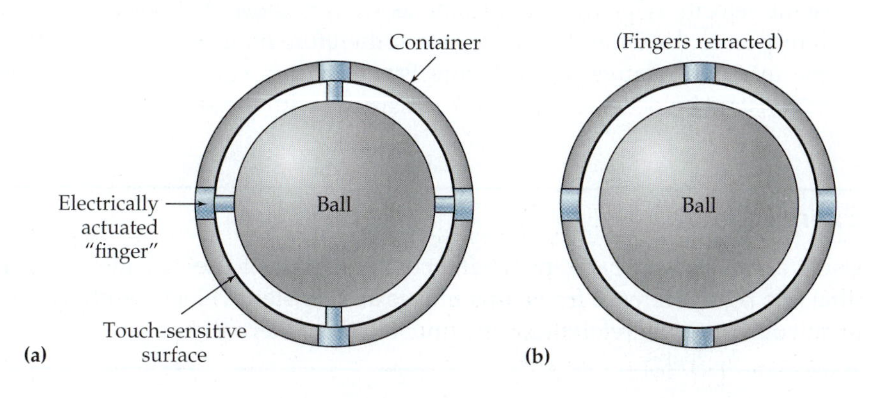

Figure R1.5
(a) A cross-sectional view of a floating-ball first-law detector. Electrically actuated "fingers" hold the ball initially at rest in the spherical container. (b) After the fingers are retracted, the ball should continue to float at rest in the container, as long as the frame to which the container is attached is inertial.

(If we want to operate this detector in a gravitational field, we need to figure out a way to cancel the gravitational force on the ball without inhibiting its freedom to move, but principle, this can be done.) If we attach such a first-law detector to the clock at each lattice location in our reference frame and *none* of these detectors registers a violation of Newton's first law, then we can say with confidence that our frame is inertial.

This definition of an inertial frame is simple enough to apply to realistic examples. For example, while a gravity-compensated detector, as shown in figure R1.5, would register no violation of the first law if it were attached to a plane at rest, we know without actually trying it that if the plane began to accelerate for takeoff, the detector's floating ball would be deflected toward the rear of the plane by the same (fictitious) force that presses us back into our chairs. Similarly, we might expect that detectors in a reference frame floating in deep space (far from any massive objects) would register no violation of Newton's first law; yet we know that detector balls in a similar frame that is rotating around its center will be deflected outward by the (fictitious) centrifugal force in that noninertial frame.

Inertial frames move with constant velocities relative to each other

The following statement is an important and useful consequence of the definition of an inertial reference frame:

> *Any* inertial reference frame will be observed to move at a *constant velocity* relative to *any other* inertial reference frame. Conversely, a rigid, nonrotating reference frame that moves at a constant velocity with respect to any other inertial reference frame must *itself* be inertial.

We first discussed this issue in chapter N8, but the *methods* we used there unfortunately employ certain assumptions about the nature of time that turn out to be inconsistent with the principle of relativity (as we will see). We can, however, prove that the statement above follows *directly* from the definition of an inertial reference frame without having to make any assumptions about the nature of time. Here is an argument for the first part of the statement above; I have left proof of the converse statement an exercise.

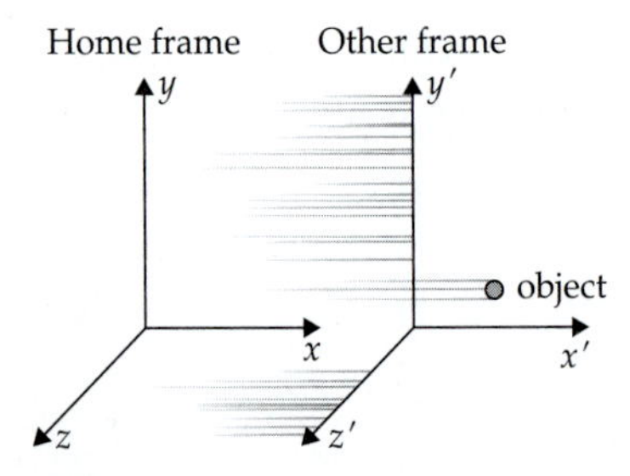

Figure R1.6
An isolated object at rest in the Other Frame must move at a constant velocity with respect to the Home Frame, so the whole Other Frame must move at the same constant velocity relative to the Home Frame.

Consider two inertial reference frames (see figure R1.6), which we will call the **Home Frame** and **Other Frame**, respectively. (*Home Frame* and *Other Frame* are phrases I will use in this text as *names* of inertial reference frames, which I will emphasize by the capitalization.) Since these are *inertial* reference frames, observers will measure an isolated object to move with a constant velocity in either frame *by definition*. Now suppose that a specific isolated object happens to be at *rest* relative to the Other Frame. Because such an object must move at a constant velocity in the Other Frame if the frame is inertial, if that object is initially at rest, it must *remain* at rest in that frame. Now let us observe the same object from the Home Frame. Since the object is isolated and the Home Frame is inertial, the object must move at a constant velocity relative to that frame as well. Because that object is at rest with respect to the Other Frame, we must therefore observe the entire Other Frame to move relative to the Home Frame with a constant velocity (the same constant velocity as the object), consistent with the statement above.

Exercise R1X.1

Using a similar approach, prove the converse part of the statement above (that is, a rigid reference frame that moves at a constant velocity with respect to any other inertial reference frame must *itself* be inertial).

R1.4 The Final Principle of Relativity

Rest has no physical meaning in relativity

Our first informal statement of the principle of relativity stated that "the laws of physics are the same in a laboratory moving at a constant velocity as they are in a laboratory at rest." We have subsequently developed the idea of a reference frame to express the essence of what we mean by a "laboratory." However, how can we physically distinguish a reference frame "moving at a constant velocity" from one "at rest"?

The short answer is that we cannot! The principle of relativity specifically states that a reference frame moving at a constant velocity is physically equivalent to a frame at rest. Therefore, there can be no physical basis for distinguishing a laboratory at rest from another frame moving at a constant velocity. Imagine you and I are in spaceships coasting at a constant velocity in deep space. You will consider yourself to be at rest, while I am moving by you at a constant velocity. I, on the other hand, will consider myself to be at rest, while you are moving by me at a constant velocity. According to the principle of relativity, there is no physical experiment that can resolve our argument about who is "really" at rest. We could, of course, agree to choose one or the other of us to be at rest, but this choice is completely arbitrary. Therefore, if the principle of relativity is true, there is no basis for assigning an absolute velocity to any reference frame: only the *relative* velocity between two frames is a physically meaningful concept.

On the other hand, it is plausible that what we really mean by a reference frame "at rest" is an inertial frame. Moreover, we have just seen that a reference frame moving at a constant velocity relative to it must also be an inertial frame. Therefore, we can remove both the vague word "laboratory" and the ambiguity of the concepts "at rest" and "moving at a constant velocity" in our original statement of the principle of relativity by restating it as follows.

Our final statement of the principle of relativity

The Principle of Relativity

The laws of physics are the same in all inertial reference frames.

This is our final polished statement of the principle of relativity. It replaces the fuzzy and ambiguous ideas in our original statement with the sharply and operationally defined idea of an *inertial reference frame*. What this principle essentially claims is that if Newton's *first* law (which describes what happens to an isolated object) is the same in two given reference frames, *then* all the laws of physics are the same in both frames. (Note that the unit's "great idea" on the front cover is a compressed version of this statement.)

R1.5 Newtonian Relativity

But what *exactly* do we mean when we say that "the laws of physics are the same" in two frames? In this section, we will discuss the Newtonian assumption about how we can synchronize clocks in an inertial reference frame. We then use this as a framework to explore what the principle of relativity means in Newtonian mechanics.

Suppose we have an inertial frame floating in deep space, ready to use. We would like to use this frame to measure the coordinates of events happening in it so as to test the laws of physics. But an important problem remains to be solved: how do we synchronize its clocks?

The Newtonian to clock synchronization

"The solution is easy," says a Newtonian physicist. "Everyone knows, as Newton himself asserted, that 'time is absolute and flows equably without regard to anything external.' *Any* good clock will therefore measure the flow

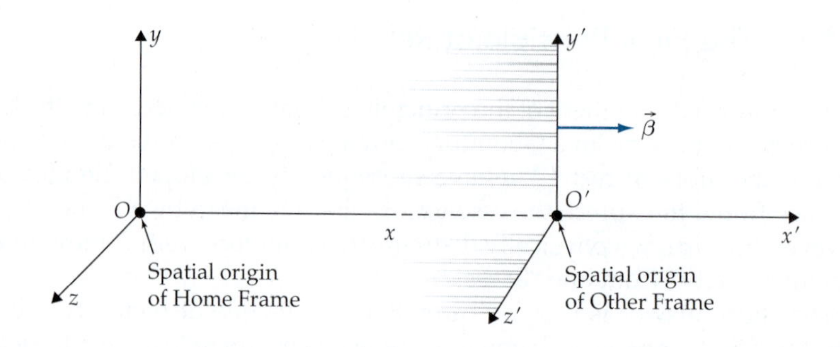

Figure R1.7
A schematic drawing of two reference frames in standard orientation. The spatial origins of the frames coincided at $t = t' = 0$ just a little while ago. (You should imagine the frame lattices intermeshing so that events can be recorded in both frames.)

of this absolute time. So we can simply designate one clock to be a master clock, carry it around to each of the lattice clocks, and synchronize that clock to the master. Since both the master and lattice clocks measure the flow of immutable absolute time, the master clock's motion as we carry it from place to place in the lattice is irrelevant. Once a lattice clock is set to agree with the master clock, it will certainly remain in agreement with it, since both clocks measure the flow of absolute time. Indeed, if the master clocks in two different reference frames are in agreement at any given event, then all the clocks in the two frames will *always* agree. It doesn't matter whether the frames are in motion with respect to each other; it doesn't even matter if they are inertial or not. This follows from the self-evident absolute nature of time."

This picture of time is straightforward, credible, and consistent with ideas about time that most of us already hold. But what are its consequences?

Again consider two inertial frames that we call the Home Frame and the Other Frame. We will often (but not always!) imagine ourselves to be in the Home Frame (so that this frame appears to *us* to be at rest). The Other Frame moves at a *constant* velocity $\vec{\beta}$ with respect to the Home Frame according to the proof given in section R1.4 ($\vec{\beta}$ is the "boost" in velocity that one needs to go from being at rest in the Home Frame to being at rest in the Other Frame).

Standard orientation for inertial reference frames

These frames might in principle have any relative orientation, but it is conventional in special relativity to use our freedom to choose the orientations to put the two frames in **standard orientation**, where the Home Frame's x, y, and z axes point in the same directions as the corresponding axes in the Other Frame. We conventionally distinguish the Home Frame and Other Frame axes by referring to the Home Frame axes as x, y, and z and the Other Frame axes as x', y', and z' (the mark is called a *prime*). It also is conventional to define the origin event (the event that defines $t = 0$ in both frames) to be the instant at which the spatial origin of one frame passes the origin of the other. We conventionally choose the common x axis so that the frames move relative to each other along this axis. Finally, we *always* choose $\vec{\beta}$ to be the velocity of the *Other* Frame relative to the Home Frame (the velocity of the Home Frame relative to the Other frame is thus $-\vec{\beta}$). Signs in *many* equations in this text depend on this convention, so it is wise to memorize this. Figure R1.7 illustrates two frames in standard orientation.

Consequences of the Newtonian view of time

Now consider an object moving in space that periodically emits blinks of light. Let the spatial position of a certain blink event as measured in the Home Frame be represented by the vector $\vec{r}\,(t)$ and the same measured in the Other Frame by $\vec{r}\,'(t)$ (we conventionally write symbols for quantities observed in the Other Frame with an attached prime). According to our assumption that time is universal and absolute, observers in both frames should agree at what time this blink event occurs: $t = t'$. The position of the spatial origin O' of the Other Frame in the Home Frame at that time is simply $\vec{\beta}t$, since the Other Frame moves at a constant velocity $\vec{\beta}$ with respect to the Home Frame, and

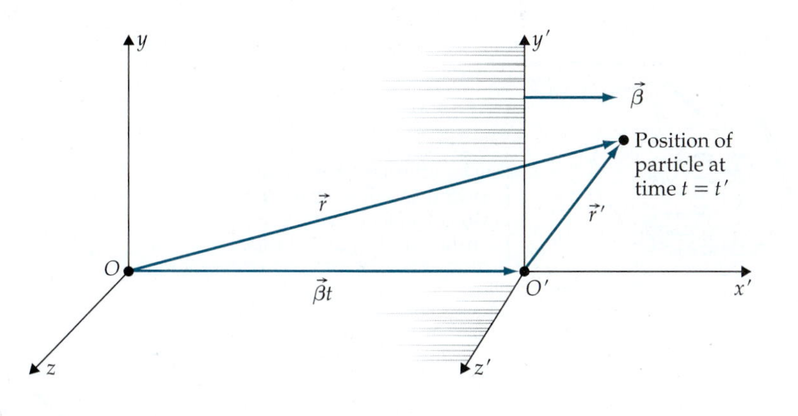

Figure R1.8
The relationship between $\vec{r}$ and $\vec{r}'$ for two inertial reference frames (assuming that time is universal and absolute).

we conventionally take both frames' origins to coincide at $t = 0$. The relationship between the object's position vectors in the two frames at the time of the blink is (as shown in figure R1.8) given by $\vec{r}(t) = \vec{r}'(t') + \vec{\beta}t$, or

$$\vec{r}'(t') = \vec{r}(t) - \vec{\beta}t \qquad \text{(R1.1)}$$

For frames in standard orientation, $\vec{\beta}$ points along the x axis, meaning that we can write equation R1.1 in component form as follows:

The Galilean transformation equations

$$t' = t \quad \text{(reminding us that time is absolute)} \qquad \text{(R1.2}a\text{)}$$
$$x' = x - \beta t \qquad \text{(R1.2}b\text{)}$$
$$y' = y \qquad \text{(R1.2}c\text{)}$$
$$z' = z \qquad \text{(R1.2}d\text{)}$$

The meaning of β without the arrow

where we define β without the arrow to be the x component of $\vec{\beta}$. It is therefore positive if the Other Frame moves in the $+x$ direction relative to the Home Frame and negative if it moves in the $-x$ direction.

Physicists call these four equations the **Galilean transformation equations**. These equations allow us to find the position of the object at a given time t' in the Other Frame if we know its position at time $t = t'$ in the Home Frame (*assuming*, of course, that time is universal and absolute).

Taking the time derivative of both sides of the last three equations yields

The Galilean *velocity* transformation equations

$$v'_x = v_x - \beta \qquad \text{(R1.3}a\text{)}$$
$$v'_y = v_y \qquad \text{(R1.3}b\text{)}$$
$$v'_z = v_z \qquad \text{(R1.3}c\text{)}$$

(Note that since $t' = t$, it really doesn't matter that we are taking a derivative with respect to t' on the left side and a derivative with respect to t on the right.) These equations tell us how to find an object's velocity in the Other Frame, given that object's velocity in the Home Frame: We call these equations the **Galilean velocity transformation equations**.

If we take the time derivative again, we get

An object's acceleration is the same in all inertial reference frames

$$a'_x = a_x \qquad \text{(R1.4}a\text{)}$$
$$a'_y = a_y \qquad \text{(R1.4}b\text{)}$$
$$a'_z = a_z \qquad \text{(R1.4}c\text{)}$$

Equations R1.4 tell us that *observers in both inertial frames agree about an object's acceleration at a given time*, even though they may well disagree about the object's position and velocity components at that time!

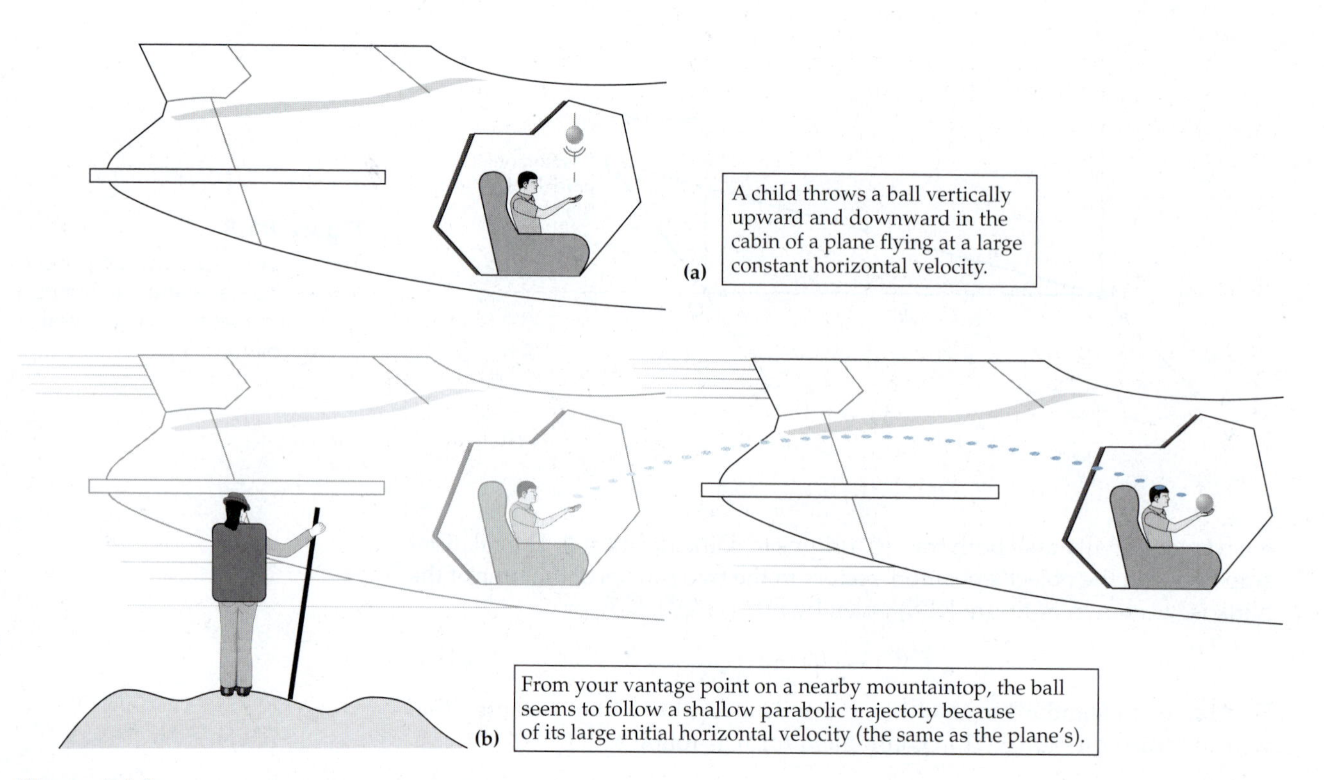

Figure R1.9
An example illustrating the application of the principle of relativity in Newtonian physics. Newton's second law describes the motion of the ball in both frames.

An illustration of how the laws of physics can be the same in different frames

Now we are in a position to discuss more fully what we might mean by "the laws of physics are the same" in every inertial reference frame in the Newtonian context at least. Consider the following example. A child on an airplane throws a ball vertically into the air and catches it again. As measured in the plane cabin (which we will take to be the Other Frame), the ball appears to travel vertically along the vertical z axis. Now imagine that you watch this process from a nearby mountaintop as the plane passes by. Instead of observing the ball travel vertically up and down, *you* will instead observe the ball to follow a shallow parabolic trajectory, because in your frame (which we will take to be the Home Frame), the ball, plane, and child all have a considerable horizontal velocity (see figure R1.9).

The ball's motion in these two reference frames looks quite different: it is vertical in the plane's cabin but parabolic (almost horizontal!) in your frame. Even so, you and an observer on the plane would *agree* that (1) the ball has a certain mass m (which you and the observer could each measure with your own balances) and thus should experience a gravitational force of magnitude $m|\vec{g}|$ acting on it, (2) this force must be the net force on the object while it is in flight (since nothing else is in contact with the ball, ignoring air friction), and (3) the ball has the same acceleration in your respective reference frames (see equations R1.4). Since you agree on the value of m, the magnitude and direction of the net force on the ball, and the acceleration that the ball experiences, you will *both* agree that Newton's second law

$$\vec{F}_{\text{net}} = m\vec{a} \quad (\text{with } \vec{F}_{\text{net}} = m\vec{g} \text{ here}) \tag{R1.5}$$

accurately predicts the ball's motion, even if you disagree about the ball's initial velocity and thus the exact character of its subsequent motion (that is, whether it is vertical or parabolic).

In a similar fashion, you might imagine observing a game of pool in the plane cabin. You on the mountaintop and your friend on the plane will totally disagree about the initial and final velocities of the balls in any given collision (since you will observe them to have a large horizontal component of velocity that your friend does not observe). Even so, you both will find that the total momentum of the balls just before a given collision is equal to their total momentum just afterward, consistent with the law of conservation of momentum (see problems R1M.7 and R1D.2).

This is what it means to say that the "laws of physics are the same" in different inertial frames. Observers in different inertial frames may disagree about the *values* of various quantities (particularly positions and velocities), but each observer will agree that if one takes the mathematical equation describing a physical law (such as Newton's second law) and substitutes in the values measured in that observer's frame, one will always find that the equation is satisfied. In other words, the same basic equations will be found to describe the laws of physics in all inertial reference frames.

R1.6 The Problem of Electromagnetic Waves

Review of Newtonian approach to relativity

In 1864, James Clerk Maxwell published a set of equations (now called **Maxwell's equations**) that summarized the laws of electromagnetism in a compact and elegant form. These equations (which are the focus of the latter part of unit E) were the culmination of decades of intensive work by many physicists, and represent one of the greatest achievements of 19th-century physics.

Among the many fascinating consequences of these equations was the prediction that one could set up *traveling waves* in an electromagnetic field, much as one can create ripples on the surface of a lake. The speed at which such electromagnetic waves travel is *completely determined* by various universal constants appearing in Maxwell's equations, constants whose values were fixed by experiments involving electrical and magnetic phenomena and were fairly well known in 1864. The predicted speed of such electromagnetic waves turns out to be about 3.00×10^8 m/s. Light was already known to have wavelike properties (as demonstrated by experiments performed by Thomas Young and Augustin-Jean Fresnel) and to travel at roughly this speed (as measured by Ole Rømer in 1675 and Jean-Bernard Leon Foucault in 1846). On the basis of this information, Maxwell concluded that light consisted of such electromagnetic waves. Later experiments confirmed Maxwell's bold assertion by showing that the value of the speed of light was indeed indistinguishable from the value predicted on the basis of the constants in Maxwell's equations. The work of Heinrich Hertz, who was able to directly generate low-frequency electromagnetic waves (that is, radio waves) and demonstrate that they had the properties predicted by Maxwell's equations, was particularly compelling.

The ether hypothesis

In short, Maxwell's equations predicted that light waves must travel at a specific speed $c = 3.00 \times 10^8$ m/s. The question is, relative to what? The consensus in the physics community at the time (one that Maxwell shared) was that electromagnetic waves were oscillations of a hypothetical medium called the **ether**, just as sound waves are oscillations in air, and water waves are oscillations in the surface of a body of water. Physicists therefore generally assumed that light waves would travel at the predicted speed c relative to this ether, and thus have this speed in a frame in which the ether is at rest.

In all *other* inertial reference frames, however, light waves must travel at a speed different from c. To see this, imagine a spaceship flying away from a space station at a velocity $\vec{\beta}$. A blinker on the space station emits a pulse of

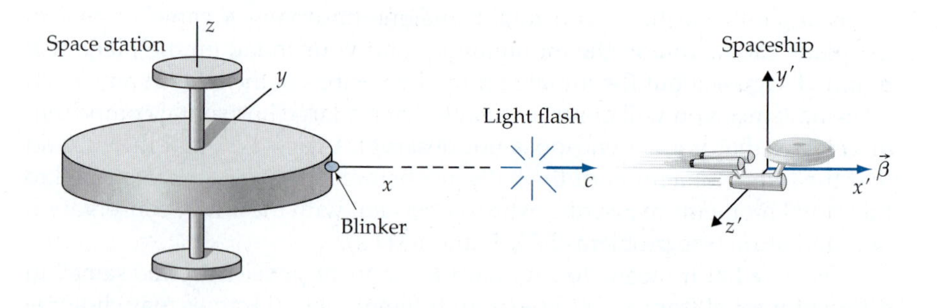

Figure R1.10
A light flash chasing a departing spaceship.

light waves toward the departing spaceship (see figure R1.10). Suppose the space station is at rest with respect to the ether and treat this as the Home Frame. Observers on the space station will then measure the emitted pulse of light to move away from the blinker at a speed of c. How rapidly would this flash of light be observed to travel in a frame fixed to the spaceship?

By construction, both the flash and the spaceship move in the $+x$ direction. The x component of the Galilean velocity transformation equation (equation R1.3*a*) in this case implies that

$$v'_{\text{light},x} = v_{\text{light},x} - \beta = c - \beta \tag{R1.6}$$

So the speed of the light flash, as measured in the frame of the spaceship (that is, the Other Frame), is $c - \beta$. This makes sense: if the spaceship happened to travel at the speed of light (so that $\beta = c$), the flash should *intuitively* appear to be motionless in the frame of the spaceship and thus never catch up with it, in agreement with equation R1.2. In any frame moving with respect to the ether, then, light waves should be measured to have a speed $\neq c$.

But this means that Maxwell's equations strictly apply only in a *certain* inertial reference frame (the frame at rest with respect to the ether), since they do specify that light waves move with a *specific* speed c. Presumably some small modifications would have to be made to these equations to make them work in frames that are *not* at rest with respect to the ether.

Physicists fail to detect the ether

Now, even its proponents admitted that this ether was peculiar stuff. It had to fill all space and permeate all objects, and yet be virtually undetectable. It had to have virtually zero density and viscosity, because it did not significantly impede anything's motion. But it also had to be extraordinarily "stiff" with regard to oscillations because the speed of waves in a medium increases with that medium's stiffness, and c is very large (for comparison, mechanical waves traveling through *solid rock* have speeds of only 6000 m/s).

In 1887, U.S. physicists Albert Michelson and Edward Morley performed a sensitive experiment designed to prove the existence of this problematic stuff. If this ether filled all space, the earth must (as a result of its orbital motion around the sun) be moving through the ether at a speed comparable to its orbital speed of about 30 km/s. This "ether wind" would make the speed of light depend on direction: a light wave traveling against the ether wind would move more slowly than a wave moving across it. So Michelson and Morley constructed a very sensitive experiment that compared the speed of two beams of light sent in perpendicular directions in a very clever way.

To the surprise of everyone involved, there was no discernible difference in the speeds of the two light waves. Michelson and Morely repeated the experiment with different orientations of the apparatus, at different times of the year (just in case the earth happened to be at rest with respect to the ether at the time of the first experiment), as did other physicists. In all cases, the result was that the speed of light seemed independent of the earth's motion.

In a modern version of the Michelson–Morley experiment performed in 1978, A. Brillet and J. L. Hall (see *Physical Review Letters*, vol. 42, p. 549, 1979) set up a laser that fed light into a Fabry-Perot cavity, which is essentially a region of space bounded by two mirrors. The laser's frequency was continually adjusted so that a specific integer number of waves fit between the mirrors. Any variation in the speed of light would cause the number of waves of a given frequency that fit between the mirrors to increase or decrease, requiring the electronics to adjust the laser's frequency to keep the number of waves fixed. The laser and cavity were mounted on a granite turntable so its orientation relative to any ether wind could be varied. While the table was rotated, Brillet and Hall observed that the fractional change in the laser frequency was $(1.5 \pm 2.5) \times 10^{-15}$. If an ether wind comparable in speed to the earth's orbital speed $|\vec{v}|$ existed, it should cause a variation in the fractional frequency on the order of magnitude of $|\vec{v}|^2/c^2$, or about 10^{-8}, more than 10 million times any possible variation consistent with experiment!

The Michelson–Morley result (and other corroborating results) caused a ruckus in the physics community, as physicists strove to explain away these results while saving the basic ether concept. Many explanations were offered but none provided a satisfactory explanation of all known experimental data.

Einstein's proposed solution to the problem

In 1905, Albert Einstein published a short paper in the European journal *Annalen der Physik* that changed the direction of physics. In that paper, Einstein proposed that since it seemed to be impossible to demonstrate the existence of the ether, the whole concept should be rejected: we should simply accept that light can move in a vacuum. But the vacuum of empty space provides no anchor for defining a special frame where the speed of light is c. (What would such a frame be attached to?) So if we accept that the speed of light in a vacuum is c, then we must accept that this speed is c in *every* inertial reference frame, in direct contradiction to equation R1.6! The assumption that *the speed of light is a frame-independent quantity* is necessary, Einstein argued, to make Maxwell's equations consistent with the principle of relativity, as they are laws of physics that predict a specific value for the speed of light.

But how can one measure a pulse of light to move at the same speed c in two different frames when those frames are not at rest with respect to each other? Einstein's bold idea, while neatly sidestepping the difficulties with the ether concept, seemed impossible to most of his contemporaries.

However, we only have three choices: either the principle of relativity is wrong, Maxwell's equations are wrong, or the Galilean velocity transformation equations are wrong. By accepting the ether concept, physicists before Einstein had opted to accept the idea that Maxwell's equations would have to be modified in frames moving with respect to the ether, thus keeping the Galilean velocity transformation but implicitly rejecting the full principle of relativity. Even as evidence against the ether hypothesis became firm and incontrovertible, rejection of the Galilean velocity transformation, so solidly based on simple and obvious ideas, seemed absurd.

On the other hand, Einstein's suggestion was elegant in its simplicity. Throw away the ether, he said. It is an unhelpful hypothesis with no experimental support. Throw away the awkward and bizarre theories that arose to explain our inability to detect the ether. Embrace instead the beautiful simplicity of the principle of relativity and Maxwell's equations. The speed of light is then the same in all inertial frames automatically, and the null results of experiments like the Michelson–Morely experiment are trivially explained.

The cost? *The Galilean transformation equations must be wrong*. But what could possibly be wrong with their derivation? It is the idea of universal and absolute time that is wrong, argued Einstein. In the next chapter, we will look at what the principle of relativity implies about the nature of time.

TWO-MINUTE PROBLEMS

R1T.1 Which of the following are (at least nearly) inertial reference frames and which are not? (Respond T if the frame is inertial, F if it is noninertial, and C if it is inertial for everyday purposes. The classification could be debatable, creating an opportunity to discuss the issues involved.)
(a) A nonrotating frame floating in deep space
(b) A frame floating in deep space that rotates at 1 rev/h
(c) A nonrotating frame attached to the sun
(d) A frame attached to the surface of the earth
(e) A frame attached to a car moving at a constant velocity
(f) A frame attached to a roller-coaster car

R1T.2 Which of the following physical occurrences fit the physical definition of an *event*?
A. The collision of two point particles
B. A point particle passing a given point in space
C. A firecracker explosion
D. A party at your dorm
E. A hurricane
F. A, B, and C
T. *Any* of the above could be an event, depending on the reference frame's scale and/or how precise the measurments need to be.

R1T.3 Since the laws of physics are the same in every inertial reference frame, there is no meaningful physical distinction between an inertial frame at rest and one moving at a constant velocity. True (T) or false (F)?

R1T.4 Since the laws of physics are the same in every reference frame, an object must have the same kinetic energy in all inertial reference frames. T or F?

R1T.5 Since the laws of physics are the same in every inertial reference frame, an interaction between objects must be observed to conserve energy in every inertial reference frame. T or F?

R1T.6 Since the laws of physics are the same in every inertial reference frame, if you perform identical experiments in two different inertial frames, you should get *exactly* the same results. T or F?

R1T.7 Imagine two boats. One travels 5.0 m/s eastward relative to the earth and the other 3.4 m/s eastward relative to the earth. We set up a reference frame on each boat with the x axis pointing eastward, and choose the first boat (arbitrarily) to be the Home Frame. The second boat is thus the Other Frame. What is the sign of β, according to the convention established in this chapter?
A. Positive
B. Negative
C. We are free to choose either sign.

R1T.8 You are in a spaceship traveling away from earth. You and Mission Control on earth agree that the $+x$ direction is the direction in which your ship is traveling relative to the earth. If you choose your own frame to be the Home Frame (so that the earth is the Other Frame), what is the sign of β, according to the convention established in this chapter?
A. Positive
B. Negative
C. We are free to choose β to have either sign.

R1T.9 Suppose you observe a collision of an isolated system of two particles. A friend observes the same collision in a reference frame moving in the $+x$ direction with respect to yours. According to the Galilean transformation equations, on which aspects of the collision will you agree with your friend? (Answer T or F.)
(a) On the value of the system's total x-momentum
(b) On the value of the system's total y-momentum
(c) On the value of the system's total z-momentum
(d) On the force $\vec{F}$ that one particle exerts on the other
(e) That the system's total momentum is conserved

R1T.10 Suppose you are in a train traveling at one-half of the speed of light relative to the earth. Assuming that photons emitted by the train's headlight travel at the speed of light relative to you, they would (according to the Galilean velocity transformation) travel at 1.5 times the speed of light relative to the earth. T or F?

R1T.11 Suppose you are in a spaceship traveling at twice the speed of light relative to the earth. Assuming that the Galilean transformation equations are true and the earth is approximately at rest relative to the ether, light from the ship's taillight will never reach the ship's bridge at its front end. T or F?

R1T.12 Suppose the Galilean transformation equations are true and your spaceship is moving at twice the speed of light relative to the ether. What odd things will you observe in your spaceship? Select all that apply. (If you are using the back of the book to communicate your answers, you can point to multiple letters with several fingers.)
A. You won't be able to see anything behind you.
B. You won't be able to see anything in front of you.
C. The beam from a laser pointer facing forward and a bit to your right will get curved toward the ship's stern.
D. Light from stars in front of you will become infinitely blue-shifted.
E. Stars a bit to the right or left of the forward direction will have their apparent positions shifted dramatically toward the ship's stern.
F. You will see none of these effects.
T. You will see all of these effects.

HOMEWORK PROBLEMS

Basic Skills

R1B.1 A train moving with a speed of 55 m/s passes through a railway station at time $t = t' = 0$. Fifteen seconds later a firecracker explodes on the track 1.0 km away from the train station in the direction the train is moving. Find the coordinates of this event in both the station frame (consider this to be the Home Frame) and the train frame. Assume the train's direction of motion relative to the station defines the $+x$ direction in both frames, and assume the Galilean transformation equations are true.

R1B.2 Suppose we select the rear end of a 120-m-long train to define the origin $x' = 0$ in the train frame, and we define a certain track signal light to define the origin $x = 0$ in the track frame. Suppose the train's rear end passes this light at $t = t' = 0$ as the train moves in the $+x$ direction at a constant speed of 25 m/s. Twelve seconds later, the engineer turns on the train's headlight. Assume the galilean transformation equations are true.
(a) Where does this event occur in the train frame?
(b) Where does this event occur in the track frame?
(Please explain your response in both cases.)

R1B.3 Suppose that boat A is moving relative to the water with a velocity of 6 m/s due east and boat B is moving with a velocity of 12 m/s due west. Assume that observers on both boats use reference frames in which the x direction points east. According to the Galilean velocity transformation equations, what is the velocity of boat A relative to boat B? (*Hint:* Draw a picture! Be sure to define which object corresponds to which frame.)

R1B.4 Suppose that spaceship A is moving relative to the earth at a speed of $0.5c$ (where c is the speed of light) in a direction we define to be the $+x$ direction. Ship B is moving at a speed of $0.9c$ in the same direction. What is the velocity of ship A relative to ship B, according to the Galilean velocity transformation equations? (*Hint:* Draw a picture! Be sure to define which object corresponds to which frame.)

R1B.5 Suppose that in an effort to attract more passengers, Amtrak trains now offer free bowling in a specially constructed "bowling alley" car. Imagine that such a train is traveling at a constant speed of 35 m/s relative to the ground. A bowling ball is hurled by a passenger on the train in the same direction as the train is traveling. Assume the Galilean transformation equations are true.
(a) The ball is measured in the ground frame to have a speed of 42 m/s. What is its speed in the frame of the train according to the Galilean velocity transformation?
(b) Suppose the ball's velocity is 8 m/s in the train frame. What is its speed relative to the ground?
(*Hints:* Draw a picture and be sure to define a $+x$ direction and specify which frame is the Home Frame.)

R1B.6 Consider a floating-ball first-law detector like the one shown in figure R1.4. If the ball is 10 cm in diameter and is placed at $t = 0$ in the center of a spherical shell whose inside diameter is 12 cm, about how long will it take the ball to hit the shell if the shell accelerates at 0.1 m/s^2?

Modeling

R1M.1 Read Galileo's 1632 presentation of the principle of relativity, as summarized on one of these websites:

en.wikipedia.org/wiki/Galileo's_ship
www.relativity.li/en/epstein2/read/a0_en/a2_en/

(or in E. F. Taylor and J. A. Wheeler, *Spacetime Physics*, 2nd ed., San Francisco: Freeman, 1992, pp. 53–55, or elsewhere online: search for "Galileo" and "Shut yourself up.") In a short paragraph, compare and contrast his presentation of the principle with the presentation in section R1.1.
(a) What for Galileo corresponds to a "laboratory moving at a constant velocity"?
(b) What for Galileo corresponds to the phrase "the laws of physics are the same" in such laboratories?

R1M.2 Imagine that Frames R Us, Inc., is constructing an economy reference frame whose price will be below every other frame on the market. Placing a clock at every point in the frame lattice is too expensive, so the company decides to place *one* clock at the origin. At all other positions, the company simply places a flag that springs up when an object goes by. The flag has the lattice location printed on it, so an observer sitting at the origin can assign spacetime coordinates to every event by noting *when* he or she sees the flag spring up (according to the clock at the origin) and noting the spatial coordinates indicated by the flag. Why *doesn't* this method yield the same spacetime coordinates as having a clock at every location would? Pinpoint the *assumption* that the engineers at Frames R Us are making that is incorrect. (See problem R1R.2 for a further exploration of the problems with this reference frame.)

R1M.3 Two firecrackers explode simultaneously 125 m apart along a railroad track, which we can take to define the x axis of an inertial reference frame (the Home Frame). A train (the Other Frame) moves at a constant 25 m/s in the $+x$ direction relative to the track frame. Assume the Galilean transformation equations are true.
(a) Do the firecrackers explode simultaneously in the train frame?
(b) How far apart are the explosions as measured in the train frame? (*Hint:* If $x_2 - x_1 = 125$ m, what is $x_2' - x_1'$?)
(c) Suppose that instead of the explosions being simultaneous, the firecracker farther ahead in the $+x$ direction explodes 3.0 s before the other in the ground frame. How far apart would the explosions be in the train frame if this were true?

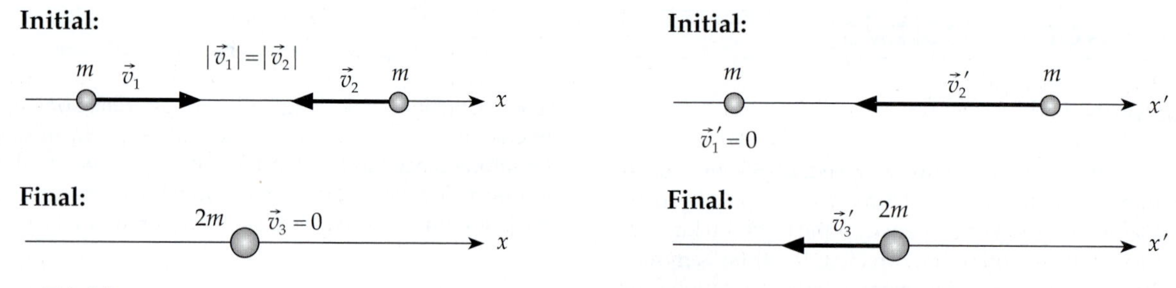

Figure R1.11
A collision between two blobs, as observed in two different reference frames.

R1M.4 Suppose that construction crews have set up a pair of blinking lights 60 m apart to mark a construction zone along a road. A car moves along the road with a speed of 30 m/s toward the lights. Let us take the road to be the Home Frame and the car to be the Other Frame. In the Home Frame, the light farther from the car (light 2) blinks 0.66 s before the other (light 1) blinks as the car approaches.

(a) What is the time interval $t_2' - t_1'$ between the blink events in the car's frame? (*Hint:* Think about the *sign.*)

(b) What is the x displacement $x_2' - x_1'$ between the events in the car frame?

R1M.5 In a certain particle accelerator experiment, two subatomic particles A and B are observed to fly away in opposite directions from a particle decay. Particle A is observed to travel with a speed of 0.6 c relative to the laboratory, and particle B is observed to travel with a speed of 0.9 c, where c is the speed of light (3.0×10^8 m/s). For simplicity's sake, let's agree to take the direction of motion of particle B to define the $+x$ direction in both the laboratory frame and particle A's frame. We would like to calculate the relative velocity of particles A and B.

(a) What object should we choose to be the Home Frame in this problem, according to the conventions established in this chapter? What object defines the Other Frame? What object is the object whose velocity we want to determine in both frames? Please explain.

(b) Use the Galilean velocity transformation equations to calculate the relative velocity of the two particles as a fraction or multiple of c. Please explain your work.

R1M.6 A highway patrol officer on the ground uses a radar gun to measure a suspect's car's speed to be 50 m/s. A patrol car is traveling at 40 m/s relative to the ground, but is moving in the opposite direction as it approaches the car. Let the direction in which the patrol car is traveling define the $+x$ direction for everyone. We would like to calculate the relative velocity of the suspect and the patrol car.

(a) Suppose we choose the patrol car to be the Home Frame. What is the sign of β in this case, according to the conventions established in this chapter? Please explain your reasoning.

(b) Use the Galilean velocity transformation equations to calculate the velocity of the suspect's car relative to the patrol car. Please explain your work.

R1M.7 Figure R1.11 above shows an inelastic collision between two blobs as observed in two different inertial reference frames. Assume the Galilean transformation equations are true.

(a) Which frame is the Home Frame, according to the convention established in this chapter? What is the sign of β? Explain your reasoning.

(b) What is v_{2x}' in terms of $|\vec{v}_1|$?

(c) What is v_{3x}' in terms of $|\vec{v}_1|$?

(d) What is the system's total x-momentum in the Home Frame, both initially and finally, in terms of $m|\vec{v}_1|$?

(e) What is the system's total x-momentum in the Other frame, both initially and finally, in terms of $m|\vec{v}_1|$?

(f) Is momentum conserved in the Home Frame?

(g) Is momentum conserved in the Other Frame?

(h) *Assuming* that energy is conserved in the Home Frame, show that it is also conserved in the Other Frame. (Ignore gravity. Note that *some* kinetic energy is converted to thermal energy U in this collision, but also that ΔU must be the *same* in both frames, because *all* observers will agree about how much hotter the final blob is compared to the initial blobs, and thus on how much the system's internal energy has increased.)

R1M.8 Some people are playing a game of shuffleboard on an ocean cruiser moving down the Hudson River at a constant speed of 17 m/s in the $+x$ direction relative to the shore. During one shot, a puck (which has a mass of 750 g and is traveling at 10 m/s in the $-x'$ direction in the boat frame) hits a puck having the same mass at rest. After the collision, the first puck comes to rest, and the other puck travels at 10 m/s in the $-x$ direction in the boat frame. (Assume that the ground frame's x axis points in the same direction as the boat frame's x' axis.)

(a) Show that the total x-momentum of the two-puck system is conserved in the boat frame. Explain carefully.

(b) Imagine that someone sitting on a bridge under which the boat is passing takes a video of this important game. What x-velocity will each puck be measured to have relative to the shore? Explain carefully.

(c) Show that in spite of the fact that the puck's x-velocities have signs and magnitudes that are *different* from those measured on the boat, the total momentum of the two-puck system is still conserved in the shore frame. Explain your work.

R1M.9 The engines on a 4000-kg jet plane accelerating for take-off exert a constant thrust of magnitude $|\vec{F}_{th}| = 20{,}000$ N on the jet as it accelerates from rest a distance $D = 1000$ meters down the runway before taking off. This take-off is observed by someone riding in a train with a constant speed of 30 m/s alongside the runway. Assume that the train and jet move in the same direction relative to the ground (which we will take to be the $+x$ direction), and assume that both the passenger and the plane are at $x = 0$ in the ground frame when the plane starts its run at $t = 0$.

(a) In unit C, we saw that if the net external force $\vec{F}_{net}$ on an object is constant, the change in the object's kinetic energy during a given interval of time should be equal to $\vec{F}_{net} \cdot \Delta\vec{r}_{CM}$, where $\Delta\vec{r}_{CM}$ is the displacement of the object's center of mass during that time. Use this to show that the jet's speed relative to the ground at take-off is 100 m/s. Explain your reasoning.

(b) Assuming that the jet's acceleration is constant, show that it takes it 20 s to reach this speed. Explain.

(c) What are the plane's initial and final x-velocities in the train frame? Explain.

(d) Assume that the passenger's position defines the origin $x' = 0$ in the train's frame. What is the jet's initial x-position in the train frame?

(e) What is its x-position at take-off in the train frame? Explain.

(f) Show that the change in the jet's kinetic energy K' in the train frame is equal to $\vec{F}'_{net} \cdot \Delta\vec{r}'_{CM}$ in that frame. (Thus, this law of physics, the momentum requirement, is the same in both frames, even though the numerical values of $K = \vec{F}_{net} \cdot \Delta\vec{r}_{CM}$ and $K' = \vec{F}'_{net} \cdot \Delta\vec{r}'_{CM}$ are *not*.)

R1M.10 A person in an elevator drops a ball of mass m from rest from a height h above the elevator floor. The elevator is moving at a constant speed $|\beta|$ downward with respect to its enclosing building.

(a) How far will the ball fall in the building frame before it hits the floor? (*Hint:* $>h$!)

(b) What is the ball's initial vertical velocity in the building frame? (*Hint:* Not 0!)

(c) Use the law of conservation of energy in the *building* frame to compute the ball's final speed (as measured in that frame) just before it hits the elevator floor.

(d) Use the Galilean velocity transformation equations and the result of part (c) to find the ball's final speed in the elevator frame.

(e) *Assume* that conservation of energy applies in the building's frame. Use the result of part (d) and the fact that the ball's acceleration is $|\vec{g}|$ to show that energy is also conserved in the elevator frame.

Derivation

R1D.1 A totally symmetric way to orient a pair of reference frames is so that their $+x$ directions point in the direction that the other frame is moving. How is this different from the "standard" orientation (draw a picture)? How would the Galilean position and velocity transformation equations be different if we were to use this convention?

R1D.2 Imagine two inertial reference frames in standard orientation, where the Other Frame moves in the $+x$ direction with x-velocity β relative to the Home Frame. Suppose an observer in the Home Frame observes the following collision: an object with mass m_1 and velocity $\vec{v}_1$ hits an object with mass m_2 traveling with velocity $\vec{v}_2$. After the collision, the objects move off with velocities $\vec{v}_3$ and $\vec{v}_4$, respectively. Do *not* assume that all or even *any* of these velocities are in the x direction. Assume, though, that total momentum is measured to be conserved in the Home Frame, that is, that

$$m_1\vec{v}_1 + m_2\vec{v}_2 = m_1\vec{v}_3 + m_2\vec{v}_4 \quad \text{(assume this!)} \qquad \text{(R1.7)}$$

Using this equation and the Galilean transformation equations, show that if the Newtonian view of time is correct, then the total momentum of the two objects will also be conserved in the Other Frame

$$m_1\vec{v}'_1 + m_2\vec{v}'_2 = m_1\vec{v}'_3 + m_2\vec{v}'_4 \quad \text{(prove this!)} \qquad \text{(R1.8)}$$

even though the velocities measured in the two frames are very different. Please show your work in detail.

Rich-Context

R1R.1 Design a first-law detector that does *not* use a floating ball as the basic active element. Your detector should primarily test Newton's first law and not some other law of physics (although it is fine if other laws of physics are involved in addition to the first law). Preferably, your detector should be reasonably practical and (if at all possible) usable in a gravitational field. (*Note:* There are *many* possible solutions to this problem. Be creative!)

R1R.2 Consider the economy reference frame described in problem R1M.2. Prove that an object that actually moves at a constant velocity close to that of light will be observed in the economy frame to move faster as it approaches the origin and slower as it departs. Also describe what happens if the object moves *faster* than light.

ANSWERS TO EXERCISES

R1X.1 Imagine that the Home Frame is an inertial frame. Consider a set of isolated objects arrayed around the Home Frame that happen to be initially at rest in that frame. Since their velocity with respect to the inertial Home Frame has to be constant, these objects will *remain* at rest relative to the Home Frame. Now, if the Home Frame moves at a constant velocity relative to the Other Frame (and the latter is rigid and nonrotating so that all parts of the Other Frame move with a constant velocity relative to the Home Frame, and vice versa), then our set of isolated objects must also move with a constant velocity relative to the Other Frame. Since these isolated objects are observed to move with a constant velocity everywhere in the Other Frame, it must be an inertial frame as well.

R2 Coordinate Time

Chapter Overview

Introduction

In this chapter, we take the first steps toward developing a conception of time that solves the problem raised by Maxwell's equations. We also start developing crucial tools for our exploration of relativity: the *spacetime diagram* and the *geometic analogy*.

Section R2.1: Relativistic Clock Synchronization

If Maxwell's equations are true laws of physics, then the speed of light c in a vacuum must be the same in all inertial reference frames. This in turn implies, as Einstein noted, that *any* method for **synchronizing** an inertial frame's clocks, if the method is to be consistent with the principle of relativity, should be equivalent to using light flashes to synchronize the clocks *assuming* that the speed of those flashes is c.

Section R2.2: SR Units

The speed of light in this approach is a frame-independent universal constant that connects time and space units. (Indeed, physicists currently *define* the meter to be the distance that light moves in exactly 1/299,792,458 s.)

In ***Système Relativistique*** **(SR) units**, we avoid the conversion factor entirely by measuring distances in **seconds**, where 1 s of distance is the distance that light moves in 1 s of time. The speed of light in this unit system is $1\text{ s}/1\text{ s} = 1$; all other speeds are similarly unitless; and energy, momentum, and mass all have units of kilograms. To convert a quantity from SI to SR units, you multiply it by whatever power of the unit operator $1 = (3 \times 10^8\text{ m}/1\text{ s})$ makes the units of meters go away; to convert from SR to SI units, you multiply by whatever power gives you the appropriate power of meters.

Section R2.3: Spacetime Diagrams

A **spacetime diagram** is an important visual aid for displaying the relationship between events. Such a diagram is simply a graph with a vertical time axis and horizontal spatial axes. *Events* in spacetime correspond to *points* on such a diagram. The connected sequence of events that describe a particle's motion in spacetime is represented by a curve on the diagram called a **worldline**, which displays how the particle's position varies with time. If the axes on the diagram have the same scale and the object moves only along the x axis, then the slope $\Delta t/\Delta x$ of its diagram worldline at any instant of time is $1/v_x$, where v_x is the particle's x-velocity at that instant. Note that the slope of any light-flash worldline will be ± 1.

Section R2.4: Spacetime Diagrams as Movies

If you view a spacetime diagram through a horizontal slit moving upward at a constant rate, you see a one-dimensional movie that displays the events and moving particles depicted on the diagram (see figure R2.5).

Section R2.5: The Radar Method

We can determine an event's spacetime coordinates using a method that is completely equivalent to that described in section R2.1 but requires only *one* clock at the frame's origin. Suppose a light flash emitted by the master clock at time t_A is reflected by an event E and returns to the master clock at time t_B. Since light moves at speed 1 in all inertial frames, event E's coordinates must be

$$t_E = \tfrac{1}{2}(t_B + t_A) \quad x_E = \tfrac{1}{2}(t_B - t_A) \tag{R2.4}$$

- **Purpose:** This equation expresses an event E's spacetime coordinates t_E and x_E in terms of the time t_A that a light flash leaves a given frame's origin and the time t_B when it returns after being reflected by the event. (Both times are measured by a master clock at the frame's origin.)
- **Limitations:** These equations assume that the event is located on the x axis, that the master clock is in an inertial reference frame, and that we are using SR units.

This approach to determining coordinates is called the **radar method**, because radar tracking of airplane positions is essentially done in this way.

Section R2.6: Coordinate Time Is Frame-Dependent

The **coordinate time** Δt between two events is the time measured in the context of an inertial reference frame. Usually, the two events will occur at different places in a given frame, and in such a case the coordinate time is the difference between the event times as recorded by two synchronized clocks in the frame, one present at each event. If the events happen to occur at the same place in the frame, then a *single* clock present at that position suffices to measure Δt.

The coordinate time between a given pair of events is a *frame-dependent* number, because although the method of clock synchronization described in section R2.1 is self-consistent *within* a given inertial frame, an observer in another inertial reference frame will *not* observe these clocks to be synchronized when comparing them to synchronized clocks in her or his own frame! This can cause, for example, an observer in one frame to think that two events are simultaneous, while an observer in a different frame concludes they are not.

This is not due to problems with the process of synchronizing multiple clocks. One can show that the radar method of defining coordinates leads to exactly the same results. The disagreement is a simple consequence of defining time so that the speed of light is the same in all inertial reference frames.

Section R2.7: A Geometric Analogy

We have no trouble with analogous frame-dependent quantities in a related but more familiar guise. Imagine superimposing two xy Cartesian coordinate systems on a town, one rotated with respect to the other. We know that surveyors using properly established but different coordinate systems will disagree about both the north–south displacement Δy and the east–west displacement Δx between two given locations in the town. The way that different inertial reference frames define spacetime coordinates for events is directly analogous to the way different coordinate axes define x and y coordinates for points on the two-dimensional plane, so we should *expect* observers in different reference frames to disagree about the value of the coordinate Δt between events (as well as about their spatial separation $|\Delta\vec{d}\,|$).

Note that we can use a tape measure to *directly* measure the separation of two points on a plane, either by measuring the *pathlength* along some arbitrary path between the points or by measuring the *distance* between them along the unique straight line connecting those points. Since we can measure these quantities directly without setting up a coordinate system, any person who calculates these quantities using coordinate-based quantities must get the same result as any other: the *pathlength* along a path and the *distance* between two points are thus coordinate-independent. We will define analogous frame-independent *time* quantities in chapter R3.

R2.1 Relativistic Clock Synchronization

At the end of chapter R1, we saw that experiments show that light has the same speed in all inertial reference frames. Einstein proposed that this was because Maxwell's equations are true laws of physics and, since they predict that electromagnetic waves must move at a certain speed c, that speed (by the principle of relativity) *must* be the same in all reference frames. The problem is that the concept of universal and absolute time (and the Galilean transformation equations that follow from it) contradicts this proposal. But if time is not universal and absolute, how can we even define what it means?

The solution, as Einstein was the first to see, is that we must define what we mean by "time" *operationally* within each inertial frame by specifying a concrete and specific procedure for synchronizing that frame's clocks that is consistent with both the principle of relativity *and* the laws of electromagnetism. But how can we synchronize clocks in such a manner?

Einstein's method of clock synchronization

Here is one method. Maxwell's equations imply that light moves through a vacuum at a certain fixed speed c. The principle of relativity requires that this speed be the same in every inertial reference frame. Therefore, *any* synchronization method consistent with the principle of relativity will lead to light being found to have speed c in any inertial reference frame.

Since the speed of light must be c in every inertial frame anyway, let us in fact synchronize the clocks in our inertial reference frame by *assuming* this is true! How do we do this? Suppose we have a master clock at the spatial origin of our reference frame. At exactly $t = 0$, we send a light flash from that clock that ripples out to the other clocks in the frame. Since we are *assuming* that light travels at a speed of $c = 299{,}792{,}458$ m/s, this light flash will reach a lattice clock exactly 1.0 meter from the master clock at exactly time $t = (1.0 \text{ m})/(299{,}792{,}458 \text{ m/s}) = 3.33564095 \times 10^{-9} \text{ s} = 3.33564095$ ns. Therefore, if we set that clock to read 3.33564095 ns exactly as the flash passes, then we know it is synchronized with the master clock. The process is similar for all the other clocks in the lattice.

So here is a first draft of a description of Einstein's method for synchronizing the clocks in an inertial reference frame:

> A light flash is emitted by clock A in an inertial frame at time t_A (as read on clock A) and received by clock B in the same frame at time t_B (as read on clock B). These clocks are defined to be **synchronized** if $c(t_B - t_A)$ is equal to the distance between the clocks. That is, the clocks are synchronized if they measure the speed of a light signal traveling between them to be c.

Exercise R2X.1

Imagine we have a clock on the earth and a clock on the moon. How can we tell if these clocks are synchronized according to this definition? Suppose we send a flash of light from the earth's clock toward the moon at exactly noon, as registered by the earth's clock. What time will the clock on the moon read when it receives the flash if the two clocks are synchronized? (The distance between the earth and the moon is 384,000,000 m.)

Note that we are *assuming* that Maxwell's equations obey the principle of relativity

"Now, wait just a minute!" I hear you cry. "Isn't all this a bit circular? You claim that Maxwell's equations predict that light will be measured to have the same speed c in every inertial reference frame. But then you go and set up the clocks so this result is *ensured*. Is this fair?"

This *is* fair. We are not trying here to *prove* that Maxwell's equations obey the principle of relativity—we are *assuming* they do, so we can determine the consequences of this assumption. To make this clear in his original paper, Einstein actually stated the frame independence of the speed of light as a separate postulate, *emphasizing* that it is an assumption. However, this is not really a *separate* assumption: it is a consequence of the assumption that the principle of relativity applies to all laws of physics, *including* Maxwell's equations. The point is that if the principle of relativity is true, the speed of light will be measured to have speed c no matter *what* valid synchronization method we use, so why not use a method based on that fact?

Moreover, there *are* other valid ways of synchronizing the clocks in a given inertial reference frame, ways that make no assumptions whatsoever about the frame independence of the speed of light.* If such a method were used to synchronize clocks in an inertial frame, such a frame could be used to verify *independently* that the speed of light is indeed frame-independent. These alternative methods yield the same consequences as one gets assuming the frame-independence of the speed of light. These methods are, however, also more complex and abstract: the definition of synchronization in terms of light is much more vivid and easy to use in practice.

R2.2 SR Units

In ordinary SI units, the speed of light c is equal to 299,792,458 m/s, a somewhat ungainly quantity. The definition of clock synchronization given in section R2.1 means that we will often need to calculate how long it would take light to cover a certain distance or how far light would travel in a certain time. You can perhaps see how messy such calculations will be.

Definition of the "second" of distance

For this reason (and many others), it will be convenient when we study relativity theory to measure distance not in the conventional unit of meters but in a new unit, called a *light-second* or just *second* for short. A **light-second** or **second of distance** is defined to be the distance light travels in 1 second of time. Since 1983, the meter has in fact been officially *defined* by international agreement as the distance light travels in 1/299,792,458 s. Therefore, 299,792,458 meter is equal to 1 light-second *by definition*.

Advantages of using this unit of distance

We can, of course, measure distance in any units we please: there is nothing magical about the meter. Choosing to measure distance in light-seconds has some important advantages. First, light travels exactly 1 second of distance in 1 second of time by definition. This allows us to talk about clock synchronization much more easily. For example, if clock A and clock B are 7.3 light-seconds apart in an inertial reference frame and a light flash leaves clock A when it reads $t_A = 4.3$ s, the flash should arrive at clock B at $t_B = (4.3 + 7.3)$ s $= 11.6$ s if the two clocks are correctly synchronized, since

*One of the simplest is described by Alan Macdonald (*Am. J. Phys.*, vol. 51, no. 9, 1983). Macdonald's method is as follows. Assume that clocks A and B emit flashes of light toward each other at (as read on each clock's *own* face). If the readings of the two clocks also agree when they receive the light signal from the other clock, they are synchronized. This method only assumes that the light flashes take the same time to travel between the clocks in each direction (that is, there is no preferred direction for light travel): it does *not* assume light has any frame-independent speed. Achin Sen (*Am. J. Phys.*, vol. 62, no. 2, 1994) presents a particularly nice example of an approach that sidesteps the synchronization issue, showing mathematically that the results of relativity follow directly from the principle of relativity. Sen's article also contains an excellent list of references.

light travels exactly 1 light-second in 1 second of time by definition. So we avoid performing ungainly unit conversions if we measure distance in this way.

Indeed, agreeing to measure distance in seconds allows us to state the definition of clock synchronization in an inertial frame in a particularly nice and concise manner (this will be our *final* draft of this description):

> We define two clocks in an inertial reference frame to be synchronized if the time interval (in seconds) registered by those clocks for a light flash to travel between them is equal to their spatial separation (in light-seconds).

In spite of the tangible advantages that measuring distance in seconds yields when it comes to talking about synchronization, this is not the most important reason for choosing to do so. In the course of working with relativity, we will uncover a deep relationship between time intervals and distance intervals, akin to the relationship between distances measured north and distances measured east on a plane. We would not think of measuring northward distances in feet and eastward distances in meters: that would obscure the fundamental similarity and relationship of these quantities. Similarly, measuring time intervals in seconds while measuring distance intervals in meters obscures the fundamental similarity in these measurements that will be illuminated by relativity theory. Choosing to measure time in the same units as distance will make this beautiful symmetry of nature more apparent.

SR units

The standard unit system used by scientists studying ordinary phenomena comprises the *Système International*, or SI, units. In this system, the units for mechanical quantities (such as velocity, momentum, force, energy, angular momentum, and pressure) are based on three fundamental units: the *meter*, the *second*, and the *kilogram*. In this text, however, we will use a slightly modified version of SI units (let's call it *Système Relativistique*, or **SR, units**) where distance is measured in *seconds* (that is, light-seconds) instead of in meters (with the other basic units being the same).

As discussed in chapter C1, we can use *unit operators* to convert a quantity from one kind of unit to another. We first write down an equation stating the basic relationship between the units in question: 1 mi = 1.609 km, for example. We then rewrite this in the form of a ratio equal to 1: either 1 = 1 mi/1.609 km or 1 = 1.609 km/1 mi. Since multiplying by 1 doesn't change a quantity, you can multiply the original quantity by whichever ratio leads you to the correct final units upon cancellation of any units that appear in both the numerator and denominator. For example, to determine what a distance of 25 km is in miles, you simply multiply 25 km by the first of the two ratios described above, as follows:

$$25\text{ km} = 25\text{ km}\cdot 1 = 25\,\cancel{\text{km}}\left(\frac{1\text{ mi}}{1.609\,\cancel{\text{km}}}\right) = \frac{25}{1.609}\text{ mi} \approx 16\text{ mi} \tag{R2.1}$$

The factors used to convert from SI distance units to SR distance units are based on the fundamental definition of the light-second: 299,792,458 m ≡ 1 s. Thus, the basic conversion factors we need are 1 = (1 s/299,792,458 m), or 1 = (299,792,458 m/1 s). For example, a distance of 25 km can be converted to a distance in light-seconds as follows:

$$\begin{aligned}25\text{ km} &= 25\text{ km}\cdot 1\cdot 1 = 25\,\cancel{\text{km}}\left(\frac{1000\,\cancel{\text{m}}}{1\,\cancel{\text{km}}}\right)\left(\frac{1\text{ s}}{2.998\times 10^{8}\,\cancel{\text{m}}}\right)\\ &= 8.3\times 10^{-5}\text{ s} = 83\ \mu\text{s}\end{aligned} \tag{R2.2}$$

meaning that 25 km is equivalent to the distance that light travels in 83 millionths of a second.

The light-second is a rather large unit of distance (the moon is only about 1.3 light-seconds away from the earth!). A more appropriate unit on the human scale is the light-nanosecond, where 1 light-nanosecond $\equiv 10^{-9}$ light-second ($= 0.2998$ m ≈ 1 ft). On the astronomical scale, the light-year (where 1 light-year $\approx 3.16 \times 10^7$ s $\approx 0.95 \times 10^{16}$ m) is an appropriate (and commonly used) distance unit. The dimensions of the solar system are conveniently measured in light-hours (it is about 10 light-hours in diameter). All these units represent extensions of the basic unit of the light-second.

Velocity is unitless in SR units

In SR units, we consider the light-second to be *equivalent* to the second of time, and both units are simply referred to as *seconds*. This means these units can be canceled if one appears in the numerator of an expression and the other in the denominator. For example, in SI units, velocity has units of meters per second, but in SR units it has units of seconds per second = unitless(!). Thus, an object that travels 0.5 light-second in 1.0 s has a speed in SR units of 0.5 s/1.0 s = 0.5 (no units!). This bare number for a speed actually represents a comparison of the object's speed to the speed of light, since light covers 1.0 s of distance in 1.0 s of time by definition. Thus, an object traveling at a speed of 0.5 (in SR units) is traveling at one-half the speed of light.

Natural SR units for other physical quantities

In a similar manner, one can find the natural units for any physical quantity in SR units. For example, a particle's kinetic energy has the same units as mass times speed squared. In SI units, these units would be kg·m²/s². Thus, the natural SI energy unit is the joule, where $1\text{ J} \equiv 1\text{ kg}\cdot\text{m}^2/\text{s}^2$. In SR units, mass times speed squared has units of $\text{kg}\cdot\text{s}^2/\text{s}^2 = \text{kg}$. Thus, the natural SR unit for energy is the kilogram (the same as the unit of mass!). How much energy is represented by an SR kilogram of energy? We can determine this by using the standard conversion factor to convert the SR distance unit of seconds to the SI distance unit of meters:

$$1\text{ kg (energy)} = 1\,\cancel{\text{kg}}\frac{\cancel{\text{s}^2}}{\cancel{\text{s}^2}}\left(\frac{2.998 \times 10^8\,\cancel{\text{m}}}{1\,\cancel{\text{s}}}\right)^2\left(\frac{1\text{ J}}{1\,\cancel{\text{kg}}\cdot\cancel{\text{m}^2}/\cancel{\text{s}^2}}\right) = 8.988 \times 10^{16}\text{ J} \quad \text{(R2.3)}$$

This unit is roughly equal to the yearly energy output of 10 full-size electric power plants!

In general, what we need to do to convert from SR units to SI units is to multiply the SR quantity by whatever power of the conversion factor $1 = (299{,}792{,}458\text{ m}/1\text{ s})$ yields the correct power of meters in the units of the final result. Similarly, to convert from SI units to SR units, we simply multiply the SI quantity by whatever power of this factor causes the units of meters to disappear. (See appendix RA for a complete discussion of how to convert units and equations from one system to the other.)

R2.3 Spacetime Diagrams

The clock synchronization method described in section R2.1 provides the last bit we needed to know to build and operate an inertial reference frame. We now know how to assign spacetime coordinates to any event occurring within that inertial frame in a manner consistent with the principle of relativity.

How to draw events on a spacetime diagram

Problems in relativity theory often involve studying how physical events relate to one another. We can conveniently depict the coordinates of events by using a special kind of graph called a **spacetime diagram**. Throughout this unit, we will find spacetime diagrams to be indispensable in helping us express and visualize the relationships between events.

Consider an event A whose spacetime coordinates are measured in some inertial frame to be t_A, x_A, y_A, z_A. To simplify our discussion somewhat,

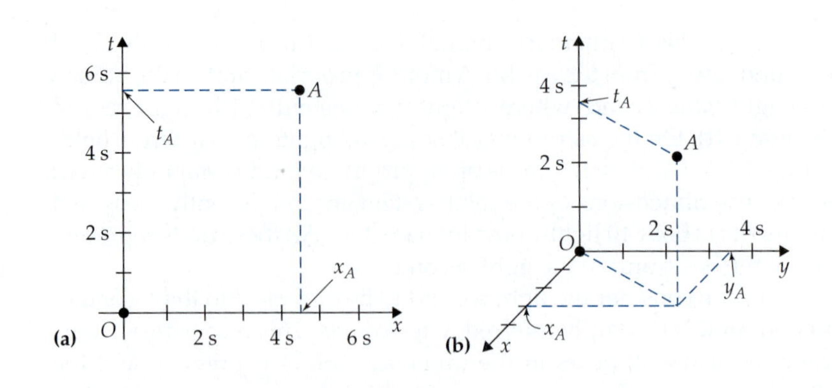

Figure R2.1
(a) How to plot an event A on a spacetime diagram. In this particular case, event A has a time coordinate $t_A \approx 5.4$ s and an x coordinate $x_A \approx 4.5$ s. (b) Plotting an event A that has nonzero x and y coordinates (but a zero z coordinate).

assume that $y_A = z_A = 0$ (that is, the event occurs somewhere along the x axis of the frame). Now imagine drawing a pair of coordinate axes on a sheet of paper. Label the vertical axis with a t and the horizontal axis with an x (it is conventional to take the t axis to be vertical in spacetime diagrams). Choose an appropriate scale for each of these axes. Now you can represent when and where event A occurs by plotting the event as a point on the diagram (in the usual manner you would use in plotting a point on a graph), as shown in figure R2.1a. We can plot any event that occurs along the x axis in space on such a diagram in a similar manner.

Note that the point marked O in figure R2.1a also represents an event. This event occurs at time $t = 0$ and at position $x = 0$. We call this event the diagram's **origin event**.

If we need to draw a spacetime diagram of an event A that occurs in space somewhere in the xy plane (that is, which has $z_A = 0$ but nonzero t_A, x_A, y_A), we must add another axis to the spacetime diagram (see figure R2.1b). The resulting diagram is less satisfactory and more difficult to draw because we are trying to represent a three-dimensional graph on a two-dimensional sheet of paper. We can't draw a spacetime diagram showing events with *three* nonzero spatial components at all, as this would involve trying to represent a *four*-dimensional graph on a two-dimensional sheet of paper. A four-dimensional diagram is hard to visualize at all, much less draw! Fortunately, one or two spatial dimensions will be sufficient for most purposes.

Worldlines

In chapter R1, we visualized describing an object's motion by imagining the object to carry a strobe light that blinks at regular intervals. If we can specify when and where each of these blink events occurs (by specifying its spacetime coordinates), we can get a pretty good idea of how the object is moving. If the time interval between these blink events is reduced, we get an even clearer picture of the object's motion. In the limit that the interval between blink events goes to zero, the object's motion can be described in unlimited detail by a list of such events. Thus, we can describe any object's motion in terms of a connected *sequence* of events. We call the set of all events occurring along the path of an object that object's **worldline**.

On a spacetime diagram, an event is represented by a point. Therefore, a worldline is represented on a spacetime diagram by an infinite set of infinitesimally separated points, which is a *curve*. This curve is nothing more than a graph of the object's position versus time (except that the time axis is conventionally taken to be vertical on a spacetime diagram). Figure R2.2 illustrates worldlines for several example objects moving in the x direction.

A worldline's slope is the *inverse* of its particle's x-velocity

Note that because the time axis is taken to be vertical, the slope of the curve on a spacetime diagram representing the worldline of an object traveling at a constant velocity in the x direction is not its x-velocity v_x (as one might expect) but rise/run $= \Delta t/\Delta x = 1/v_x$! Thus, the slope of the curve

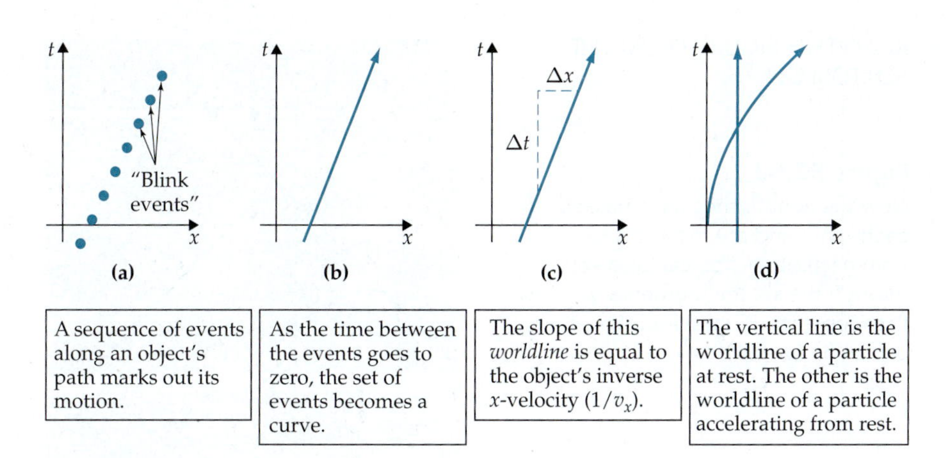

Figure R2.2
A sequence of spacetime diagrams illustrating various important things to know about worldlines.

Figure R2.3
The worldline of a particle traveling in a circular path in the *xy* plane.

representing the worldline of an object at rest is infinity and *decreases* as v_x increases. The worldline of an object traveling at a constant x-velocity has a constant slope.

Occasionally, we need to draw a spacetime diagram of an object moving in two spatial dimensions. The spacetime diagram in such a case is necessarily three-dimensional, which is hard to draw on two-dimensional paper. Figure R2.3 shows an example of such a spacetime diagram.

Light-flash worldlines have slopes of ±1

In drawing spacetime diagrams, we also conventionally use the *same-size* scale on both axes. If we do this, then the worldline of a flash (that is, a very brief pulse) of light always has a slope of either 1 (if the flash is moving in the $+x$ direction) or -1 (if the flash is moving in the $-x$ direction), since light travels 1.0 second of distance in 1.0 second of time by definition in every inertial reference frame. We also conventionally draw the worldline of a flash of light with a dashed line instead of a solid line (see figure R2.4).

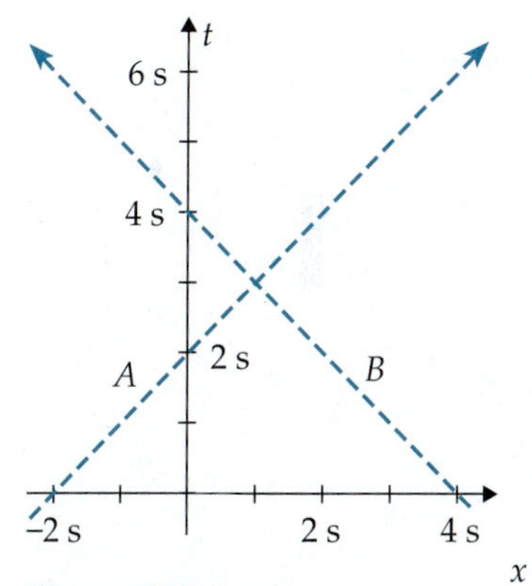

Figure R2.4
Light-flash worldlines on a spacetime diagram. Worldline *A* represents a light flash moving in the $+x$ direction. Worldline *B* represents a flash traveling in the opposite direction.

Exercise R2X.2

On figure R2.4, draw the worldline of a particle moving in the $-x$ direction through the origin event with a speed of 0.2 (1 light-second per 5 s).

R2.4 Spacetime Diagrams as Movies

One can easily get confused about what a spacetime diagram really represents. For example, in the spacetime diagram shown in figure R2.4, one can easily forget that the light flashes shown are moving in only one dimension (along the x axis), not in two. Their velocity vectors therefore point opposite to each other, not perpendicular to each other.

Making a movie viewer for a spacetime diagram

Here is a technique you can use to make the meaning of any spacetime diagram clear and vivid: turn it into a movie! Here's how. Take a 3 × 5 index card and cut a slit about $\frac{1}{16}$ inch wide and about 4 inches long, using a knife or a razor blade. This slit represents the spatial x axis at a given instant of time. Now place the slit over the x axis of the spacetime diagram. What you see through the slit is what is happening along the spatial x axis at time $t = 0$. Now slowly move the slit up the diagram, keeping it horizontal. You will see

READ THIS FIGURE FROM THE BOTTOM UP!

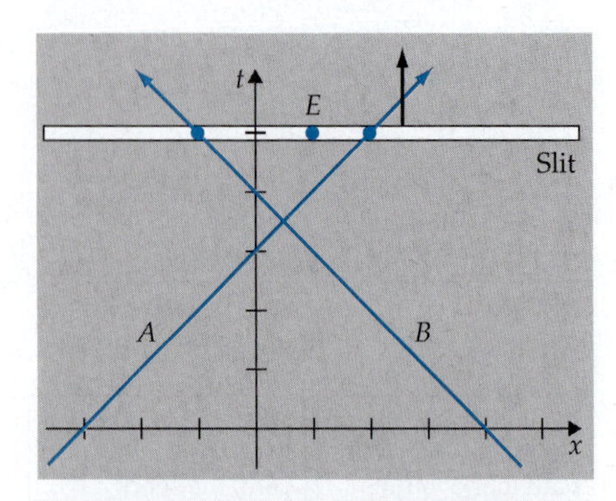

Figure R2.5d
Now the light flashes have passed each other and are moving away from each other. You can also see through the slit the momentary flash representing the firecracker explosion (event *E*).

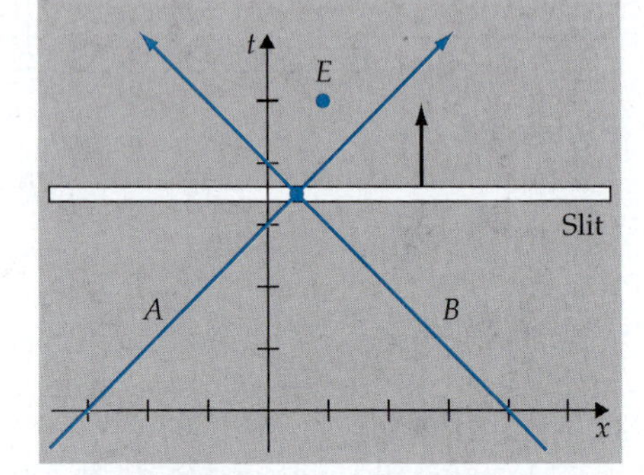

Figure R2.5c
At this instant, the light flashes pass through each other at a position of about $x = +\frac{1}{2}$ s.

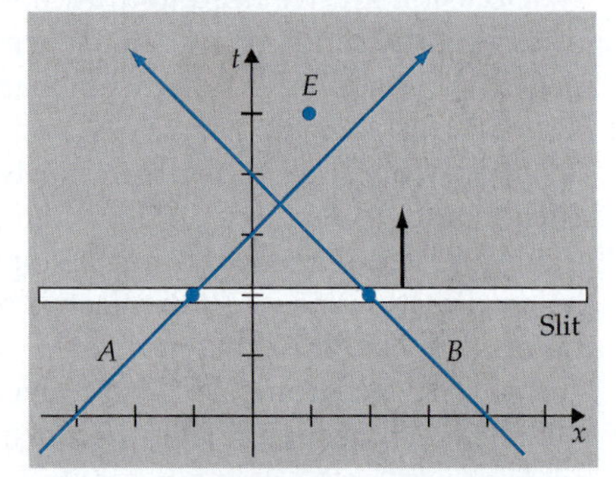

Figure R2.5b
As time passes (and you move the slit up the diagram), the dots representing the light flashes approach each other.

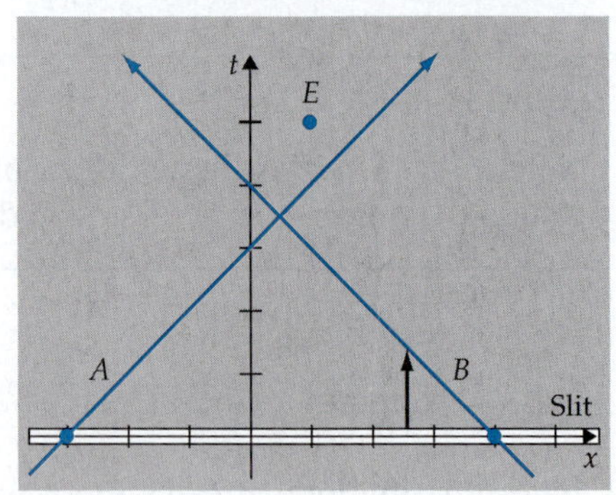

Figure R2.5a
The spacetime diagram is basically the same as figure R2.4 with a firecracker explosion (event *E*) thrown in to make things more interesting. At time $t = 0$, the light flashes are represented by black dots that you can see through the slit at $x = -3$ s and $x = +4$ s.

through the slit what is happening along the spatial x axis at successively later times. You can watch the objects whose worldlines are shown on the diagram move to the left or right as the slit exposes different parts of their worldlines. Events drawn as dots on the diagram will show up as flashes as you move the slit past them. What you see through the slit as you move it up the diagram is essentially a movie of what happens along the x axis as time passes. Figure R2.5 illustrates the process.

If you employ this technique, you cannot fail to interpret a spacetime diagram correctly. After a bit of practice with the card, you will be able to convert diagrams to movies in your head.

R2.5 The Radar Method

If we are willing to confine our attention to events occurring only along the x axis (and thus to objects moving only along that axis), we can determine the spacetime coordinates of an event with a single master clock and some light flashes: we don't need to construct a lattice at all! The method is analogous to locating an airplane by using radar.

Suppose that at the spatial origin of our reference frame (that is, at $x = 0$), we have a master clock that periodically sends flashes of light in the $\pm x$ directions. Imagine a certain flash emitted by the master clock at t_A happens to illuminate an event E of interest that occurs somewhere down the x axis. The reflected light from the event travels back along the x axis to the master clock, which registers the reception of the reflected flash at time t_B (see the spacetime diagram of figure R2.6).

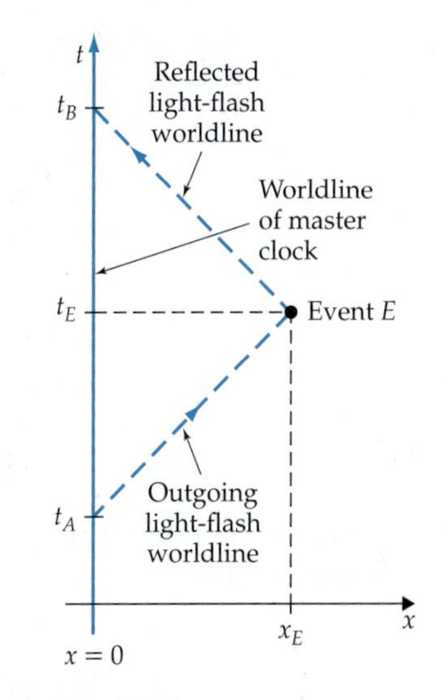

Figure R2.6
At time t_A, the master clock at rest at $x = 0$ in the Home Frame sends out a flash of light, which reflects from something at event E and returns to the master clock at time t_B.

The values of the emission and reception times t_A and t_B are sufficient to determine both the location and the time that event E occurred! Consider first how we can determine the location. The light flash's round-trip time is $t_B - t_A$. Since in this time the light covered the distance from $x = 0$ to event E and back, and since the light flash travels 1 second of distance in 1 second of time by definition, the distance to event E (in seconds) must be one-half of the round-trip time (in seconds). The x coordinate of event E is therefore $x_E = \pm\frac{1}{2}(t_B - t_A)$. We can determine the sign of x_E by noting whether the reflected flash comes from the $-x$ or $+x$ direction. (In this case, the reflected flash comes from the $+x$ direction, so we select the plus sign.)

We can determine the event's time coordinate as follows. Since the light flash traveled the same distance to the event as back from the event, and since the speed of light is a constant, the event must have occurred exactly halfway between times t_A and t_B. The midpoint in time between times t_A and t_B can be found by computing the average, so $t_E = \frac{1}{2}(t_B + t_A)$.

In summary, therefore, the spacetime coordinates of event E are

$$t_E = \tfrac{1}{2}(t_B + t_A) \qquad x_E = \tfrac{1}{2}(t_B - t_A) \qquad \text{(R2.4)}$$

- **Purpose:** This equation expresses an event E's spacetime coordinates t_E and x_E in terms of the time t_A at which a light flash leaves a given frame's origin and the time t_B when it returns after being reflected by the event. (Both times are measured by a master clock at the frame's origin.)
- **Limitations:** This equation assumes that the event is located on the x axis, that the master clock is in an inertial reference frame, and that we are using SR units.

This method yields the same result as the clock lattice method

Equations R2.4 represent a method of determining the spacetime coordinates of an event that does *not* require the use of a complete lattice of synchronized clocks. But you should be able to convince yourself that *this method produces exactly the same coordinate values as you would get from a clock lattice*. For example, imagine you actually had a lattice clock at x_E (the location of event E). The distance between that clock and the master clock at the origin must be equal to one-half the time it would take a flash of light to go from one clock to the other and back, since light travels 1 second of distance in 1 second of time by definition. The lattice clock at x_E at the time of event E must read t_A + (the light travel time between the two clocks) $= t_A + \frac{1}{2}(t_B - t_A) = \frac{1}{2}(t_B + t_A)$, since we are assuming that the lattice clocks are synchronized, which means (by definition) that they measure a light flash to travel between them at the speed of light (1 second of distance in 1 second of time).

Using this method to determine spacetime coordinates is therefore equivalent to using a lattice of synchronized clocks. In some cases in this text, we will find it clearer or more convenient to use one method, and in some cases the other. The important thing to realize is that *either* the radar method or the clock lattice method provides specific and well-defined method for assigning time coordinates to events, and they are equivalent because both express the assumption that the speed of light is a frame-independent constant. These methods essentially define what time means in special relativity, and thus will provide the foundation for most of the arguments in the remainder of the text. *Make sure you thoroughly understand both methods.*

Adapting the method for three spatial directions

We call this section's method the **radar method** because radar installations actually use a three-dimensional generalization of this approach to track the trajectories of aircraft (the impracticality of using a clock lattice to do the same is obvious!). We can precisely locate an object in three spatial dimensions if we record not only the time that the outgoing pulse was sent and the time that the reflected pulse was received but also the *direction* from which the reflected pulse was received. The analysis is a bit more complicated (see problems R2M.7 and R2M.8), but the basic idea is the same.

R2.6 Coordinate Time Is Frame-Dependent

Once we have satisfactorily synchronized the clocks in an inertial reference frame, we can use them to measure the time coordinates of various events that occur in that frame. In particular, we can measure the time between two events A and B in our reference frame by subtracting the time read by the clock nearest event A when it happened from the time read by the clock nearest event B when it happened: $\Delta t_{BA} \equiv t_B - t_A$. Note that this method of measuring the time difference between two events requires the use of a pair of synchronized clocks in an established inertial reference frame. Such a measurement therefore *cannot be performed* in the absence of an inertial frame.

So define the **coordinate time** between two events as follows:

The definition of *coordinate time* between events

> The coordinate time Δt between two events either by a pair of synchronized clocks at rest *in a given inertial reference frame* (one clock present at each event) or by a single clock at rest *in that inertial frame* (if both events happen to occur at that clock in that frame).

The coordinate time between events depends on your choice of frame

Now, suppose the observer in some inertial reference frame (let's call this frame the Other Frame: we'll talk about a Home Frame in a bit) sets out to synchronize its clocks. In particular, let us focus on two clocks in that frame that lie on the x axis an equal distance to the left and right of the master clock at $x' = 0$. At $t' = 0$, the observer causes the center clock to emit two flashes of light, one traveling in the $+x'$ direction and the other in the $-x'$ direction. Let's call the emission of these flashes from $x' = 0$ at $t' = 0$ the origin event O.

As both of the other clocks are the *same distance* from the center clock and since the speed of light is 1 (light-)second/second in *every* inertial reference frame, the left-hand clock will receive the left-going light flash (call the event of reception event A) at the *same* time as the right-hand clock receives the right-going flash (event B). By the definition of synchronization, both clocks should therefore be set to read the *same* time at events A and B (a time in seconds equal to their common distance from the center clock).

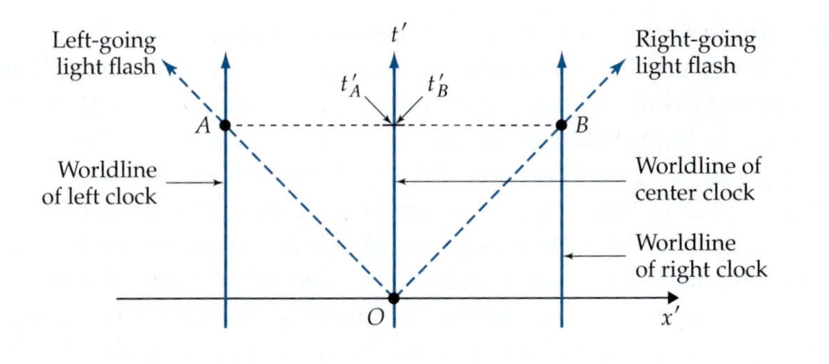

Figure R2.7
The synchronization of two clocks equally spaced from a center clock, as observed in the Other Frame. If the right and left clocks are set to agree at events *A* and *B*, they will be synchronized with each other.

The spacetime diagram in figure R2.7 illustrates this process. Note that since all three clocks are at rest in this frame, their worldlines on the spacetime diagram are vertical. Moreover, since the speed of light is 1 s/s in this (and every other inertial) frame, the worldlines of the light flashes will have slopes of ± 1 on the spacetime diagram (that is, they make a 45° angle with each axis) as long as the axes have the same scale. On this diagram, it is clear that events A and B really do occur at the same time in the Other Frame.

Now consider a different inertial reference frame (the Home Frame), within which the Other Frame moves in the $+x$ direction at an x-velocity β. How will an observer in this frame interpret these *same events*? For convenience, let us take the event of the emission of the flashes to be the origin event in this frame also (so event O occurs at $t = x = 0$ in the Home Frame).

The observer in the Home Frame will agree that the right and left clocks in the Other Frame are always equidistant from the center clock in the Other Frame. Moreover, at $t = 0$, when the center clock passes the point $x = 0$ in the Home Frame as it emits its flashes, the right and left clocks are equidistant from the emission event. But as the light flashes are moving to the outer clocks, the Home Frame observer observes the left clock to move up the x axis *toward* the flash coming toward it, and the right clock to move up the x axis *away* from the flash coming toward it. The left-going light flash therefore has less distance to travel to meet the left clock than the right-going flash does to meet the right clock. Since the speed of light is 1 in the Home Frame as well as in the Other Frame, this means that the left clock receives its flash first. *Therefore, the Home Frame observer observes event A to occur* ***before*** *event B.*

Figure R2.8 shows a spacetime diagram of the process as observed in the Home Frame. Note that the clocks are *not* at rest in the Home Frame, so their worldlines on a Home Frame spacetime diagram will be equally spaced lines with slopes of $1/\beta$ indicating that the clocks are moving to the right at a speed β. The light flashes have a speed of 1 s/s in the Home Frame (as they do in *any* inertial frame), so we must draw their worldlines with a slope of ± 1 on the spacetime diagram.

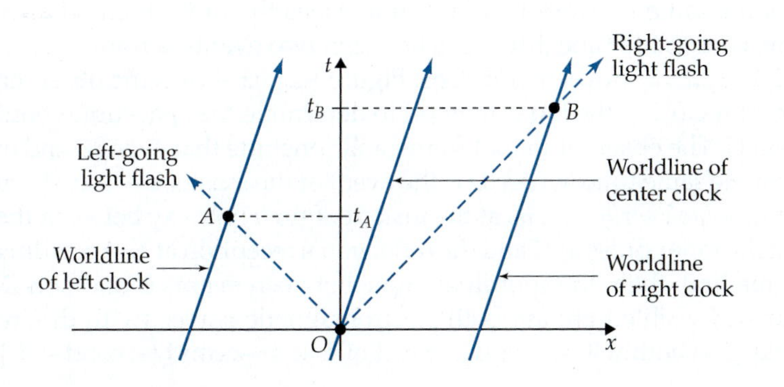

Figure R2.8
The same events as observed in the Home Frame. In this frame, event *B* is measured to occur *after* event *A*.

In summary, the coordinate time between events A and B as measured in the Other Frame is $\Delta t' = 0$ (by construction here), but the coordinate time between these events as measured in the Home Frame is $\Delta t \neq 0$. We see that the coordinate times between the *same two events* measured in different reference frames are *not* generally equal. We say that the coordinate time differences are **relative** (that is, they depend on one's choice of inertial reference frame).

Why? If each observer synchronizes the clocks in his or her own reference frame according to our definition, *each will conclude that the clocks in the other's frame are not synchronized.* Notice that the Other Frame observer has set the right and left clocks to read the *same* time at events A and B. Yet these events do *not* occur at the same time in the Home Frame. Therefore, the Home Frame observer will claim that the clocks in the Other Frame are not synchronized. (Of course, the Other Frame observer will say the same thing about the clocks in the Home Frame.) The definition of synchronization that we are using makes perfect sense *within* any inertial reference frame, but it does not allow us to synchronize clocks in *different* inertial frames. In fact, the definition *requires* that observers in different inertial frames measure *different* time intervals between the same two events, as we have just seen.

Spatial coordinate differences are relative also

In general, two observers in different frames will also disagree about the *spatial* coordinate separation between the events. Consider events C and D that both occur at the center clock in the Other Frame, but at different times. Since the center clock defines the location $x' = 0$ in the Other Frame, the events have the same x' coordinate in that frame, so $\Delta x' = 0$. But in the Home Frame, the center clock is measured to move in the time between the events, and so the two events do *not* occur at the same place: $\Delta x \neq 0$ (see figure R2.9).

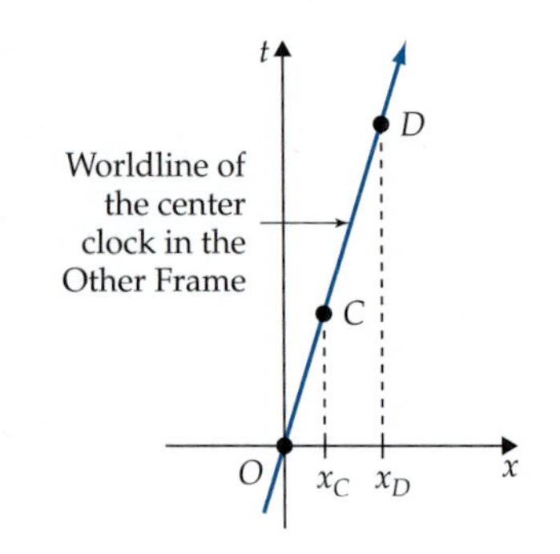

Figure R2.9
Events C and D both occur at the same place in the Other Frame ($\Delta x' = 0$) but not in the Home Frame ($\Delta x \neq 0$). (This diagram is drawn by an observer in the Home Frame.)

Exercise R2X.3

Note that the frame dependence of the *spatial* coordinate difference between two events has nothing to do with clock synchronization or relativity: this would be true even if time were universal and absolute. Show, using the Galilean transformation equations, that if the separation between two events in the Other Frame is $\Delta x' = 0$ but $\Delta t' \neq 0$, then the separation between these events in the Home Frame is *not* zero ($\Delta x \neq 0$).

The reason why observers in different inertial frames disagree about whether the clocks in a given frame are synchronized is that *synchronization is defined so that light flashes are measured to have a speed of* 1 *in every inertial frame*: the frame-dependence of coordinate time differences is a logical consequence of this. This can be illustrated by considering the radar method of assigning spacetime coordinates. Although the radar method does not involve the use of synchronized clocks, it does depend on the assumption that the speed of light is the same in every inertial frame. Does the radar method also imply that the coordinate time difference between two events is frame-dependent?

The radar method yields the same conclusion

Figure R2.10 shows that it does. Figure R2.10a shows the observer in the Other Frame using the radar method to determine the spacetime coordinates of event C. The observer in that frame will conclude that event C and event D occur at the same time, where D is the event of the master clock at $x' = 0$ registering $t'_D = \frac{1}{2}(t'_A + t'_B)$, that is, at the instant of time halfway between the emission of the radar pulse at t'_A and the reflection's reception at t'_B. According to the radar method, then, the coordinate time between events C and D is $\Delta t' = 0$. [Radar and visible light are both electromagnetic waves (with different frequencies), so both will move at a speed of 1 light-second/second = 1.]

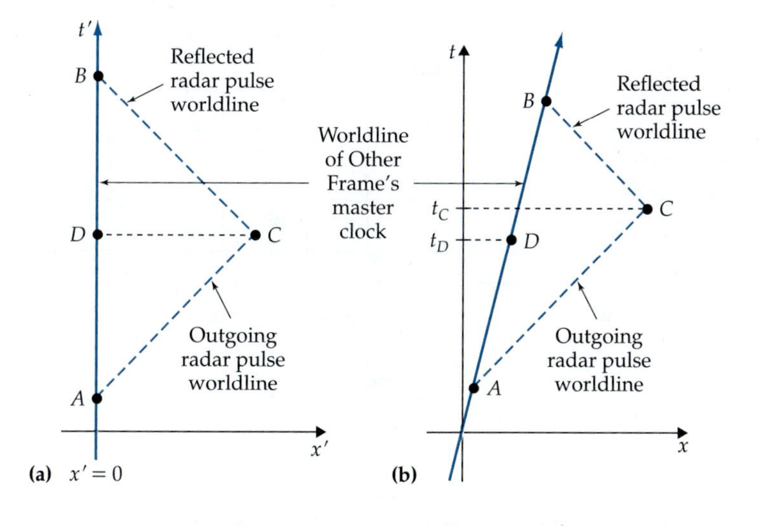

Figure R2.10

(a) In the Other Frame, events *C* and *D* are defined to be simultaneous if *D* occurs at the master clock at a time halfway between the emission event *A* and the reception event *B*. The coordinate time difference between events *C* and *D* in the Other Frame is thus $\Delta t' = 0$. (b) In the Home Frame, the Other Frame's master clock moves to the right as time passes, so its worldline is slanted. On the other hand, radar pulse worldlines still have slope ± 1, as shown. This means that an observer in the Home Frame will conclude that event *C* happens after event *D*, so the coordinate time difference between the events is $\Delta t \neq 0$.

When the same sequence of events is viewed from the Home Frame, though, a different conclusion emerges (see figure R2.10b). According to observers in the Home Frame, the Other Frame's master clock is moving along the x axis with some speed β, so in a spacetime diagram based on Home-Frame measurements, that clock's worldline will appear as a slanted line (with slope $1/\beta$) instead of being vertical. Radar pulse worldlines, on the other hand, still have slopes of ± 1, just as they did in the Other Frame spacetime diagram. The inevitable result (as you can see from the diagram) is that observers in the Home Frame are forced to conclude that event C occurs *after* event D does, and thus that the time difference between events C and D in the Home Frame is $\Delta t \neq 0$.

Frame dependence of coordinate time follows from principle of relativity

The point is that the relativity of the coordinate time interval between events is a direct consequence of the fact that we are *defining* coordinate time by assuming that the speed of light is 1 in every inertial reference frame. Remember, though, that we *must* make this assumption if the laws of electromagnetism are to be consistent with the principle of relativity!

R2.7 A Geometric Analogy

You may find it troubling that coordinate differences between events are not absolute but are instead frame-dependent. This is particularly true of the time coordinate separation: it is not easy to let go of the Newtonian notion of absolute time! The fact is, though, *we have no trouble at all with these ideas when they appear in a related but more familiar guise.*

An illustration of alternative coordinate systems in plane geometry

Consider Askew, a hypothetical town somewhere in the western United States (its embarrassed residents wish its precise location to remain secret). Most towns in the rural United States have streets that run directly north/south or east/west. The surveyor who laid out the streets of Askew in 1882, however, *tried* to calibrate his compass against the North Star the night before, but in fact had forgotten exactly where it was (it was a long time since he had this stuff in high school, after all) and ended up choosing a star that turned out to be 28° east of the true North Star. Therefore, all the streets of Askew are twisted 28° from the standard directions.

Now, if we would like to assign x and y coordinates to points of interest in this town (or any town), we need to set up a Cartesian coordinate system. We conventionally to orient coordinate axes on a plot of land so that the x and y

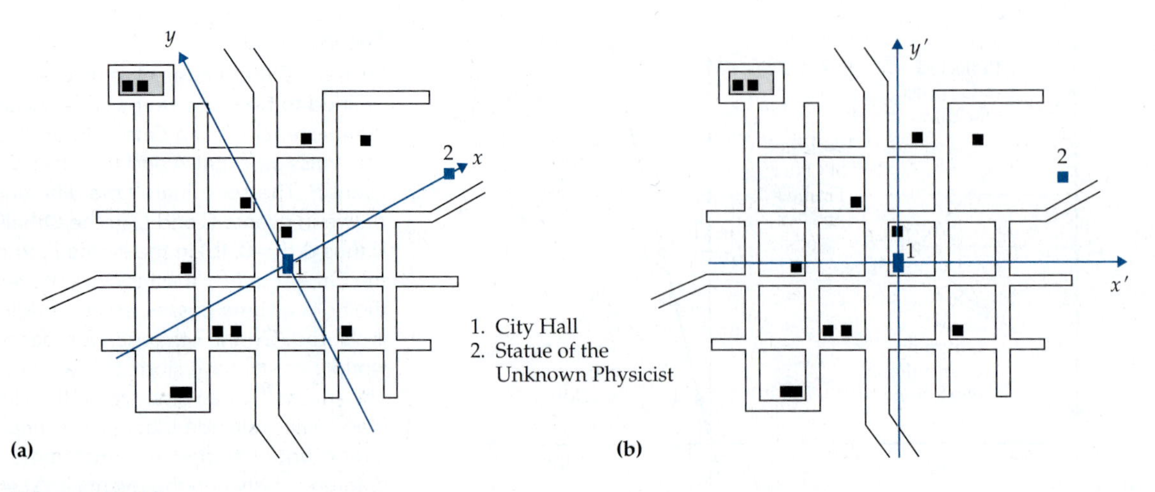

Figure R2.11
(a) A standard northward-oriented Cartesian coordinate system superimposed on the town of Askew.
(b) A more convenient coordinate system oriented 28° clockwise.

axes point north and east, respectively (see figure R2.11a). This is usually also convenient, since the streets in *most* towns are parallel to those axes. There is no reason why this *must* be done, though, and in Askew's case, it is actually more convenient to use a coordinate system tilted 28° clockwise (figure R2.11b). Note that City Hall is the origin of both coordinate systems.

We can (of course) use any coordinate system we like to quantify the positions of points of interest in the town, since coordinate systems are arbitrary human artifacts that we impose for our convenience on the physical world. However, the coordinates we actually *obtain* for various points certainly *do* depend on the coordinate system used. For example, the coordinate differences between City Hall and the Statue of the Unknown Physicist in Memorial Park might be $\Delta y = 0$, $\Delta x = 852.0$ m in the standard coordinate system, but $\Delta y' = 399.9$ m, $\Delta x' = 752.3$ m in the conveniently oriented coordinate system.

We are not surprised by frame dependence of coordinates in this case!

Is it surprising that the results are different? Do the differences in the results cause us to suspect that one or the other coordinate system has been set up incorrectly? Hardly! We already *know and expect* that differently oriented coordinate systems on a plane will yield different coordinate measurements. This causes no discomfort, because we understand this is the way things are.

In an entirely analogous way, we have carefully and unambiguously defined a procedure for setting up an inertial reference frame and synchronizing its clocks. This definition happens to imply that spacetime coordinate measurements in different frames yield different results. This should be no more troubling to us than the fact that Askew residents who use different sets of coordinate axes will assign different coordinates to various points in town. *Coordinates have meaning only in the context of the coordinate system or inertial frame we use to observe them.*

The only reason that the relativity of time coordinate differences is a difficult idea is because we don't have *common experience* with inertial reference frames moving with high enough relative speeds to display the difference. The frames that we typically experience in daily life have relative speeds below 300 m/s, or about one-millionth of the speed of light. If for some reason we could only construct Cartesian coordinate systems on the surface of the earth that differed in orientation by no more than one-millionth of a radian, then we might also consider Cartesian coordinate differences as being "universal and absolute" as well!

So, to summarize, the coordinate differences between points on a plane (or events in spacetime) are "relative" because coordinate systems (or inertial reference frames) are human artifacts that we *impose* on the land (or on spacetime) to help us quantify that physical reality by assigning coordinate numbers to points on the plane (or events in spacetime). Because we are free to set up coordinate systems (or reference frames) in different ways, the coordinate differences between two points (or events) reflect not only something about their real physical separation, but also something about the artificial choice of coordinate system (or reference frame) we have made.

So, is it true, then, that *everything* is relative? Is there nothing we can measure about the physical separation of the points (or events) that is absolute (that is, independent of reference frame)?

Distance and pathlength, on the other hand, are coordinate-*independent*

The *distance* between two points is such a quantity. For example, the distance between Askew's City Hall and the Statue of the Unknown Physicist is $|\Delta\vec{d}| = (\Delta x^2 + \Delta y^2)^{1/2} = [(852.0\text{ m})^2 + 0]^{1/2} = 852.0$ m in the north-oriented coordinate system and $|\Delta\vec{d}'| = [(\Delta x')^2 + (\Delta y')^2]^{1/2} = [(399.9\text{ m})^2 + (752.3\text{ m})^2]^{1/2} = 852.0$ m in the convenient coordinate system. It doesn't matter what coordinate system one uses to calculate $|\Delta\vec{d}|$: you always get the same answer.

The distance between two points on a plot of ground thus reflects something deeply real about the nature of the plot of ground itself, independent of the human coordinate systems we impose on it. This distance is independent of coordinate systems because we can, in fact, determine it directly *without using a coordinate system at all* simply by laying a tape measure between the points! Since this method yields an unambiguous distance, valid calculations of this distance in *any* coordinate system should yield the same value.

Of course, there are many *ways* that one could lay a tape measure between City Hall and the Statue of the Unknown Physicist. One could lay the tape measure along a straight path between the two points: this would measure the distance "as the crow flies," which is what is usually meant by the phrase "the distance between two points." But there are other possibilities. One might, for example, lay the tape measure two blocks down Elm Street from City Hall, then one block over along Grove Avenue, then up Maple Street, and so on. This would measure a different kind of distance between the two points that we might call a *pathlength*.

Both the straight-line distance and the more general pathlength between two points can be measured directly with a tape measure, and thus are quantities independent of any coordinate system. But the distance and the pathlength between two points may not be the same. In general, the pathlength between two points will depend on the path chosen, and will always be greater than (or at best equal to) the straight-line distance.

The three kinds of spatial separation

To summarize, we can quantify the separation of two points on a plane three totally different ways. We can measure the *coordinate separations* between the points, using a coordinate system. (The results will depend on our choice of coordinate system.) We can measure the *pathlength* between them with a tape measure laid along a specified path. (The result here will depend on the path chosen, but is independent of coordinate system.) Or we can measure the *distance* between the points with a tape measure laid along the unique path that is the straight line between the points. Because in this last case the tape's path is unique, the distance between two points is a unique number that quantifies in a basic way the separation of those points in space.

Analogously, we can measure the time between two events in spacetime in three different ways. *The coordinate time* between events is analogous to the coordinate separation of points and so depends on one's choice of reference frame. In the next chapter, we will see how we can define frame-*independent* times analogous to *pathlength* and *distance*.

TWO-MINUTE PROBLEMS

R2T.1 Imagine that in the distant future you (on earth) are watching a transmission from Pluto, which at the time is 5.0 light-hours from earth. You notice that a clock on the wall behind the person speaking in the video reads 12:10 p.m. You note that your watch reads exactly the same time. Is the station clock synchronized with your watch?
A. Yes, it is.
B. No, it isn't.
C. The problem doesn't give enough information to tell.

R2T.2 Suppose you receive a message from a starbase that is 13.0 light-years from earth. The message is dated July 15, 2127. What year does your calendar indicate at the time of reception if your calendar and the station's calendar are correctly synchronized?
A. 2127
B. 2114
C. 2140
D. Other (specify)

R2T.3 The speed of a typical car on the freeway expressed in SR units is most nearly
A. 10^{-7}
B. 10^{-10}
C. 10^{-8}
D. 10^{-6}
E. 10^{-4}
F. Other (specify)
T. None of these answers is right: we must state units!

R2T.4 Suppose you are sitting at the origin of an inertial reference frame. You *see* (that is, you receive the light from) an event E occurring near a clock at $x = -30$ ns at a time $t = 80$ ns. When do you *observe* that event to occur?
A. $t_E = 0$
B. $t_E = 30$ ns
C. $t_E = 50$ ns
D. $t_E = 80$ ns, of course
E. $t_E = 110$ ns
F. Some other time (specify)

R2T.5 The spacetime diagram in figure R2.12 shows the worldlines of various objects. Which object has the largest speed at time $t = 1$ s?

R2T.6 The spacetime diagram in figure R2.12 shows the worldlines of various objects. Which object has the largest speed at time $t = 4$ s?

R2T.7 The spacetime diagram in figure R2.12 shows the worldlines of various objects. Which worldline cannot possibly be correct? (Explain why.)

R2T.8 In figure R2.12, the object whose worldline is labeled B is moving along the x axis. T or F?

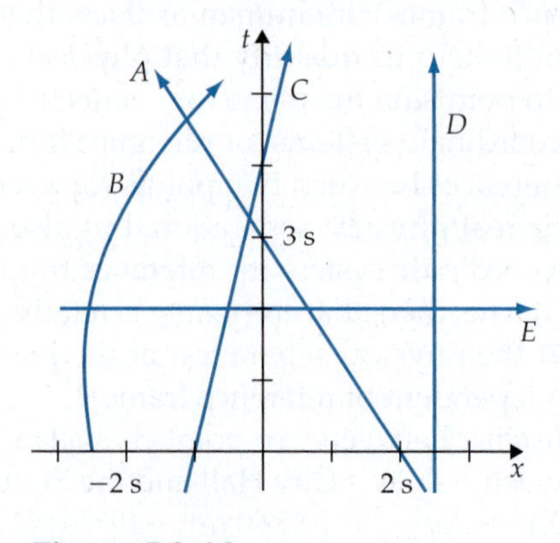

Figure R2.12
Worldlines of various objects.

R2T.9 A light flash leaves a master clock at $x = 0$ at time $t = -12$ s, is reflected from an object a certain distance in the $-x$ direction from the origin, and then returns to the origin at $t = +8$ s. From this information, we can infer that the spacetime coordinates of the reflection event are $[t, x] =$
A. [4 s, 20 s]
B. [−4 s, −20 s]
C. [10 s, −2 s]
D. [2 s, −10 s]
E. [−2 s, −10 s]
F. Other (specify)

R2T.10 Coordinate time would be frame-independent if the Newtonian concept of time were valid. T or F?

R2T.11 Consider a Home Frame and an Other Frame that moves in the $+x$ direction with respect to the Home Frame.
(a) Observers in the Home Frame will conclude that the clocks in an Other Frame will be out of synchronization, even if the observers in the Other Frame have carefully synchronized clocks using the Einstein prescription. T or F?
(b) Specifically, Home Frame observers will see that for events farther and farther up the common $+x$ axis, the times registered by Other Frame clocks at the events
A. Become further and further ahead.
B. Become further and further behind.
C. Remain the same.
D. Have no clear relationship to the values that Home Frame clocks register for the same events.

R2T.12 In the geometric analogy, the coordinate time difference Δt between two events in spacetime corresponds to
A. The north–south separation between points on a plane.
B. The distance between points on a plane.
C. A certain pathlength between points on a plane.
D. The separation between the events in spacetime.
E. Something else (specify).

HOMEWORK PROBLEMS

Basic Skills

R2B.1 (Practice with SR units.)

(a) What is the earth's diameter in seconds?

(b) A highway sign reads "Speed Limit 6×10^{-8}," meaning speed in SR units. What is this in miles per hour?

(c) Argue that the SR unit of acceleration is s^{-1}. What is $1\ s^{-1}$ expressed as a multiple of $|\vec{g}|$?

R2B.2 (Practice with SR units.)

(a) A sign on a hiking trail reads "Viewpoint: 5.5 μs." About how long would it take you to walk to this viewpoint at a typical walking speed of 1 m/s?

(b) Section R1.2 mentions that the Voyager 2 spacecraft achieved speeds in excess of 72,000 km/h. What is this speed in SR units?

(c) In SI units, power is measured in watts (where $1\ W \equiv 1\ J/s = 1\ kg \cdot m^2/s^3$). Argue that the natural SR units of power are kg/s. Let's define 1 SR-watt = 1 SRW ≡ 1 kg/s. A large electric power plant produces energy at a rate of about 10^9 W. What is this in SRW?

R2B.3 (Practice with SR units.)

(a) Argue that the SR units of mass, momentum, and kinetic energy are simply kilograms.

(b) Imagine that a truck with a mass of 25 metric tons (that is, 25,000 kg) is barreling down a highway at a speed of 59 mi/h. What is the truck's momentum in kilograms?

(c) What is the truck's kinetic energy in kilograms?

R2B.4 (Practice with SR units.)

(a) Argue that force has SR units of kg/s.

(b) What is 1 kg/s in newtons?

(c) What are the SR units of pressure?

(d) What are the SR units of density?

R2B.5 For each of the worldlines shown in figure R2.13, describe in words what the particles are doing, giving numerical values for velocities when possible. For example, you might say, "Particle *A* is traveling in the $+x$ direction with a constant speed of $\frac{1}{3}$."

R2B.6 Suppose you send out a light flash at $t = 3.0$ s, as registered by your clock, and you receive a return reflection showing your kid brother making a silly face at $t = 11$ s, as registered by your clock.

(a) At what time did your brother actually make this face?

(b) How far is your brother away from you (express your result in seconds and kilometers)?

(c) Is this far enough away that he can't really be a nuisance?

R2B.7 Suppose you send out a radar pulse at $t = -22$ h, as registered by your clock, and receive a reflection from an alien spacecraft at $t = +12$ h as registered by your clock.

(a) Is the spaceship inside or outside the solar system?

(b) When did the spaceship reflect the radar pulse?

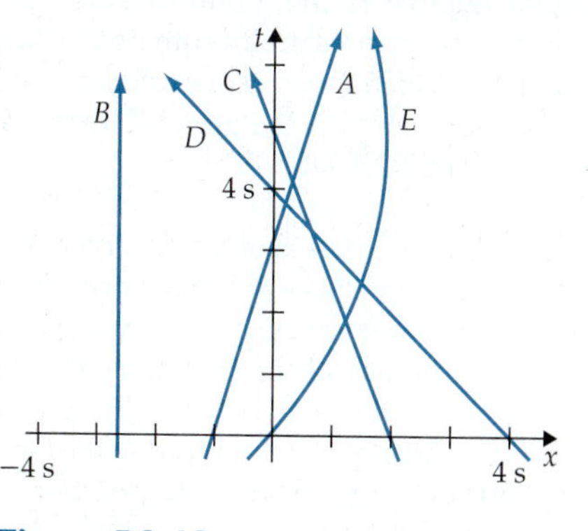

Figure R2.13
Wordlines of various objects.

R2B.8 A firecracker explodes 30 km away from an observer sitting next to a certain clock *A*. The light from the firecracker explosion reaches the observer at exactly $t = 0$ according to clock *A*. Suppose the firecracker's flash illuminates the face of another clock *B* that is sitting next to the firecracker. What time will clock *B* register at the moment of illumination if it is correctly synchronized with clock *A*? Express your answer in milliseconds.

R2B.9 Suppose you are in an inertial frame in empty space with a clock, a telescope, and a powerful strobe light. A friend is sitting in the same frame at a very large (unknown) distance from your clock. At precisely 12:00:00 noon according to your clock, you set off the strobe lamp. Precisely 30.0 s later, you see in your telescope the face of your friend's clock illuminated by your strobe flash.

(a) How far away is your friend from you (in seconds)?

(b) What should you see on the face of your friend's clock if that clock is synchronized with yours?

Describe your reasoning in a few short sentences.

R2B.10 Draw a spacetime diagram that displays worldlines for the following particles.

(a) Particle *A* travels at a constant speed of $\frac{3}{4}$ in the $+x$ direction and passes the point $x = 0$ at time $t = 2$ s.

(b) Particle *B*, which at time $t = 0$ is at position $x = +2$ s is traveling at a speed of $\frac{1}{2}$ in the $-x$ direction, is slowing down as time passes, and eventually comes to rest at $x = 0$ at time $t = 8$ s.

(c) Light flash *C* passes the position $x = 0$ at time $t = -2$ s as it travels in the $+x$ direction.

R2B.11 Draw a spacetime diagram that shows the following worldlines.

(a) Particle *A* travels at a constant speed of $\frac{3}{5}$ in the $-x$ direction and passes the point $x = 0$ at time $t = -2$ s.

(b) Particle *B*, which at time $t = 0$ is at rest at position $x = 0$, accelerates in the $+x$ direction asymptotically toward the speed of light as time passes.

(c) Light flash *C* passes the position $x = 0$ at time $t = +3$ s as it travels in the $-x$ direction.

R2B.12 Two firecrackers A and B are placed at $x' = 0$ and $x' = 100$ ns, respectively, on a train that is moving in the $+x$ direction relative to the ground frame. According to synchronized clocks on the train, both firecrackers explode simultaneously. Which firecracker explodes first according to synchronized clocks on the ground? Explain carefully. (*Hint:* Study figure R2.8 carefully.)

R2B.13 Figure R2.8 implies that an observer in the Home Frame concludes that clocks that have been properly synchronized in the Other Frame are *not* synchronized in the Home Frame. Would an observer in the Other Frame conclude that clocks that have been carefully synchronized in the Home Frame are not synchronized in the Other Frame? Draw a spacetime diagram from the point of view of the Other Frame to justify your response.

R2B.14 Redraw figure R2.8, assuming that the Newtonian concept of time is true. How does your redrawn diagram differ from the original, and how is this difference related to the behavior of light according to the Newtonian and relativistic models?

Modeling

R2M.1 The spacetime diagram below shows the worldline of a rocket as it leaves the earth, travels for a certain amount of time, comes to rest, and then explodes.

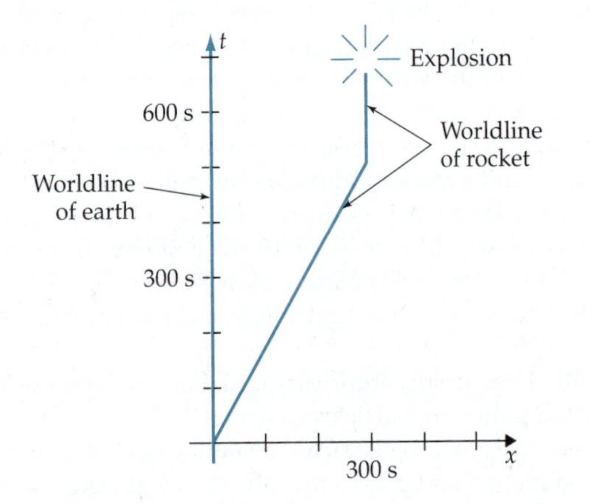

(a) The rocket leaves the earth; the rocket comes to rest in deep space; the rocket explodes. What are the coordinates of each of these three events?

(b) What is the rocket's constant speed relative to the earth before it comes to rest?

(c) A light signal from the earth reaches the rocket just as it explodes. Indicate on the diagram exactly where and when this light signal was emitted.

R2M.2 Suppose a spaceship leaves the earth at event A and travels in the $+x$ direction, accelerating at a constant rate from rest at time $t = 0$ to a final speed of $\frac{4}{5}$ at time $t = 1$ h. It remains at that speed thereafter. Just as the spaceship reaches that speed, it emits a laser signal back toward the earth (call this event B), which reaches the earth at event C. Draw a *quantitatively* accurate spacetime diagram of this situation (as observed in a frame attached to the earth, with the earth at $x = 0$) that shows the worldlines of the earth, the spaceship, the returning laser signal, and events A, B, and C.

R2M.3 An alien spaceship traveling at a constant velocity of $\frac{1}{3}$ in the $+x$ direction passes the earth (call this event A) at time $t = 0$. Just as the spaceship passes, people on the earth launch a probe, which accelerates from rest toward the spaceship at such a rate that it catches up to and passes the alien spaceship (call this event B) when both are 10 min of distance from the earth. As it passes the alien spaceship, the probe takes a photo and sends it back to the earth as an encoded radio message that travels at the speed of light. The message reaches the earth at event C. Draw a *quantitatively* accurate spacetime diagram of this situation (as observed in a frame attached to the earth, with the earth at $x = 0$) that clearly shows the worldlines of the earth, the alien spaceship, the probe, the returning radio message, and events A, B, and C. In particular, clearly indicate when the people on earth receive the photo.

R2M.4 A rocket launched from the moon travels away from it at a speed of $\frac{2}{5}$. Call the event of the rocket's launching event A. After 125 s, as measured in the reference frame of the moon, the rocket explodes: call this event B. The light from the explosion travels back to the moon: call its reception event C. Let the moon be located at $x = 0$ in its own reference frame, and let event A define $t = 0$. Assume that the rocket moves along the $+x$ direction.

(a) Draw a spacetime diagram of the situation, drawing and labeling the worldlines of the moon, the rocket, and the light flash emitted by the explosion and received on the moon.

(b) Draw and label events A, B, and C as points at the appropriate places on the diagram. Write down the t and x coordinates of these events.

R2M.5 A spaceship in deep space is approaching a space station at a constant speed of $|\vec{v}| = \frac{2}{3}$. Let the space station define the position $x = 0$ in its own reference frame. At time $t = 0$, the spaceship is 10.0 light-hours away. At that time and place (call this event A), the spaceship sends a laser pulse of light toward the station, signaling its intention to dock. The station receives the signal at its position (call this event B), and after a pause of 100 min (everyone was at lunch), emits another laser pulse signaling permission to dock (call this event C). The spaceship receives this pulse (call this event D) and immediately begins to decelerate at a constant rate. It arrives at rest at the space station (call this event E) 4.0 h after event D, according to station clocks.

(a) Carefully construct a spacetime diagram that shows the worldlines of the space station, the approaching spaceship, and the two light pulses. Also indicate the time and place (in the station's frame) of events A through E by labeling the corresponding points on the diagram. Scale your axes using the hour as the basic time and distance unit (you might subdivide each hour into units of $\frac{1}{3}$ h).

(b) In particular, exactly when and where does event D occur? Event E? Write down the coordinates of these events in the station frame, and explain how you *calculated* them (reading them from the diagram is a useful check, but is not enough).

(c) Compute the magnitude of the spaceship's average acceleration between events D and E in SR units (s^{-1}) and as a multiple of $|\vec{g}|$. Note that a shockproof watch can typically tolerate an acceleration of about $50|\vec{g}|$.

R2M.6 Suppose a spaceship is docked at a space station floating in deep space. Assume the space station defines the origin in its own frame. At $t = 0$ (call this event A) the spaceship starts accelerating in the $+x$ direction away from the space station at a constant rate (as measured in the station frame). The spaceship reaches a cruising speed of 0.5 after 8 h as measured in the station frame (call this event B). [*Hint:* You will find it helpful to construct a spacetime diagram as you go through the parts of this problem: such a diagram is required in part (e) anyway.]

(a) *Where* does event B occur? Explain your reasoning.

(b) Find the magnitude of the ship's acceleration in SR units (s^{-1}) and as a multiple of $|\vec{g}|$. Note that a shockproof watch can tolerate an acceleration of $\approx 50|\vec{g}|$.

(c) At event B, the spaceship sends a radio signal to the station, reporting it has reached cruising speed. This signal reaches the station at event C. When and where does this event occur? (*Hint:* Radio signals are electromagnetic waves that travel at the speed of light.)

(d) The technician responds to this message 0.5 h later after returning from a coffee break: call this event D. When and where does the ship receive this acknowledgment (event E)?

(e) Draw a careful spacetime diagram of this situation, showing the worldlines of the space station, the spaceship, the radio signals, and events A through E. Be sure to label all these items appropriately.

R2M.7 An air traffic control radar installation receives a radar pulse reflected from a certain jet plane 280 μs after the pulse was sent. The signal comes from a direction that is 35° north of west and 5.5° up from horizontal. If the sending of the outgoing pulse defines $t = 0$, in a frame fixed to the earth's surface and oriented in the usual way with the installation at the spatial origin of the frame, then what are the spacetime coordinates $[t, x, y, z]$ of the plane at the instant it reflects the pulse? (*Hint:* Radar pulses are electromagnetic waves, so they travel at the speed of light.)

R2M.8 Imagine you are in a spaceship prospecting for asteroids. Your radar installation receives a pulse reflected from a nice large asteroid 1.24 s after it was sent, and the returning signal comes from a direction 25° to the right and 18° up from the direction your ship is facing. Assuming that the direction your ship is facing defines the $+x$ direction, and the up direction is the $+z$ direction, and that we define $t = 0$ to be when the pulse is sent, what are the spacetime coordinates of the asteroid at the time the pulse is reflected? (*Hint:* Radar pulses are electromagnetic waves, so they travel at the speed of light.)

R2M.9 (*Seeing* is not the same as *observing*!) Suppose at time $t = 0$ you (on earth) simultaneously *receive* a message (sent via a laser transmission) emitted (event A) by the outpost on Venus (which is 180 Gm away at the time) and a message emitted (event B) by the outpost on Mars (which is 270 Gm away). Each message requires an urgent response that must be received no more than 40 min after it was sent. Which message do you respond to first?

R2M.10 (*Seeing* is not the same as *observing!*) Imagine that a bullet-train running at a *very* high speed passes two track-side signs (A and B), as shown in the aerial view below.

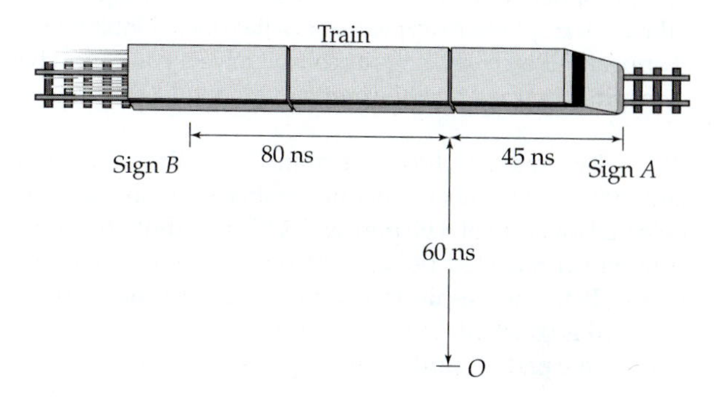

Let event A be the passing of the front end of the train by sign A, and let event B be the passing of the rear end of the train by sign B. An observer is located at the cross marked by an O in the diagram.

(a) This observer *sees* (that is, receives light with her eyes) event A to occur at time $t = 0$ and *sees* event B to occur at time $t = 25$ ns. When does she *observe* these events to occur? That is, what would a clock present at sign A read at event A, and what would a clock present at sign B read at event B if these clocks were correctly synchronized with the clock at O?

(b) In what way is the diagram misleading about the implied time relationship between events A and B? (*Hint:* Remember that the clocks at signs A and B must be synchronized with the clock at O in such a way that they would read the speed of a light signal traveling between them and O to be 1 s/s. Given this, the distance between O and A, and the time that light from event A reached O, can you infer when event A must have happened?)

R2M.11 After reading this chapter, your roommate exclaims, "Relativity cannot be right! This chapter claims that events that are simultaneous in one inertial reference frame are not simultaneous in another. Yet it is clear that two events are really simultaneous or really not simultaneous! This is not something that different observers could disagree about; or if they do, one has to be right and the other wrong!" Carefully and politely argue to your roommate that relativity *could* be right even so, and pinpoint the assumption that your roommate makes that could be debated. (*Hint:* You might be able to use the geometric analogy to good effect. Two different surveyors set up differently oriented coordinate systems on a plot of land. In one system, two rocks both lie along the x axis; in the other, they do not. Is this a problem?)

Rich-Context

R2R.1 Imagine that an advanced alien race, bent on keeping humans from escaping into the galaxy, places an opaque spherical force field around the solar system. The force field is 6 light-hours in radius, is centered on the sun, and is formed in a single instant of time as measured by synchronized clocks in an inertial frame attached to the sun. That instant corresponds to 9 p.m. on a certain night in your time zone. When does the opaque sphere appear to start blocking light from the stars from your vantage point on earth (8.33 light-minutes from the sun)? Does the opaque sphere appear all at once? If not, how long does it take for the sphere to appear, and what does it look like as it appears? Describe things as completely as you can. (This is inspired by the novel *Quarantine* by Greg Egan.)

R2R.2 Two radar pulses sent from the earth at 6:00 a.m. and 8:00 a.m. one day bounce off an alien spaceship and are detected on earth at 3:00 p.m. and 4:00 p.m. (but you aren't sure which reflected pulse corresponds to which emitted pulse). Is the spaceship moving toward earth or away? If its speed is constant (but less than c), when will it (or did it) pass by the earth? (*Hint:* Draw a spacetime diagram.)

R2R.3 (*Seeing* is not the same as *observing!*) Imagine you are sitting immediately adjacent to a set of train tracks. A certain bullet-train running at a speed of 0.5 (in SR units) on these tracks has three blinking lights, one at each end, and one in the middle. The end lights are 200 ns of distance from the middle light according to measurements in your ground frame. As the train's center passes you (a negligible distance away), you *see* all three lights blink simultaneously. Do you *observe* them to blink simultaneously? Explain. If you do *not* observe them to blink simultaneously, which light blinks first, and how much in advance of the center light does it blink? Explain carefully.

R2R.4 A train is moving due east at a large constant speed on a straight track. Suppose that Harry is riding on the train exactly halfway between its ends. Sally is sitting by the tracks only a few feet from the train. Let the event of Harry passing Sally be the origin event O in both frames. At this same instant, both Harry and Sally receive the light from lightning flashes that have struck both ends of the train, leaving scorch marks on both the train and the track. Harry concludes that since he is in the middle of the train and he received the light from the strikes at the same time, the lightning strikes must have occurred at the same time in his reference frame. Is he right? If not, which strike (the one at the front of the train or the one at the rear) really happened first? Can Sally conclude from her seeing the flashes at the same time that the strikes happened at the same time in the ground frame? Why or why not? If not, which strike happened first in her frame? (This problem is adapted from one of Einstein's own illustrations of the implications of the frame-independent speed of light.)

Advanced

R2A.1 A meter stick moves at a speed of 0.5 (in SR units) along a line parallel to its length that passes within 1 m of a camera. The camera shutter opens for an incredibly brief instant just as the meter stick's center passes closest to the camera. Explain why the marks on the meter stick do *not* look equally spaced in the resulting picture, and describe what they look like. (Ignore length contraction.)

ANSWERS TO EXERCISES

R2X.1 The distance between the earth and the moon in seconds is

$$384{,}000\,\cancel{\text{km}}\left(\frac{1000\,\cancel{\text{m}}}{1\,\cancel{\text{km}}}\right)\left(\frac{1\text{ s}}{3.0\times10^8\,\cancel{\text{m}}}\right)=1.28\text{ s} \qquad \text{(R2.5)}$$

So the clock on the moon should read 1.28 s after noon if it is synchronized with the earth clock.

R2.6 With the new worldline (shown in color), figure R2.4 becomes figure R2.14. Note that the particle worldline has a slope of −5 on the diagram, since it moves 1 s of distance in the $-x$ direction per 5 s of time.

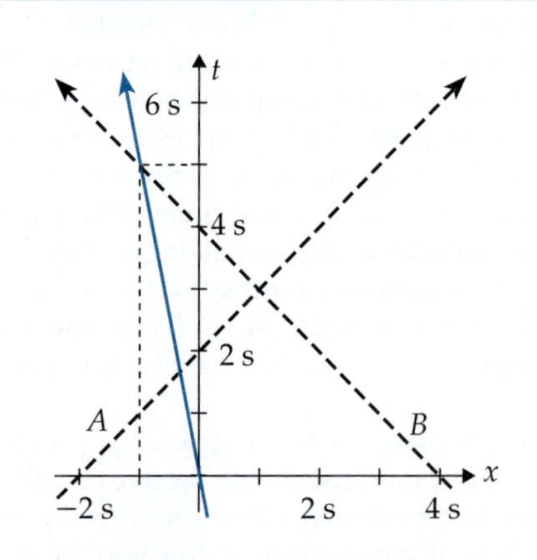

Figure R2.14
Figure R2.4 with the new worldline.

R2X.3 Consider events C and D with x coordinates x_C and x_D, respectively, in the Home Frame and x'_C and x'_D, respectively, in the Other Frame. According to the Galilean transformation equations (equations CA), we have

$$\begin{aligned} x'_D - x'_C &= (x_D - \beta t_D) - (x_C - \beta t_C) \\ &= (x_D - x_C) - \beta(t_D - t_C) \end{aligned} \qquad \text{(R2.6)}$$

If $\Delta x' = x'_D - x'_C = 0$, then

$$0 = (x_D - x_C) - \beta(t_D - t_C) \;\Rightarrow\; x_D - x_C = \beta(t_D - t_C) \qquad \text{(R2.7)}$$

So if $\Delta t' = \Delta t = t_D - t_C \neq 0$, then $\Delta x = x_D - x_C \neq 0$ as well.

R3 The Spacetime Interval

Chapter Overview

Introduction

Chapter R2 provided conceptual foundations for our discussion of time in the theory of relativity. In this chapter, we begin building quantitative links between various methods of measuring time by introducing the *metric equation*, an equation that is essentially the Pythagorean theorem for spacetime.

Section R3.1: The Three Kinds of Time

The analogy to a tape measure in spacetime is a single clock that moves in such a way as to be present at both events. Any clock that is present at both of two events measures a frame-independent **proper time** $\Delta\tau$ (short for *proprietary time*) along its particular worldline. If the worldline happens to be the unique constant-velocity worldline between the events (so that the clock is inertial), the clock also measures the frame-independent **spacetime interval** Δs between the events. *Proper time* along a worldline is analogous to the pathlength along a path on a plane, and the *spacetime interval* between events to the *distance* between points on a plane. Table R3.1 in the chapter summarizes the definitions of the three kinds of time and table R3.2 summarizes the geometric analogy between spacetime and plane geometry.

Note that the spacetime interval Δs between two events is *always* also a coordinate time Δt between those events (as measured in the particular inertial frame where the events happen at the same place) as well as a particular proper time $\Delta\tau$ (measured between the events along a worldline that happens to be straight).

If the analogy is true, it should not be any more surprising that Δt, $\Delta\tau$, and Δs between two given events have different values than that Δy, the pathlength, and the distance $|\Delta\vec{d}|$ have different values between two given points. Experimentally, these times *are* in fact different.

Section R3.2: The Metric Equation

Given the coordinate-dependent coordinate differences Δx and Δy between two points on a plane, we can use the Pythagorean theorem to compute the coordinate-independent *distance* $|\Delta\vec{d}| = (\Delta x^2 + \Delta y^2)^{1/2}$ between those points. Our goal in this section is to find the analogous equation that links the frame-dependent coordinate differences Δt and $|\Delta\vec{d}|$ between two events in a given inertial reference frame with the frame-independent *spacetime interval* between those events.

A **light clock** measures time by counting round trips of a flash of light bouncing between two mirrors a given distance L apart. In the clock's rest frame, a round trip takes a time $2L$. Now imagine a light clock that moves at a constant velocity in the Home Frame with its beam path perpendicular to its velocity. Let event A be the event of the flash bouncing off one of the mirrors, and let event B be the event when the flash hits the same mirror again. The light clock itself is present at both events and is inertial, so the time $2L$ it measures between the events is the spacetime interval Δs between these events. In the Home Frame, however, the light flash follows a zigzag path of length $2\left[L^2 + \left(\frac{1}{2}|\Delta\vec{d}|\right)^2\right]^{1/2}$, where $|\Delta\vec{d}|$ is the distance between the events in that frame. Since the speed of light is 1 in the Home Frame, this is equal to the coordinate time Δt between the events in that frame, so $\Delta t = 2\left[L^2 + \left(\frac{1}{2}|\Delta\vec{d}\,|\right)^2\right]^{1/2} > \Delta s$. If we rearrange terms in this equation, we arrive at the **metric equation** for spacetime

$$\Delta s^2 = \Delta t^2 - |\Delta\vec{d}\,|^2 = \Delta t^2 - \Delta x^2 - \Delta y^2 - \Delta z^2 \qquad \text{(R3.5)}$$

- **Purpose:** This equation specifies the frame-independent spacetime interval Δs between two events, given their coordinate separations Δt, Δx, Δy, and Δz in any given inertial frame.
- **Limitations:** This equation applies only in an inertial reference frame.
- **Notes:** This equation is to spacetime what the Pythagorean theorem is to space.

We have proved this equation only for cases where $\Delta t > |\Delta\vec{d}\,|$, but in fact it yields a frame-independent value of Δs^2 for *all* pairs of events, as we will see in chapter R7.

Section R3.3: About Perpendicular Displacements

The proof of the metric equation *assumes* that the distance between the light clock mirrors is L in both the clock frame and the Home Frame. An argument by contradiction presented in this section shows that the principle of relativity *requires* that the magnitude of any displacement measured in two different inertial reference frames have the same value in both frames as long as the displacement is perpendicular to the direction of the frames' relative motion.

Section R3.4: Evidence Supporting the Metric Equation

Muons are subatomic particles that decay with a characteristic half-life of 1.52 μs as measured in their rest frame, as if they had an internal clock telling them when to decay. We can measure the time that the internal clocks of a batch of muons measure as they move between two detectors by observing the fraction of muons that decay. Since these muons are present at both detection events, their clocks measure the spacetime interval Δs between the events. Observers in the detector frame can measure the coordinate time Δt and distance $|\Delta\vec{d}\,|$ between the detection events. This section describes a muon experiment of this type that unambiguously supports the predictions of the metric equation.

Section R3.5: Spacetime Is *Not* Euclidean

The most important difference between the metric equation and the Pythagorean equation is that the former has minus signs between terms, while the latter has only plus signs. This implies, for example, that $\Delta s = 0$ if $\Delta t = |\Delta\vec{d}\,|$ for a given pair of events, even though the events might look like two distinct points on a spacetime diagram. This is analogous to two points at 90° north latitude that look distinct on a flat map of the earth, but are in fact a single point (the north pole) on the real earth. Just as a flat map cannot accurately represent the non-Euclidian geometry of the earth, so a flat spacetime diagram cannot accurately represent the non-Euclidian geometry of spacetime: a pair of events whose separation on a spacetime diagram may look larger than that of another pair may in fact have a *smaller* spacetime interval between them.

Section R3.6: More About the Geometric Analogy

In two-dimensional plane geometry, the set of all points equidistant from a given point is a *circle*. But the minus signs in the metric equation mean that the set of all events that are the same spacetime interval from a given event is a *hyperbola*.

Section R3.7: Some Examples

This section presents examples that illustrate applications of the metric equation.

R3.1 The Three Kinds of Time

In chapter R2, we saw that we can measure the separation between two points on a plane in three different ways: (1) We can measure the north–south *coordinate* separation Δy between those points in a suitable Cartesian coordinate system; (2) we can measure the *pathlength* $\Delta\ell$ between those points along some defined path connecting them; or (3) we can measure the *distance* $|\Delta\vec{d}|$ between those points along the unique straight-line path between them. That these completely different methods of measuring the separation between the points yield numerically different results should not be, and is not, surprising: it is part of our common experience.

In this chapter, we will see that, analogously, we can measure the time separation between two events in spacetime in three fundamentally different ways. We have already seen that the *coordinate time* Δt between two events in spacetime, as registered by synchronized clocks in a given inertial reference frame, is analogous to the north–south coordinate separation Δy between two points. In particular, Δy and Δt are both *frame-dependent* quantities whose numerical values depend on our free choice of coordinate-axis orientation in the case of Δy or inertial reference frame in the case of Δt. On the plane, in contrast, the pathlength and distance are coordinate-*independent* quantities because we can measure them *directly* (using a tape measure) without referring to any particular coordinate system. How might we measure the time separation between two events in spacetime in an analogous way?

A tape measure stretched between two points marks off the distance between those points and presents a scale that can be laid right next to the two points for easy and unambiguous reading. Every observer, no matter what coordinate system they prefer, will look at the tape measure and see the same result as every other observer. The analogy for time is a clock that travels between the two events in such a way so as to be *physically present* at each event. Like the tape measure, this clock marks off the time between those events, and since the clock's face is right there at each event, *everyone* looking at that clock will agree as to the value it displays as each event happens, and thus will agree as to what this particular clock has registered as the time interval between those events: this quantity is therefore **frame-independent**. We call a time interval measured in this manner a *proper time:*

> Any time between two events measured by a clock present at both events is a **proper time** $\Delta\tau$ between those events. The numerical value of a proper time measured by a given clock between two given events is a *frame-independent* quantity.

(Note that the adjective "proper" can be misleading here. In English, this word has fairly recently come almost exclusively to mean "appropriate," or "correct in manners." But the word used to mean "proprietary," and that is the meaning intended here: proper time is the time between the events measured specifically by the *particular* clock in question. *Path time* might be a more appropriate phrase in current English.)

Proper time depends on the clock worldline

There is, however, one thing that the proper time between two events *might* well depend on other than the events themselves. It might depend on the *worldline* that the clock follows in traveling from one event to the other, just as a pathlength measured by a tape measure depends on the path along which it is laid (see figure R3.1). We will see in chapter R4 that the worldline-dependence of proper time is a straightforward consequence of the principle of relativity and is indeed an experimental *fact*. For now, it is enough to see that this path dependence is a *possibility* suggested by the geometric analogy.

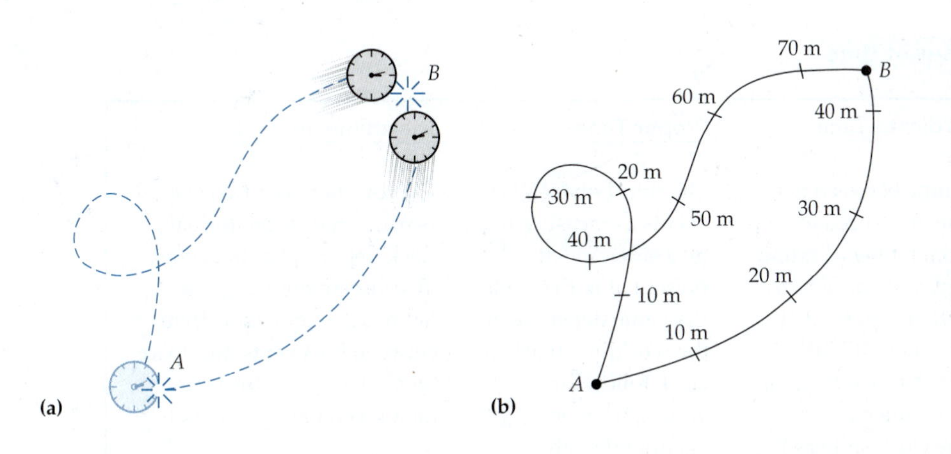

Figure R3.1
(a) A clock traveling along a given path in such a way that it is present at both events measures the proper time between them. (b) The value of this proper time may depend on the clock's worldline through spacetime, just as the analogous pathlength between two points depends on the path along which it is measured.

What is the spacetime analogue of distance?

Now, when measuring distances on a plane, we distinguish the *pathlength* between two points measured along a certain path and the *distance* between those points as follows. "Distance" is the pathlength measured along a special path: the unique straight-line path between the two points. Because the straight-line path is unique, the distance between two points along a straight line is a unique number reflecting something definite about the separation of those points in space. With this in mind, consider a clock present at two events in spacetime that is *inertial* (that is, an attached first-law detector registers no violation of Newton's first law). Such a clock follows a *unique worldline* through spacetime between the events: observed in any inertial frame, the clock travels between the events in a straight line, and since there is only one value of a constant velocity that will be just right to get the clock from one event to the other, the clock's velocity along that line is also uniquely determined.

Definition of the *spacetime interval* between events

The **spacetime interval** Δs between two events is the proper time measured by an *inertial* clock present at both events. This quantity is a unique, frame-independent number that depends on the separation of the events in space and time and nothing else.

Spacetime interval is a proper time and also a coordinate time

It is important to note that the definitions of *coordinate time*, *proper time*, and the *spacetime interval* between two events overlap in certain special cases. The spacetime interval between two events is a special case of a proper time between two events, just as the distance between two points is a special case of the pathlength between two points. An inertial clock present at both events also measures the *coordinate time* between those events in the clock's own reference frame, since the time interval measured between two events by a clock or clocks at rest in any inertial reference frame is a coordinate time by definition. So the spacetime interval between two events is a special case of a proper time *and* a special case of a coordinate time (see figure R3.2).

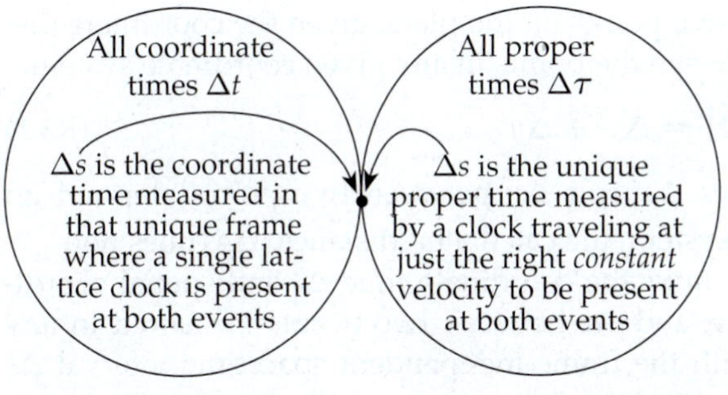

All coordinate times Δt

Δs is the coordinate time measured in that unique frame where a single lattice clock is present at both events

All proper times $\Delta \tau$

Δs is the unique proper time measured by a clock traveling at just the right *constant* velocity to be present at both events

Figure R3.2
Let points in the left circle represent the set of all possible coordinate times Δt that observers in inertial frames moving at various different relative velocities might measure between two given events. Let points in the right circle represent the set of all possible proper times $\Delta \tau$ measured between the same events by clocks present at both events but moving between them along various different worldlines. The single point in common between these sets is the spacetime interval Δs between those two events.

Table R3.1 Three kinds of time

	Coordinate Time	Proper Time	Spacetime Interval
Definition	The time between two events measured in an inertial reference frame by a *pair of synchronized* clocks, one present at each event. (If both events happen to occur at the same place, a single clock suffices.)†	The time between two events as measured by a single clock present at both events. (Its value depends on the worldline that the clock follows in getting from one event to the other.)	The time between two events as measured by an inertial clock present at both events. (Because an inertial clock follows a unique worldline between the events, the spacetime interval's value is unique for a given pair of events.)
Conventional symbol	Δt	$\Delta\tau$	Δs
Frame-independent?	No	Yes	Yes
Geometric analogy	Spatial coordinate differences	Pathlength	Distance

†*Note:* Alternatively, the coordinate time difference between two events might be inferred from measurements of the spacetime coordinates of these events using the radar method.

Table R3.2 The geometric analogy

Plane Geometry		Spacetime Geometry
Map	↔	Spacetime diagram
Points	↔	Events
Paths or curves	↔	Worldlines
Coordinate systems	↔	Inertial reference frames
Relative rotation of coordinate systems	↔	Relative velocity of inertial reference frames
Differences between spatial coordinate values	↔	Differences between spacetime coordinate values
Pathlength along a path	↔	Proper time along a worldline
Distance between two points	↔	Spacetime interval between two events

Tables R3.1 and R3.2 summarize the three kinds of time and the now-complete geometric analogy, respectively.

R3.2 The Metric Equation

Even though we can measure the distance between two points on a plane using a tape measure, we can also use the Pythagorean theorem to *calculate* the distance $|\Delta\vec{d}\,|$ between two points on the plane given the coordinate displacements Δx and Δy between the points in any given coordinate system:

$$|\Delta\vec{d}\,|^2 = \Delta x^2 + \Delta y^2 \quad \text{(R3.1)}$$

Note that while the values Δx and Δy between two points depend on one's choice of coordinate system, the calculated distance $|\Delta\vec{d}\,|$ does not.

An analogous formula links the coordinate time Δt and spatial coordinate displacements Δx, Δy, and Δz between two events measured in any inertial reference frame with the frame-independent spacetime interval Δs

between those events. This equation, which we call the *metric equation*, enables us to escape the "relativity" of inertial reference frames and quantify the separation of the events in *absolute* (frame-independent) terms. Our goal in this section is to *derive* this equation from the principle of relativity.

The following derivation is the very heart of the special theory of relativity. The metric equation is the key to understanding all the unusual and interesting consequences of the theory of relativity. You should make a special effort to understand this argument thoroughly.

To make the argument easier, consider a special kind of clock we call a **light clock** (figure R3.3 shows an idealized version). A light clock consists of two mirrors a fixed distance L apart and a flash of light that bounces back and forth between the mirrors. As the light flash bounces off the bottom mirror, a detector in that mirror sends a signal to an electronic counter. The clock dial thus essentially registers the number of round trips that the light flash has completed. Since the speed of light is *defined* to be 1 second of distance per second of time in any inertial frame, we should calibrate the clock's face to register a time interval of $2L$ (where L is expressed in seconds) for each "tick" of the clock (that is, each time the light flash bounces off the bottom mirror): the clock will then read the correct time as long as it is inertial.

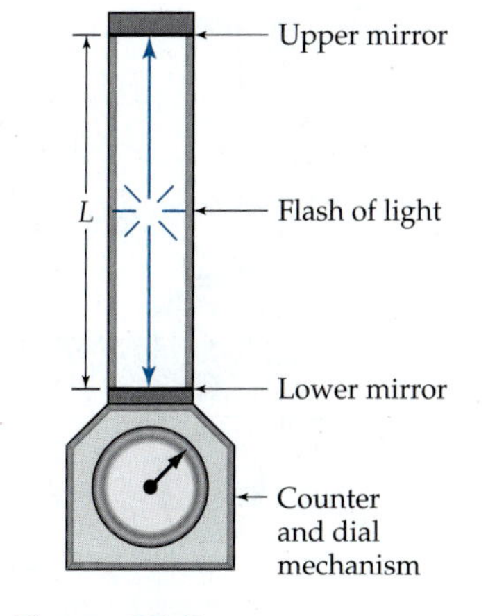

Figure R3.3
Schematic diagram of a light clock. Each "tick" of the light clock represents the passage of a time interval equal to $2L$ (in SR units).

Derivation of the metric equation

Now consider an arbitrary pair of events A and B. Let the coordinate time interval and spatial distance between these events (as measured in the Home Frame) be Δt and $|\Delta\vec{d}| = (\Delta x^2 + \Delta y^2 + \Delta z^2)^{1/2}$, respectively. Suppose we have a light clock moving between these events (with its beam path oriented perpendicular to its direction of motion) at just the right constant velocity to be present at both events. To simplify our argument, let us also suppose that the length L between the light clock mirrors has just the right value so that events A and B happen to coincide with successive ticks of the light clock (in principle, we could always adjust L to make this true for the two given events). The situation is illustrated in figure R3.4 below (the figure is repeated on the next page for your convenience).

In the inertial frame of the light clock, both events occur at the clock face, and the clock's light flash completes exactly one round trip. The time interval recorded by this clock between events A and B is thus exactly $2L$. Since this inertial clock is present at both events, it registers the spacetime interval between these events, so $\Delta s = 2L$.

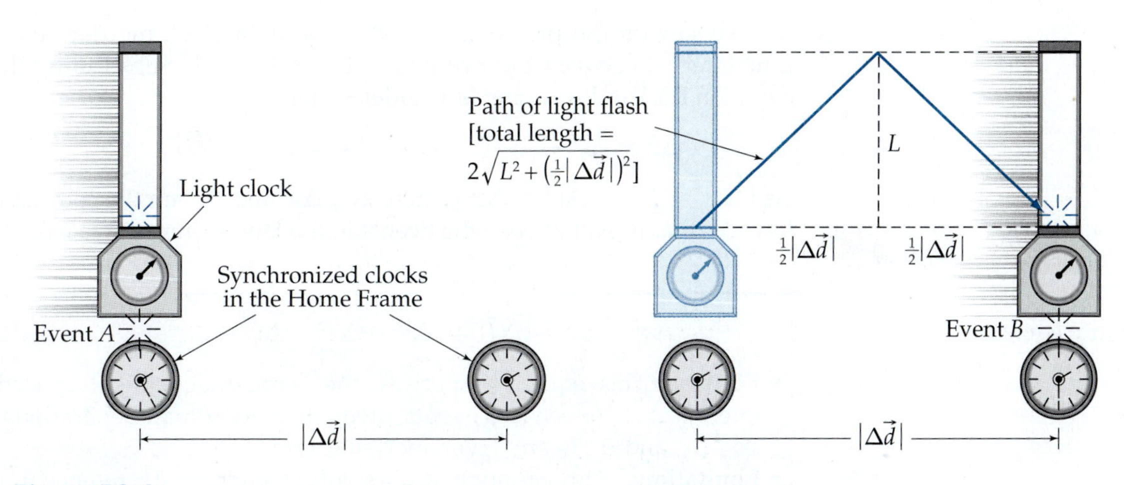

Figure R3.4
As the light clock moves from event A to event B in the Home Frame, its internal light flash will be observed to follow the zigzag path shown.

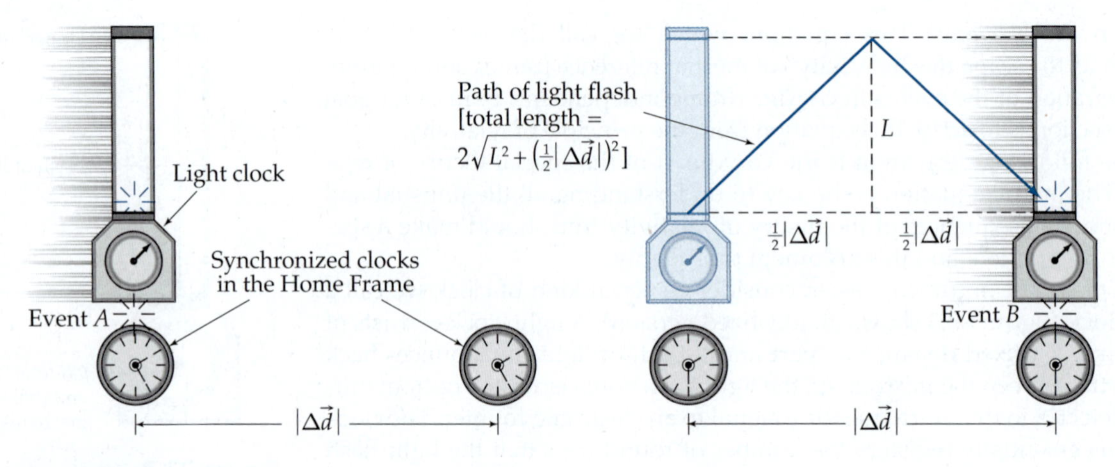

Figure R3.4
(Repeated here for the reader's convenience.) As the light clock moves from event A to event B in the Home Frame, its internal light flash will be observed to follow the zigzag path shown.

On the other hand, in the Home Frame, each event's time is registered by the clock nearest the event. Since the events occur at different places, we determine the coordinate time interval Δt between the events by comparing readings from a *pair* of clocks. In this frame, the light clock is observed to move a distance $|\Delta\vec{d}|$ in the time interval Δt. An observer in the Home Frame observes the light flash follow the zigzag path shown in figure R3.4. As you can see from the figure, the total distance that the light flash travels in the Home Frame is (according to the Pythagorean theorem)

$$2\sqrt{L^2+(\tfrac{1}{2}|\Delta\vec{d}|)^2} = \sqrt{4L^2+|\Delta\vec{d}|^2} = \sqrt{(2L)^2+|\Delta\vec{d}|^2} \quad \text{(R3.2)}$$

Since the synchronized clocks in the Home Frame must (by definition of *synchronization*) measure the speed of light to be 1, the coordinate time interval Δt registered on the pair of synchronized clocks in the Home Frame must be equal to the distance that the light flash traveled between the events:

$$\Delta t = \sqrt{(2L)^2+|\Delta\vec{d}|^2} \quad \text{(R3.3)}$$

But we saw on the previous page that the light clock registers the spacetime interval between the two events to be $\Delta s = 2L$. Substituting this into equation R3.3 and squaring both sides yields

$$\Delta t^2 = \Delta s^2 + |\Delta\vec{d}|^2 \quad \text{or} \quad \Delta s^2 = \Delta t^2 - |\Delta\vec{d}|^2 \quad \text{(R3.4)}$$

As $|\Delta\vec{d}|^2 = \Delta x^2 + \Delta y^2 + \Delta z^2$ (where Δx, Δy, and Δz are the coordinate differences measured between the events in the Home Frame), we have finally

The metric equation

$$\Delta s^2 = \Delta t^2 - |\Delta\vec{d}|^2 = \Delta t^2 - \Delta x^2 - \Delta y^2 - \Delta z^2 \quad \text{(R3.5)}$$

- **Purpose:** This equation specifies the frame-independent spacetime interval Δs between two events, given their coordinate separations Δt, Δx, Δy, and Δz in any given inertial frame.
- **Limitations:** This equation applies only in an inertial reference frame.

This extremely important equation links the frame-*independent* spacetime interval Δs between any two events to the frame-*dependent* coordinate separations Δt, Δx, Δy, and Δz measured between those events *in any arbitrary inertial reference frame!* Note that we have not sacrificed anything by using a light clock in this argument: since the speed of light is defined to be 1 in any inertial frame, *any* decent clock that we construct must agree with what the light clock says. The only real limitation to our argument is that Δt must be greater than $|\Delta \vec{d}\,|$ for the two events in question, so that it is possible for a light flash to travel between the events. (Note that if $\Delta t < |\Delta \vec{d}\,|$, then equation R3.5 yields an imaginary value for Δs, an absurd result indicating that the conditions of the proof have been violated.)

Since the spacetime interval Δs between two events in spacetime is analogous to the distance $|\Delta \vec{d}\,|^2$ between two points on a plane, the formula $\Delta s^2 = \Delta t^2 - \Delta x^2 - \Delta y^2 - \Delta z^2$ is directly analogous to the Pythagorean theorem $|\Delta \vec{d}\,|^2 = \Delta x^2 + \Delta y^2$. Note that the Pythagorean theorem also relates a coordinate-independent quantity (the distance $|\Delta \vec{d}\,|$ between two points) with quantities whose values depend on the choice of coordinate system (the coordinate differences Δx and Δy). Indeed, the formula for the spacetime interval would be just like a four-dimensional version of the Pythagorean theorem if it were not for the minus signs that appear. We will see that these minus signs have a variety of interesting and unusual consequences.

We call equation R3.5 the **metric equation**. It is the link between our human-constructed reference frames and the absolute physical reality of the separation between two events in space and time. It is difficult to overemphasize this equation's importance: virtually all the rest of our study of the theory of relativity will be devoted to exploring its implications!

R3.3 About Perpendicular Displacements

What if L is not the same in both frames?

The previous argument *assumes* that the vertical separation L between the light clock's mirrors is the same in both the light clock frame (where we used it to compute the spacetime interval) and the Home Frame (where we used it to compute the coordinate time). But how do we *know* this is true? Since coordinate differences between events are generally frame-dependent, what gives us the right to assume that mirror's separation has the same value in both frames? This is not a trivial issue, because in chapter R6 we will see that observers in two different frames will disagree about the length of displacements measured parallel to the line of the frames' relative motion.

In this section, however, I will argue that the principle of relativity directly implies that if we have two inertial reference frames in relative motion along a given line, any displacement measured *perpendicular* to that direction of motion *must* have the same value in both reference frames.

The proof presented here will be a proof by contradiction. This kind of argument is a bit tricky to follow, so pay attention. Here's how it works. We will assume that there *is* a contraction (or expansion) effect that applies to perpendicular lengths and then show that the existence of such an effect contradicts the principle of relativity. Turned around, this argument then implies that if the principle of relativity is true, no such effect can exist.

Proof that distances measured perpendicular to the line of relative motion of two frames is the *same* in both frames

Consider two inertial reference frames (a Home Frame and an Other Frame) in standard orientation, so that the line of relative motion is along the frames' common x and x' axes. In each frame, we set up a measuring stick along the y or y' direction with spray-paint nozzles set 1.00 meter apart

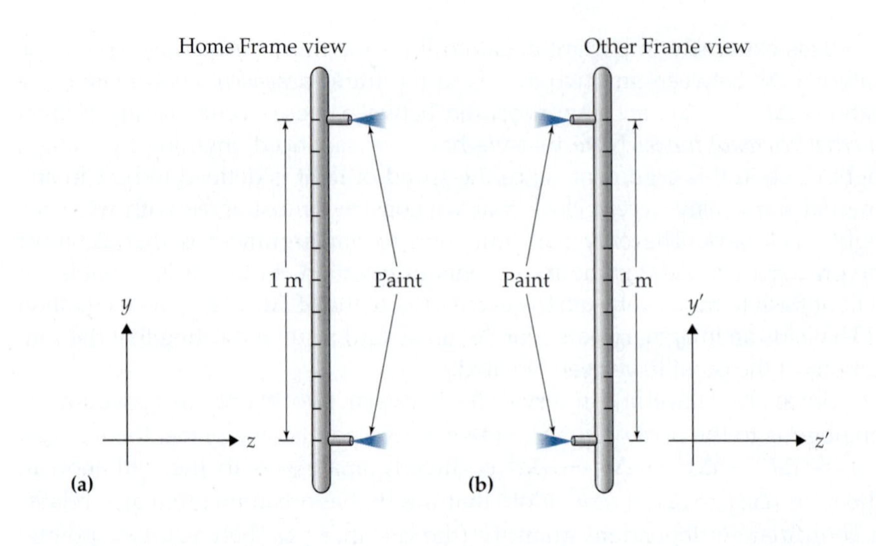

Figure R3.5
(a) The spray paint nozzles on the Home Frame measuring stick are 1.00 meter apart in the *y* direction as measured in that frame. They point directly at their counterparts in the Other Frame so that stripes are painted on the Other Frame's measuring stick as it moves by. The *x* axis points directly into the plane of the paper here.
(b) Similarly, the paint nozzles on the Other Frame stick are 1.00 meter apart in that frame and are pointed to paint stripes on the Home Frame measuring stick as it moves by.

(as shown in figure R3.5). Note that the common x and x' axes (which lie along the line of the frames' relative motion) are perpendicular to the plane of the diagram: as the frames move relative to each other, one measuring stick moves into the paper and the other out. The paint nozzles in each frame are pointed toward the other frame's measuring stick, so as the two measuring sticks pass, they will spray-paint stripes on each other.

Now suppose that some kind of frame-dependent contraction occurs so that an observer in the Home Frame observes the Other Frame's measuring stick (which is moving relative to the Home Frame) to be vertically contracted, meaning that an observer in the Home Frame measures the spray-paint nozzles on that stick to be *less* than 1.00 meter apart. This in turn means that the stripes painted by these nozzles will be less than 1.00 meter apart in the Home Frame: they will be painted *inside* the nozzles on the Home Frame's measuring stick. This also means that the nozzles on the measuring stick at rest in the Home Frame must paint stripes on the stick in the Other Frame which are *outside* the latter stick's nozzles (see figure R3.6a).

Now, the principle of relativity requires that the laws of physics be exactly the same in any inertial reference frame. This specifically means that if you perform exactly the same experiment in two inertial reference frames, you should get exactly the same result. There should be *no* way of experimentally distinguishing the two frames. How does this principle apply here?

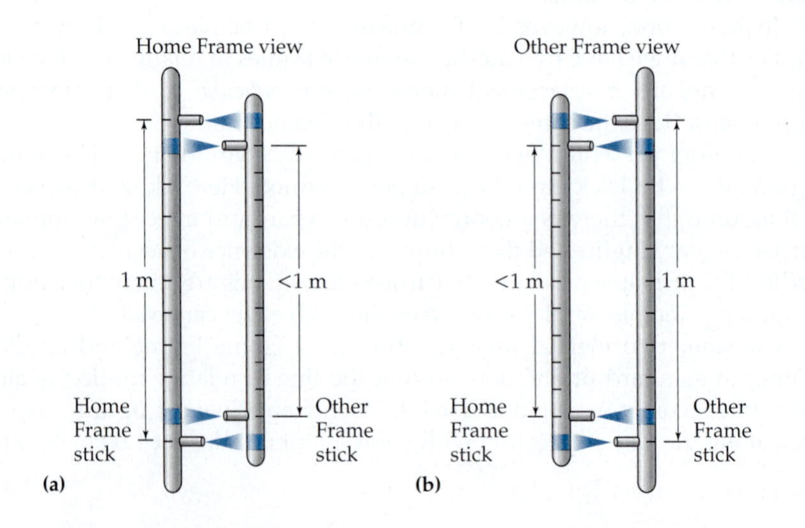

Figure R3.6
If a contraction effect exists, then the principle of relativity implies that observers in each frame must observe the other's stick to be contracted. (a) So an observer in the Home Frame observes the Home Frame stick to paint stripes *outside* the Other Frame's nozzles as the latter moves by (into the plane of the drawing here).
(b) Similarly, an observer in the Other Frame observes her or his stick to paint stripes *outside* the nozzles on the Home Frame's stick as the latter moves by (*out* of the plane of the paper here).

In the Other Frame, it is the Home Frame stick that is moving. Therefore, the principle of relativity requires that if a frame-dependent contraction effect exists, an observer in the Other Frame *must* measure the Home Frame stick to be contracted, just as the Home Frame observer measured the Other Frame stick to be contracted. This in turn means that the stripes painted by the Other Frame stick will be *outside* the Home Frame stick's nozzles, and the stripes painted by the Home Frame stick will be *inside* the Other Frame stick's nozzles, as shown in figure R3.6b.

Now, figure R3.6a and R3.6b describe a logical contradiction. In figure R3.6a, stripes get painted on the Home Frame stick *inside* its nozzles. In figure R3.6b, stripes get painted on the Home Frame stick *outside* its nozzles. These cannot be simultaneously true! The paint marks on the Home Frame stick are permanent and unambiguously visible to all observers in *every* reference frame. They cannot be "inside" the nozzles according to some observers and "outside" to others. So *either* figure R3.6a or figure R3.6b can be true, but *not both*. But the principle of relativity *requires* that both be true!

How can we resolve this conundrum? The only way is to reject the hypothesis that got us into this trouble in the first place—that is, the hypothesis that distances measured perpendicular to the line of relative motion of the frames have different values in the two frames. If we assume that there is no contraction (or expansion) effect operating between the frames, then there is no problem with the principle of relativity. As shown in figure R3.7, both sticks will paint stripes across each other's nozzles. The situation is exactly the same in both frames, and the contradiction disappears.

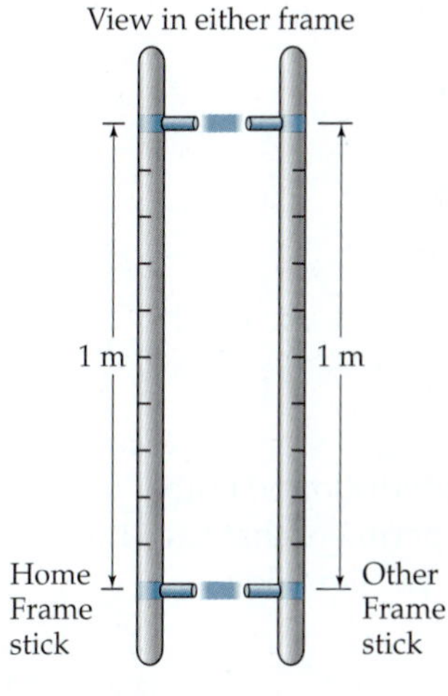

Figure R3.7
If we assume that no contraction effect exists, then the contradiction in figure R3.6 disappears.

This argument forces us to conclude that

Any displacement measured perpendicular to the line of relative motion of two inertial frames must have the same value in both frames.

This means that the distance L between the mirrors used in the derivation of the metric equation does in fact have the same value in the light clock frame as it does in the Home Frame, so our derivation should be correct.

Implication: L *is* the same in both frames in the proof of the metric equation

Exercise R3X.1

We will see in chapter R6 that a measuring stick is observed in a given reference frame to be contracted *parallel* to its direction of motion in that frame. Explain why the argument above *cannot* exclude contractions or expansions parallel to the line of motion. (*Hint:* What kind of stripes would the sticks paint on each other if they moved relative to each other in a direction parallel to their lengths?)

R3.4 Evidence Supporting the Metric Equation

Careful and compelling as the derivation of the metric equation in section R3.2 may be, we as physicists should not simply accept such an equation without some experimental confirmation. One of the classic experiments testing the validity of the metric equation involves **muons**. A muon is an elementary particle that is a more massive version of the electron (see appendix CA). Muons are continually generated in the upper atmosphere (at heights of approximately 60 km) by the interaction of cosmic rays with atmospheric gas molecules. Some of these muons stream downward toward the earth with speeds in excess of 0.99 (that is, 99% of c).

Now, muons are unstable, decaying after a short time into lighter particles. Muons at rest in a laboratory have a half-life of about 1.52 μs, which means that if you have N muons at a certain time, after 1.52 μs you will have $\frac{1}{2}N$ left; after another 1.52 μs, you will have $\frac{1}{4}N$ left, and so on. A batch of muons moving together can thus serve as a clock: to determine how much time has passed in the muon frame, all we need to do is measure the number of remaining muons in the bunch. Imagine that each muon contains a built-in clock, and that each time the clock "ticks," the muon has a certain small probability of decaying, a probability such that after 1.52 μs of ticks have passed, one-half of the muons in a bunch will have decayed.

Description of muon experiment that tests the metric equation

One can build a muon detector that counts the number of muons reaching it from a particular direction and traveling at a particular speed. Suppose we set up two detectors that register only muons traveling vertically downward at a speed of roughly 0.994 as measured in the earth's reference frame. We place one at the top of a mountain, another at the foot of the mountain 1907 meter (≈6.36 μs of distance) lower, and in each case we count the number of muons the detector sees per unit time.

Let's follow a single muon that happens to go through both detectors. Let event A be the event of this muon passing through the upper detector, and event B be this muon passing through the bottom detector. The distance $|\Delta\vec{d}|$ between these events in the earth's frame of reference is 6.36 μs in SR units. The coordinate time interval Δt between these events measured in the earth's frame is simply the time required for a muon traveling at a speed of 0.994 to traverse this distance: $\Delta t = |\Delta\vec{d}|/|\vec{v}| = 6.36\ \mu\text{s}/0.994 = 6.40\ \mu\text{s}$.

Since our muon is present at each of these events by definition and moves between them at a constant velocity of 0.994 downward, the clock inside this muon measures the spacetime interval Δs between the events. If the Newtonian conception of time were true, the muon clock and earth clocks would agree: $\Delta s = \Delta t = 6.40\ \mu\text{s}$. This is $6.40\ \mu\text{s}/1.52\ \mu\text{s} \approx 4.21$ muon half-lives, so most of the muon's co-moving siblings that make it through the top detector would decay before reaching the bottom detector. Specifically, if N muons go through the top detector, then we expect to see N times $\left(\frac{1}{2}\right)^{4.21} \approx N/18.5$ make it to the bottom detector (if the Newtonian assumption about time is true).

But if the *metric* equation is true, the spacetime interval between the two events is $\Delta s = \left(\Delta t^2 - |\Delta\vec{d}|^2\right)^{1/2} \approx [(6.40\ \mu\text{s})^2 - (6.36\ \mu\text{s})^2]^{1/2} \approx 0.714\ \mu\text{s}$. This time, which is the time that our muon and its siblings measure between the events, is only $0.714\ \mu\text{s}/1.52\ \mu\text{s} \approx 0.47$ of a muon half-life, so most of the muons' internal clocks will *not* signal that it is time to decay before they reach the bottom detector. Specifically, if N muons pass through the upper detector, you can show that about $N/1.38$ should make it to the bottom detector.

Exercise R3X.2

Verify that $\left(\frac{1}{2}\right)^{4.21} = 1/18.5$ and $\left(\frac{1}{2}\right)^{0.47} = 1/1.38$.

So the Newtonian conception of time predicts that the ratio of the number of muons passing through the upper detector to the number passing through the lower detector should be 18.5, while the metric equation predicts that it should be 1.38. This is a substantial difference that can be easily measured.

Actual results support the metric equation

This experiment was done in the early 1960s by D. H. Frisch and J. B. Smith (*Am. J. Phys.*, vol. 31, p. 342, 1963). They reported observing the ratio to be 1.38 (within experimental uncertainties), thus confirming the metric equation (and clearly excluding the Newtonian conception of time).

R3.5 Spacetime Is *Not* Euclidean

We have found the analogy between ordinary Euclidean plane geometry and spacetime geometry to be illuminating, and this basic analogy will remain quite helpful. Nevertheless, it is important at this point to describe some of the important *differences* between Euclidean geometry and spacetime geometry due to the minus signs in the metric equation $\Delta s^2 = \Delta t^2 - \Delta x^2 - \Delta y^2 - \Delta z^2$ that do not appear in the corresponding Pythagorean theorem $|\Delta \vec{d}\,|^2 = \Delta x^2 + \Delta y^2$.

The distance between two points on a map of a flat space is proportional to the actual distance in space

One important difference concerns the representation of distances on a map and spacetime intervals on a spacetime diagram. If one prepares a scale drawing (for example, a map) of a town, the distance between points on the map is *proportional* to the actual distance between those points in space. That is, distances on the drawing directly correspond to distances in the physical reality being represented. In figure R3.8a, for example, to determine the distance between City Hall and the Statue of the Unknown Physicist, one need merely measure the distance (in inches) between the two sites on the map shown there and multiply by the conversion factor (1000 m = 1 in.). It doesn't matter how the line between the two sites is oriented or where the sites are located on the drawing: the distance in the physical space being represented by the map is always proportional to the distance measured on that map.

The distance between events on a spacetime diagram is *not* proportional to Δs

However, it is *not* true that the displacement between two points on a spacetime diagram is proportional to the spacetime interval between the corresponding events. In fact, the spacetime interval between two events separated in space can even be zero (see figure R3.8b)!

A spacetime diagram thus may accurately display the spacetime *coordinates* of various events, but the distances between the points representing those events on the diagram are *not* proportional to the actual spacetime intervals between those events in spacetime.

Analogy: distances on a flat map of curved earth

This is very strange, and it may seem particularly strange that two events (such as A and C in figure R3.8b) can occur at different places and times and yet have zero spacetime interval between them. Nonetheless, there's a useful analogy with something you may have seen before. Consider a map of the world where the lines of longitude and latitude are drawn as equally spaced straight lines (see figure R3.9). Have you ever noticed how the continents' shapes and sizes appear very warped near the north and south poles on such maps? For example, look at Antarctica. It looks huge and seems to be shaped like a strip. But in fact it is not so large, and it has a nearly *circular* shape. Its size and shape are quite distorted by the the map. The shapes of Greenland and northern Canada are quite distorted as well. Indeed, the two points

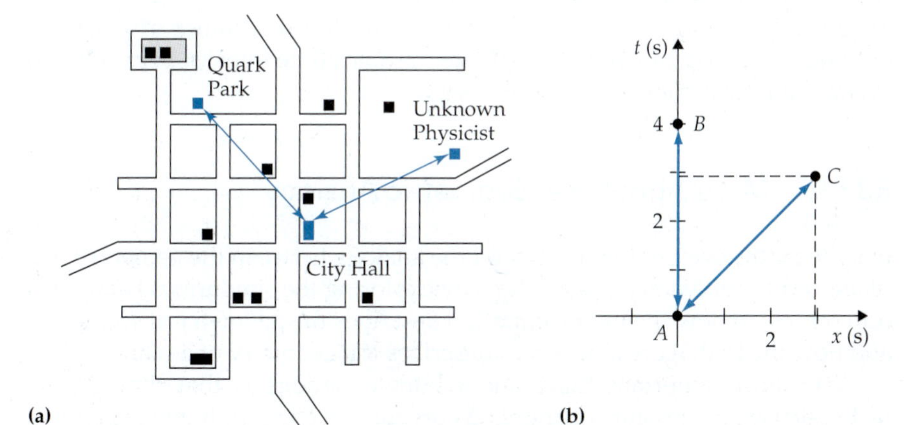

Figure R3.8
(a) A map of Askew (1 in = 1 km). The actual distance between City Hall and the Statue of the Unknown Physicist is 852 m. The lengths of both colored arrows on the map is thus 0.852 in.
(b) Both events B and C are the same distance from event A *on the spacetime diagram*, but the spacetime interval between A and B is actually 4 s, while the spacetime interval between A and C is *zero* (since $\Delta t = \Delta d$ here).

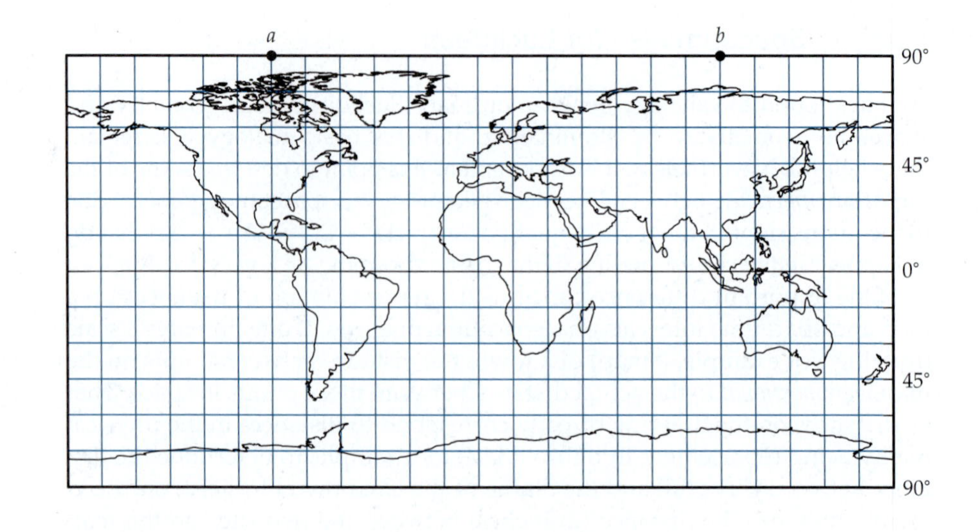

Figure R3.9
A Platte Carre projection map of the world, where the lines of longitude are represented as equally spaced straight lines.

marked a and b on the map are both at 90° north latitude—that is, at the *north pole*. Though these points are separated by a large distance on the map, the physical distance between these points on the earth's surface is actually zero!

The problem is that the geometry of the earth is not the same as that of the map

Why does this map not accurately represent the distances between points on the earth's surface? The problem is that the earth's surface as a whole is that of a *sphere*, which has a very different geometry from the Euclidean geometry of a flat sheet of paper. For example, on a sheet of paper the interior angles of a triangle always add to 180° and parallel lines never intersect. But on the earth's surface, the interior angles of a triangle add to *more* than 180° (consider a triangle with one vertex at the north pole and two vertices at the equator), and initially parallel lines may converge or diverge (consider lines of longitude, which are parallel at the equator!). Because of these fundamental geometric differences between the surface of the earth and the sheet of paper, any flat map of the earth will *necessarily* be distorted: one cannot make a map of the surface of the earth on a flat sheet of paper such that distances on the sheet correspond to actual distances on the earth.

Similarly, the geometries of spacetime and a spacetime diagram are different

Similarly, one cannot draw a spacetime diagram so distances between points on the drawing are proportional to the spacetime intervals between the corresponding events. Like the earth's surface, spacetime's geometry differs from that of the flat paper on which a spacetime diagram is drawn. The minus signs in the metric equation are symptomatic of this difference.

Just as you would not expect a flat map of the earth to accurately represent distances on the earth's surface, don't expect a spacetime diagram to accurately represent the spacetime intervals between events. A spacetime diagram displays the *coordinates* of events and the worldlines of particles, nothing more. You can always calculate the spacetime interval between two events from their coordinates if necessary.

R3.6 More About the Geometric Analogy

In spite of this, we can further extend the analogy between the geometry of a plane and the geometry of spacetime by exploring the similarities (as well as differences) in how the metric equation describes the geometry of spacetime and how the Pythagorean theorem describes the geometry of a plane.

The most important thing about both equations is that they enable us to calculate an absolute quantity (Δs or $|\Delta\vec{d}\,|$) in terms of frame-dependent

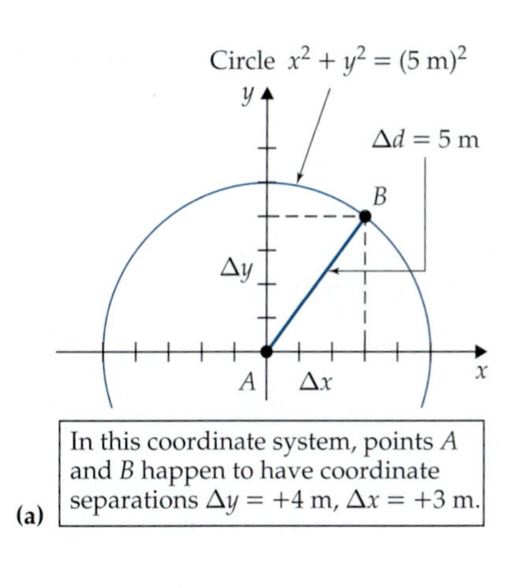

(a) In this coordinate system, points A and B happen to have coordinate separations $\Delta y = +4$ m, $\Delta x = +3$ m.

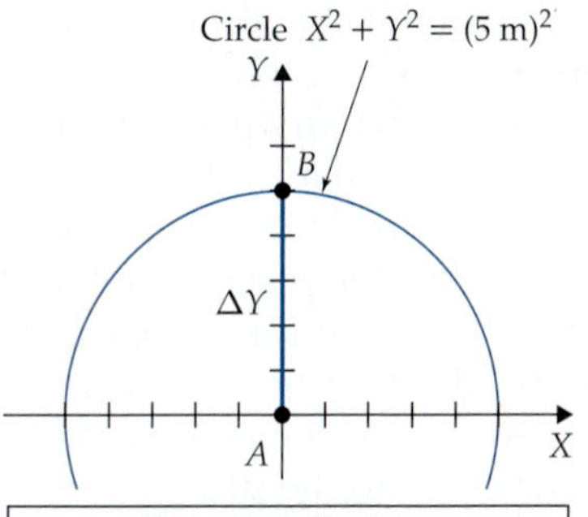

(b) We can find a coordinate system in which A and B lie along the vertical axis (that is, $\Delta X = 0$). In this unique system, the coordinate separation ΔY is equal to the distance between the points.

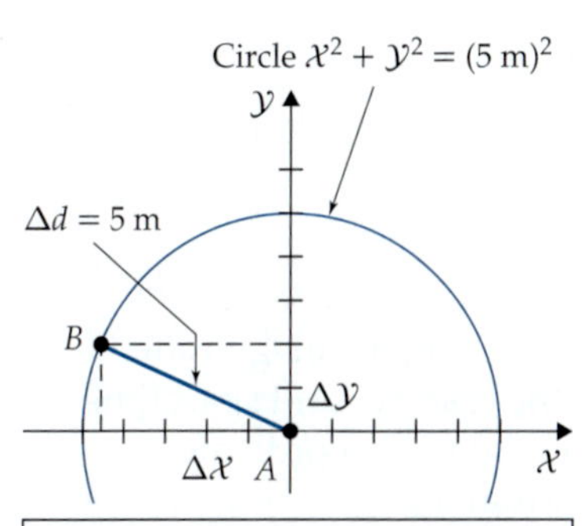

(c) If we twist the axes further clockwise relative to the underlying space, we can find a coordinate system where $\Delta \mathcal{Y} = 2$ m and $\Delta \mathcal{X} = -4.6$ m. In *any* system, B will lie *somewhere* along the circle shown.

Figure R3.10

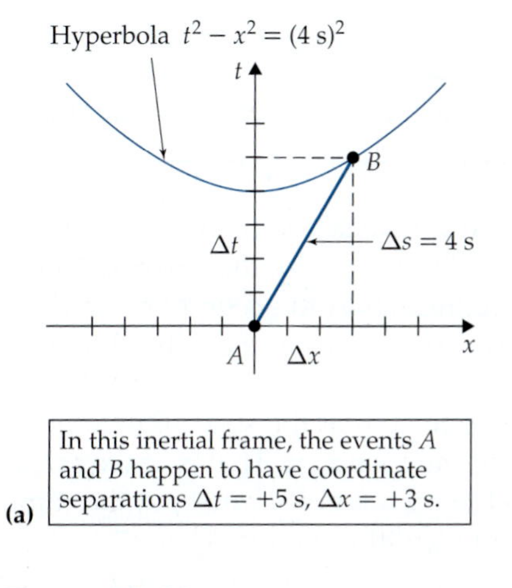

(a) In this inertial frame, the events A and B happen to have coordinate separations $\Delta t = +5$ s, $\Delta x = +3$ s.

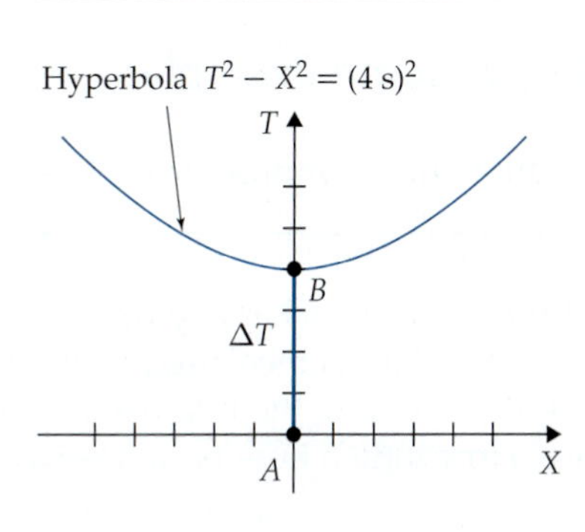

(b) We can find an inertial frame in which A and B occur at the same place (that is, $\Delta X = 0$). In this unique system, the coordinate time ΔT is equal to the space-time interval Δs between the points.

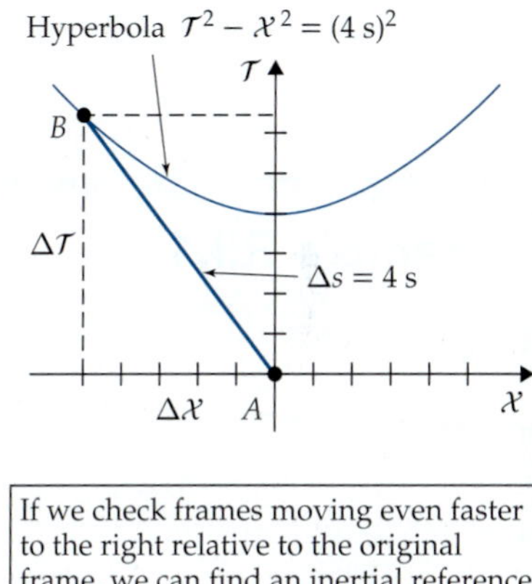

(c) If we check frames moving even faster to the right relative to the original frame, we can find an inertial reference frame where $\Delta \mathcal{T} = +6.4$ s and $\Delta \mathcal{X} = -5$ s. In any frame, B will lie *somewhere* along the hyperbola shown.

Figure R3.11

coordinate differences measured in an arbitrary inertial frame or coordinate system. This similarity is illustrated in figures R3.10 and R3.11. Figure R3.10 shows the same pair of points on the plane (A and B) that are 5 meters apart, plotted in various coordinate systems having different orientations with respect to "north." Note that if we set up the coordinate systems so that point A is at the origin, then point B in each coordinate system lies somewhere on the circle defined by the equation $x^2 + y^2 = \text{constant } |\Delta\vec{d}|^2$, where $|\Delta\vec{d}|^2$ is the squared distance between the points (since $|\Delta\vec{d}|$ is the distance between the points in *all* coordinate systems). In these drawings, I have kept the axes of each coordinate system vertical and horizontal, and rotated the space containing the points A and B "underneath" these coordinate axes: the points A and B are meant to be *the same physical points* in all the diagrams.

When viewed in different coordinate systems, point B always lies somewhere on a *circle* around point A

Similarly, figure R3.11 shows a pair of events (A and B) separated by a spacetime interval of 4 s, plotted on spacetime diagrams drawn by observers in different inertial frames. If we choose A to be the origin event in these frames, then event B lies somewhere on the curve defined by $t^2 - x^2 = \text{constant} = \Delta s^2$, where Δs^2 is the frame-independent squared spacetime interval between the events. Such a curve is a *hyperbola*, as shown. (Note that we are assuming $\Delta y = \Delta z = 0$ for these two events.) Again, remember, that these spacetime diagrams are meant to show how different observers would plot the *same* physical events A and B on their various diagrams.

When viewed in different reference frames, event B lies somewhere on a *hyperbola* about event A

The point is that the set of all points a given distance from the origin on the plane form a circle, but the set of all events a given spacetime interval from the origin event in spacetime form a hyperbola. The reason both curves aren't circles is that the metric equation has minus signs that doesn't appear in the Pythagorean relation. But there is a nice one-to-one correspondence between circles in plane geometry and hyperbolas in spacetime geometry.

Comparing the magnitudes of the distance and spacetime interval with coordinate separations

Note that one consequence of the difference between the metric equation and the Pythagorean relation is that in figure R3.10, we see that the north–south coordinate separation between a pair of points is always *less* than or equal to the distance between the points (the "hypotenuse" on the diagram): $\Delta y \leq |\Delta\vec{d}\,|$. In figure R3.11, though, we see that the coordinate time Δt between a pair of events is always *greater* than or equal to the spacetime interval Δs between them: $\Delta t \geq \Delta s$ even though the "hypotenuse" that represents Δs on the diagram *looks* larger.

R3.7 Some Examples

The following examples illustrate some applications of the metric equation.

Example R3.1

Problem: A firecracker explodes. A second firecracker explodes 25 ns away and 52 ns later, as measured in the Home Frame. In another inertial frame (the Other Frame), the two explosions are measured to occur 42 ns apart in space. How long a time passes between the explosions in the Other Frame?

Solution The key in this problem is to recognize that the spacetime interval between the two explosion events is frame-independent. That is, if we calculate it using the metric equation in the Home Frame, we must get the same answer we would get if we calculated it in the Other Frame. That is,

$$\Delta t^2 - |\Delta\vec{d}\,|^2 = \Delta s^2 = (\Delta t')^2 - |\Delta\vec{d}\,'|^2 \tag{R3.6}$$

Solving this equation for the unknown $\Delta t'$ yields

$$(\Delta t')^2 = \Delta t^2 - |\Delta\vec{d}\,|^2 + |\Delta\vec{d}\,'|^2 = (52\text{ ns})^2 - (25\text{ ns})^2 + (42\text{ ns})^2 = 3800\text{ ns}^2$$

$$\Delta t' = \sqrt{3800\text{ ns}^2} \approx 62\text{ ns} \tag{R3.7}$$

Exercise R3X.3

Suppose two events that are separated by 30 ns of distance in the Home Frame are also simultaneous in that frame. If in the Other Frame, the events are separated by 10 ns of time, what is their spatial separation in the Other Frame according to the metric equation?

Example R3.2

Problem: A certain physics professor fleeing the wrath of a set of irate students covers the length of the physics department hallway (a distance of about 120 ns) in a miraculous time of 150 ns as measured in the frame of the earth. Assuming the professor moves at a constant velocity, how much time does the professor's watch measure during the trip from one end of the hallway to the other?

Solution Part of the trick in many relativity problems is to rephrase a word problem in terms of *events*. In this case, let event A be the professor entering the hallway and event B be the professor's expeditious departure from the other end. In the reference frame of the earth, these events occur a time $\Delta t = 150$ ns apart and a distance $|\Delta\vec{d}| = 120$ ns apart. The professor's watch, however, is present at each of the events, so that watch registers the *spacetime interval* between these two events. Therefore, by the metric equation, the professor's watch reads

$$\Delta s^2 = \Delta t^2 - |\Delta\vec{d}|^2 = (150 \text{ ns})^2 - (120 \text{ ns})^2 = 8100 \text{ ns}^2$$

$$\Delta s = \sqrt{8100 \text{ ns}^2} = 90 \text{ ns} \qquad \text{(R3.8)}$$

Example R3.3

Problem: (A first glance at the "twin paradox".) A spaceship departs from the solar system and travels at a constant speed to the star Alpha Centauri 4.3 light-years away, then instantaneously turns around (never mind about the impossible acceleration!) and returns at the same constant speed. Assume the trip takes 13 years as measured by clocks here on earth. How long does the trip take as measured by clocks on the spaceship?

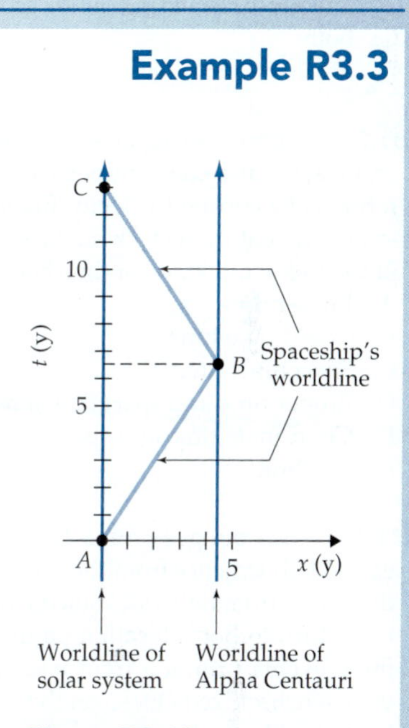

Figure R3.12
Spacetime diagram of a trip to Alpha Centauri and back.

Solution Again, we need to translate the word problem into a problem about measuring the time between events. Let event A represent the ship's departure from the solar system, event B its arrival at Alpha Centauri, and event C its return to the solar system (see figure R3.12). A clock in the spaceship does *not* measure the spacetime interval between events A (departure from the solar system) and C (return to the solar system) even though the clock is present at both events. This is because the clock is accelerated when the spaceship turns around, and so the clock is not inertial. To find the total elapsed time registered on the ship clock, we can, however, consider each leg of the trip separately. The ship's clock *does* measure the spacetime interval between events A and B, and it also measures the spacetime interval between events B and C, as it is inertial during each leg of the trip and is present at the events in question. The total time registered by the ship's clock is thus the sum of the spacetime intervals between A and B and between B and C.

We can use the metric equation to compute these spacetime intervals from the coordinate differences for these events measured in the earth's frame. Events A and B occur $\Delta t = 6.5$ y apart in time and a distance $|\Delta\vec{d}| = 4.3$ y apart in space. The spacetime interval between these events is

$$\Delta s_{AB} = \sqrt{(6.5 \text{ y})^2 - (4.3 \text{ y})^2} \approx 4.9 \text{ y} \qquad \text{(R3.9)}$$

The spacetime interval between events B and C is the same. The total elapsed time for the trip as measured by a clock on the ship is thus 2(4.9 y) = 9.8 y, which is somewhat shorter than the time of 13 y measured by clocks on earth.

Note that the line on the diagram connecting points A and B looks *longer* than 6.5 y, but the spacetime interval that this line represents is actually *shorter* than 6.5 y. This is an illustration of the issue discussed in section R3.6.

TWO-MINUTE PROBLEMS

R3T.1 A person riding a merry-go-round passes very close to a person standing on the ground once (event A) and then again (event B). Assume the ground is an inertial frame and that the rider moves at a constant speed.

(a) Which person's watch measures a *proper time* $\Delta\tau$ between events A and B?

(b) Which person's watch measures the spacetime interval Δs between those events?

(c) Which person's watch (if any) measures the coordinate time Δt between those events in some inertial frame?

A. The rider in the merry-go-round
B. The person standing on the ground
C. Both
D. Neither

R3T.2 A spaceship departs from the solar system (event A) and travels at a constant velocity to a distant star. It then returns at a constant velocity, finally returning to the solar system (event B). A clock on the spaceship registers which of the following kinds of time between these events?

A. Proper time
B. Coordinate time
C. Spacetime interval
D. Proper time and spacetime interval
E. Coordinate time and spacetime interval
F. All three

R3T.3 Alice bungee-jumps from a bridge above a deep gorge. Bob watches from the bridge. Let event D be Alice's departure from Bob's location on the bridge, and event R be her return to Bob's location on the bridge. Carol observes these events from a a train passing over the bridge, and uses synchronized clocks on the train to measure the time between Alice's departure and return.

(a) Which person's watch or clocks register(s) a proper time between events D and R?

(b) Which person's watch or clocks register(s) the spacetime interval between those events?

(c) Which person's watch or clocks register(s) a coordinate time between those events in some inertial frame?

A. Alice
B. Bob
C. Carol
D. Alice and Bob
E. Bob and Carol
F. Alice and Carol
T. All three observers

R3T.4 The spacetime interval Δs between two events can never be larger than the coordinate time Δt between those events as measured in *any* inertial reference frame. T or F?

R3T.5 Two events occur 5.0 s apart in time and 3.0 s apart in space. A clock traveling at a speed of 0.60 can be present at both these events. What time interval will such a clock measure between the events?

A. 8.0 s
B. 5.8 s
C. 5.0 s
D. 4.0 s
E. 2.0 s
F. Other (specify)

R3T.6 Consider the events A, B, C, and D shown in the spacetime diagram below.

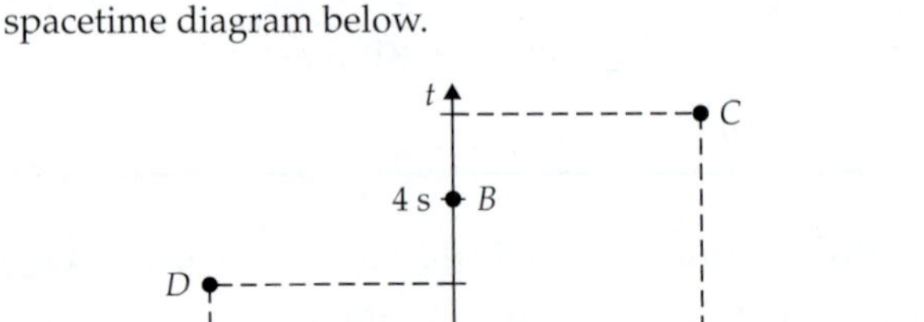

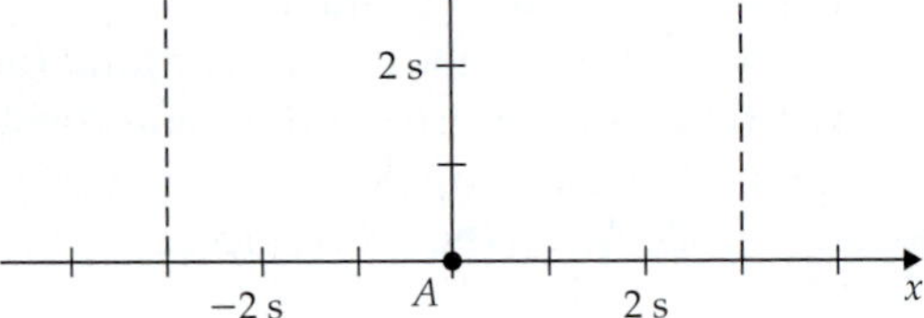

(a) What is the spacetime interval between events A and B?

(b) Between A and C?

(c) Between A and D?

A. 0 s
B. 2 s
C. 3 s
D. 4 s
E. 5 s
F. Other (specify)

R3T.7 Consider the spacetime diagram below. Let the spacetime interval between events O and A be Δs_{OA}, and let the spacetime interval between events O and B be Δs_{OB} Which of these two spacetime intervals is larger? (Assume that the y and z coordinates of all these events are zero.)

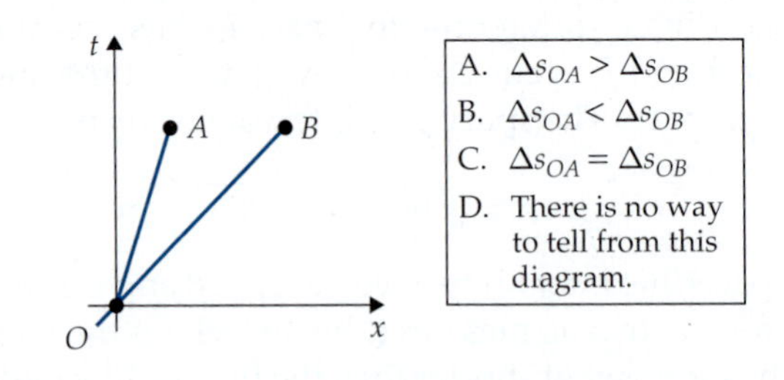

R3T.8 An inertial clock present at two events always measures a *shorter* time than a pair of synchronized clocks in *any* inertial reference frame would register between the same two events (as long as the events don't occur at the same place in that frame). T or F?

R3T.9 Consider a train moving at a speed of 0.5 relative to the ground. A light in one of its windows blinks repeatedly. An observer on the ground will necessarily *see* (not *observe*) those blinks to be separated by a larger time interval than a person on the train would. T or F?

HOMEWORK PROBLEMS

Basic Skills

R3B.1 Clock P is at rest alongside a racetrack. A jockey on horseback checks her watch against clock P as she passes it during the first lap (event A) and then checks her watch again as she passes clock P the second time (event B).

(a) Which clock (clock P or the watch) measures the spacetime interval between events A and B?

(b) Which measures proper time? (Be careful!)

(c) Do either of the clocks measure coordinate time between the events in the ground frame? Discuss.

R3B.2 Alyssa is a passenger on a train moving at a constant velocity relative to the ground. She synchronizes her watch with the station clock as she passes through the Banning town station, and then compares her watch with the station clock as she passes through the Centerville town station farther down the line. Assume the ground is an inertial frame, and assume the Banning and Centerville clocks are synchronized in that frame.

(a) Is the time she measures between the events of passing through these towns a proper time? Is it a coordinate time in some inertial reference frame? Is it the spacetime interval between the events?

(b) If one subtracts the Centerville station clock reading from the Banning station clock reading, what kind of time interval between the events does one obtain? Defend your answers carefully.

R3B.3 Alice is driving a race car around an essentially circular track. Brian, who is sitting at a fixed position at the edge of the track, measures the time Alice takes to complete a lap by starting his watch when Alice passes by his position (call this event E) and stopping it when Alice passes his position again (call this event F). Figure R3.13 illustrates these events. Cara and Dave, who are passengers in a train that passes very close to Brian, also observe these events. Cara happens to be passing Brian just as Alice passes Brian the first time, and Dave happens to pass Brian just as Alice passes Brian the second time. Assume that the clocks used by Alice, Brian, and Cara are close enough together that we can consider them all to be "present" at event E, and similarly that those used by Alice, Brian, and Dave are "present" at event F. Assume the ground frame is an inertial reference frame, and the train travels at a constant velocity.

(a) Who measures a proper time between events E and F?

(b) Who (if anyone) measures a coordinate time between events E and F?

(c) Who (if anyone) measures the spacetime interval between events E and F?

Carefully explain your reasoning in each case.

R3B.4 In a certain inertial reference frame, two events are separated in time by $\Delta t = 25$ ns and by $|\Delta\vec{d}| = 15$ ns in space. What is the spacetime interval between these events?

R3B.5 In the reference frame of the solar system, two events are separated by 5.0 h of time and 4.0 h of distance. What is the spacetime interval between these events?

R3B.6 In the Home Frame, two events are observed to occur with a spatial separation of 12 ns and a time coordinate separation of 24 ns.

(a) An inertial clock travels between these events in such a manner as to be present at both events. What time interval does this clock read between the events?

(b) What is the speed of this clock, as measured in the Home Frame?

R3B.7 An alien spaceship moving at a constant velocity goes from one end of the solar system to the other (a distance of 10.5 h) in 13.2 h as measured by clocks on earth. What time does a clock on the spaceship read for the passage? (*Hint:* Rephrase in terms of events.)

R3B.8 A space probe journeys at a constant velocity from earth to Tau Ceti (a distance of 11.9 y) in a time of 1.0 y as measured by the probe's internal clock. How long did the trip take according to clocks on earth?

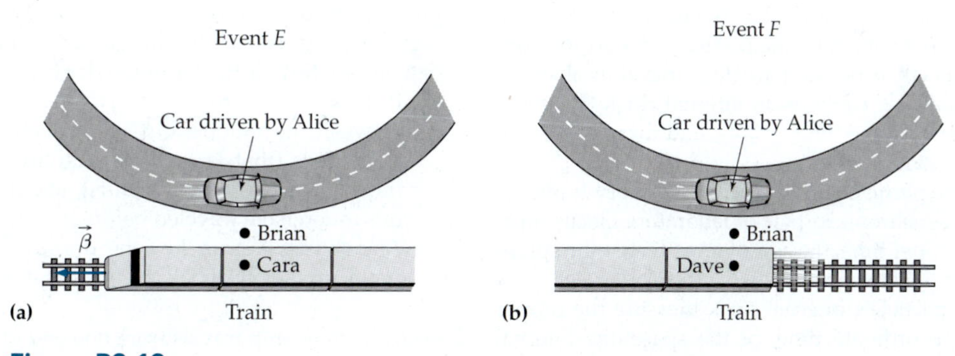

Figure R3.13
(a) Event E in the situation described in problem R3B.1. (b) Event F in that situation.

R3B.9 In the Home Frame, we observe two events to occur 500 ns apart in time and 300 ns apart in space. In an Other Frame, these events occur 400 ns apart in time. What is the spatial separation of these events in the Other Frame?

R3B.10 In the solar system frame, two events are measured to occur 3.0 h apart in time and 1.5 h apart in space. Observers in an alien spaceship measure the two events to be separated by only 0.5 h in space. What is the time separation between the events in the alien's frame?

Modeling

R3M.1 Imagine that in the year 2065, you are watching a live broadcast from the space station at the planet Neptune, which is 4.0 light-hours from earth at the time. (Assume that the TV signal from Neptune is sent to earth via a laser light communication system.) At exactly 6:17 p.m. (as registered by the clock on your desk), you see a technician on the TV screen suddenly exclaim, "Hey! We've just detected an alien spacecraft passing by here." Let this be event *A*. Exactly 1 h later, the alien spaceship is detected passing by earth: let this be event *B*. Assume that the earth and Neptune stations can be considered parts of the inertial reference frame of the solar system, and assume that the spaceship travels at a constant velocity.

(a) During the broadcast, you can see on your TV screen the face of a clock sitting on the technician's desk. What time should you see on this clock face at 6:17 p.m. your time if that clock is synchronized with yours?

(b) What is the coordinate time between events *A* and *B* in the solar system frame? Carefully explain.

(c) What is the speed of the alien spaceship, as measured in the solar system frame?

(d) What kind(s) of time would the spaceship's clock measure between events *A* and *B*?

R3M.2 A *particle accelerator* is a device that boosts subatomic particles to speeds close to that of light. Such an accelerator is typically shaped like a ring (which may be several kilometers in diameter): the particles are constrained by magnetic fields to travel inside the ring. Imagine such an accelerator having a radius of 2.998 km. Assume there are two synchronized clocks (*P* and *Q*) located on opposite sides of the ring. A certain particle in the ring is measured to travel from clock *P* to clock *Q* in 34.9 μs, as registered by those clocks. Let event *A* be the particle's departure from clock *P* and event *B* be the particle's arrival at clock *Q*. Assume the particle contains an internal clock that measures the time between these events, and that the particle travels at a constant speed.

(a) What is the particle's speed in the laboratory frame?

(b) Does the synchronized pair of laboratory clocks measure the proper time, the coordinate time, or the spacetime interval between events *A* and *B*?

(c) Does the particle's internal clock measure the proper time, the coordinate time, or the spacetime interval between events *A* and *B*?

(d) What is the spacetime interval between these events?

R3M.3 At $t = 0$, an alien spaceship passes by the earth: let this be event *A*. At $t = 13$ min (according to synchronized clocks on earth and Mars), the spaceship passes by Mars, which is 5 light-minutes from earth at the time: let this be event *B*. Radar tracking indicates that the spaceship moves at a constant velocity between earth and Mars. Just after the ship passes earth, people on earth launch a probe whose purpose is to catch up with and investigate the spaceship. This probe accelerates away from earth, moving slowly at first, but moving faster and faster as time passes, eventually catching up with and passing the alien ship just as it passes Mars. In all parts of this problem, you can ignore the effects of gravity and the relative motion of earth and Mars (which are small) and treat earth and Mars as if they were both at rest in the inertial reference frame of the solar system. Also assume that both the probe and the alien spacecraft carry clocks.

(a) Draw a quantitatively accurate spacetime diagram of the situation, including labeled worldlines for the earth, Mars, the alien spacecraft, and the probe. Also label events *A* and *B*.

(b) Whose clocks measure coordinate times between events *A* and *B*? Explain carefully.

(c) Whose clocks measure proper times between these events? Explain.

(d) Does any clock in this problem measure the spacetime interval between the events? If so, which one and why? If not, why not?

R3M.4 Suppose you and a friend are riding in trains that are moving relative to each other at relativistic speeds. As you pass each other, you both measure the time separation and spatial separation of two firecracker explosions that occur on the tracks between you. (You can measure the latter by measuring the distance between the scorch marks that the explosions leave on the side of your train.) You find the firecracker explosions to be separated by 1.0 μs of time and 0.40 μs of distance in your frame. By radio, your friend reports that the explosions were separated by only 0.60 μs of time in your friend's frame? Is this possible? If it is, find the spatial separation of the events in your friend's frame. If not, explain why not.

R3M.5 A muon is created by a cosmic-ray interaction at an altitude of 60 km. Imagine that, after its creation, the muon hurtles downward at a speed of 0.998, as measured by a ground-based observer. After the muon's "internal clock" registers 2.0 μs (which is a bit longer than the average life of a muon), this particular muon decays.

(a) If clocks on the ground were to measure the same time between the muon's birth and death as the muon's clock does (that is, special relativity is not true and time is universal and absolute), about how far would this muon have traveled before it decayed?

(b) As relativity *is* true, how far does this muon actually travel (in the ground frame) before it decays?

R3M.6 A spaceship travels from one end of the Milky Way galaxy to the other (a distance of about 100,000 y) at a constant velocity of magnitude $|\vec{v}| = 0.999$, as measured in the frame of the galaxy. How much time does a clock in the spaceship register for this trip? (*Hint:* Rephrase this problem in terms of events.)

R3M.7 In one inertial frame (the Home Frame), we observe two events to occur at the same place but $\Delta t = 32$ ns apart in time. In another inertial frame (the Other Frame), the same two events are observed to occur 45 ns apart in space.

(a) What is the coordinate time interval between the events in the Other Frame?

(b) Compute the speed of the Home Frame as measured by observers in the Other Frame. (*Hint:* The events occur at the same place in the Home Frame. So how far does the Home Frame move in the time between the events as seen in the Other Frame? What is the time between the events in the Other Frame?)

R3M.8 The new earth–Pluto shuttle line boasts that it can take you between the two planets (which are about 5.0 h of distance apart) in 2.5 h (according to a rider's watch). Assume that acceleration and deceleration periods are very brief so that you spend essentially all the trip traveling at a constant velocity.

(a) What time interval must synchronized clocks in the solar system's reference frame register between the shuttle's departure from earth and its arrival at Pluto if the advertisement is true?

(b) What is the shuttle's cruising speed?

R3M.9 Suppose a round trip to Alpha Centauri takes 2.0 y as measured by clocks on the spaceship making that trip. Assuming the ship accelerates and decelerates essentially instantly when necessary, how long does the trip take in the frame of the earth? (See example R3.3.)

R3M.10 The following spacetime diagram shows the worldline of a rocket as it leaves the earth, travels for a certain time, comes to a stop, and then explodes.

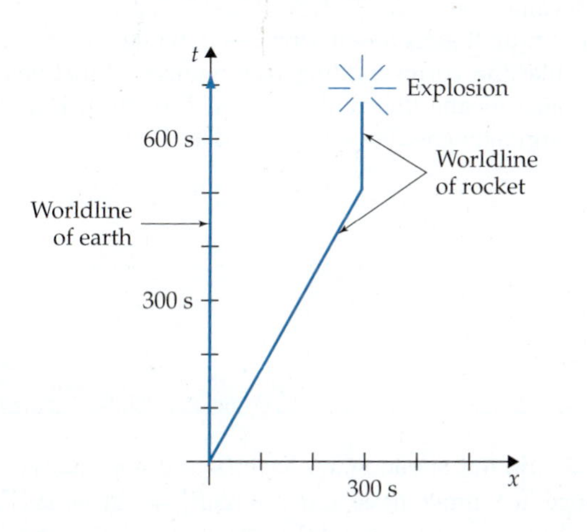

(a) How much time elapses between the rocket's departure and its unfortunately explosive demise as measured by that rocket's flight recorder?

(b) What is the spacetime interval between these two events?

R3M.11 Suppose a certain kind of unstable subatomic particle decays with a half-life at rest of about 2.0 μs (that is, if at a certain time you have N such particles at rest, then 2.0 μs later you will have $\frac{1}{2}N$ remaining). We can consider a batch of such particles as being a clock whose decreasing numbers register the passage of time. Now suppose that with the help of a particle accelerator, we manage to produce in the laboratory a beam of these particles traveling at a speed $|\vec{v}| = 0.866$ in SR units (as measured in the laboratory frame). This beam passes through a detector A, which counts the number of particles passing through it each second. The beam then travels a distance of about 2.08 km to detector B, which also counts the number of particles passing through it, as shown below.

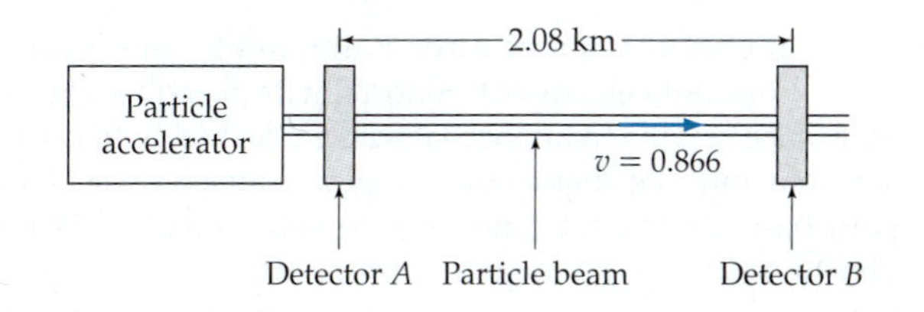

(a) Let event A be the passing of a given particle through detector A, and let event B be the passing of the same particle through detector B. How much time will a laboratory observer measure between these events? (*Hint:* You don't need to know anything about relativity to answer this question!)

(b) How much time passes between these events as measured by the clock inside the particle, according to relativity theory? If relativity is true, about what fraction of the particles that pass through detector A survive to pass through detector B?

(c) According to the Newtonian concept of time, the time measured by a particle clock between the events would be the same as the time measured by laboratory clocks. If this were so, what fraction of the particles passing through detector A survive to detector B?

Rich-Context

R3R.1 In 2095, a message arrives at earth from the growing colony at Tau Ceti (11.9 y from earth). The message asks for help in combating a virus that is making people seriously ill (the message includes a complete description of the viral genome). Using advanced technology available on earth, scientists are quickly able to construct a drug that prevents the virus from reproducing. You have to decide how much of the drug can be sent to Tau Ceti. The space probes available on short notice could either boost 200 g of the drug (in a standard enclosure) to a speed of 0.95, 1 kg to a speed of 0.90, 5 kg to a speed of 0.80, or 20 kg to a speed of 0.60 relative to the earth. The only problem is that a sample of the drug in a standard enclosure at rest in the laboratory is observed to degrade due to internal chemical processes at a rate that will make it useless after 5.0 y.

(a) Explain why it is possible to send the drug to Tau Ceti, even though the ship must travel for more than 11.9 y.

(b) How much can you send?

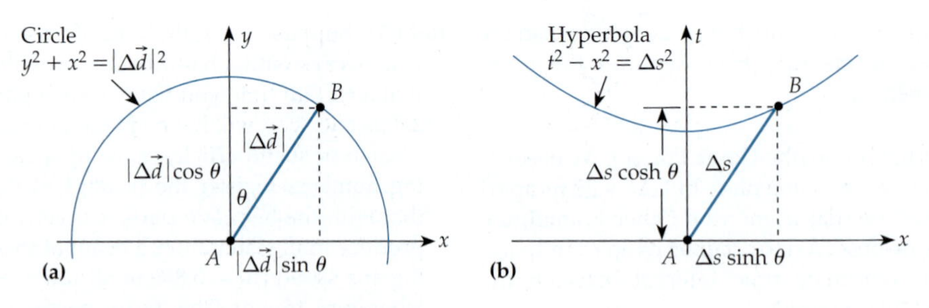

Figure R3.14
(a) Plane trigonometry. (b) Spacetime trigonometry. (Note that $\Delta s \cosh\theta > \Delta s$, even though the "hypotenuse" representing Δs *looks* bigger on the diagram.)

R3R.2 Your boss is on the earth–Pluto shuttle, which travels at a constant velocity of 0.60 straight from earth to Pluto, a distance of 5.0 h in an inertial frame attached to the sun. An hour into the flight (according to your boss's watch) your boss sends a laser message to you on earth, asking you to send a wake-up call appropriately timed so that your boss can catch a 1-h nap (as measured on your boss's watch). You *immediately* reply with the wake-up call and an apology that the call is late, claiming in your defense that the laws of physics prevented a timely response.

(a) Why was your boss's request impossible?

(b) How long did your boss sleep (according to your boss's watch) before your message was received at the shuttle? Explain carefully.

(*Hint:* Draw a spacetime diagram of the situation.)

Advanced

R3A.1 Just as we can describe the relationship between the hypotenuse of a triangle and the coordinate lengths of its sides by using the sine and cosine functions, it turns out that we can describe the relationship between the spacetime interval Δs between two events in terms of the coordinate separations Δt and Δx between those events in terms of the hyperbolic sine and cosine functions. The *hyperbolic sine* and *hyperbolic cosine* functions of a quantity θ are defined as follows:

$$\sinh\theta = \tfrac{1}{2}(e^{\theta} - e^{-\theta}) \qquad \cosh\theta = \tfrac{1}{2}(e^{\theta} + e^{-\theta}) \qquad \text{(R3.10)}$$

(a) Prove that $\cosh^2\theta - \sinh^2\theta = 1$. This means that if the spacetime interval between two events occurring along the spatial x axis is Δs, then the coordinate separations Δt and Δx between these events can be written $\Delta t = \Delta s \cosh\theta$ and $\Delta x = \Delta s \sinh\theta$ for some appropriately chosen value of θ, just as in plane geometry $\Delta x = \Delta d \cos\theta$ and $\Delta y = \Delta d \sin\theta$ (see figure R3.14).

(b) Argue that θ in the hyperbolic case is *not* the angle that line AB makes with the t axis in the spacetime diagram of figure R3.14b. Argue in fact that as $\theta \to \infty$, the angle that AB makes with the t axis approaches 45°.

(c) Argue that if $|\vec{v}|$ is the speed of an object that goes from event A to event B at a constant velocity, the "angle" θ is in fact $\tanh^{-1}|\vec{v}|$.

(d) When $|\vec{v}| = 0.80$, $\theta = 1.10$ (if your calculator can do inverse hyperbolic functions, verify this). What are the values of $\cosh\theta$ and $\sinh\theta$ for this value of θ? (Use the definitions of these functions given above if your calculator cannot evaluate hyperbolic functions).

(e) When $|\vec{v}| = 0.99$, we have $\theta = 2.65$ (again, verify if you can). What are the values of $\cosh\theta$ and $\sinh\theta$ for this value of θ?

(f) Argue that as $\theta \to 0$, $\sinh\theta \to 0$, while $\cosh\theta \to 1$, just like the corresponding trigonometric functions. This also means that $\tanh\theta \to 0$ in this limit. Use this to argue that as $|\vec{v}| \to 0$, $\Delta s \to \Delta t$ and $\Delta x \to 0$.

ANSWERS TO EXERCISES

R3X.1 Measuring sticks placed parallel to the line of relative motion will simply paint stripes down each other's length. There is no way to extract information about the length of a given stick from such stripes. One *could* paint very brief pulses, but one would need to use synchronized clocks to do this, which involves that whole issue.

R3X.2 (Just type it into your calculator.)

R3X.3 In this situation, we have $\Delta t = 0$ and $|\Delta\vec{d}| = 30$ ns. Since we must have $(\Delta t')^2 - |\Delta\vec{d}'|^2 = \Delta t^2 - |\Delta\vec{d}|^2$, and since we are given that $\Delta t' = 10$ ns, we can find the spatial separation $|\Delta\vec{d}'|$ between the events in the Other Frame by solving the equation above for $|\Delta\vec{d}'|$:

$$|\Delta\vec{d}'| = \sqrt{(\Delta t')^2 - \Delta t^2 + |\Delta\vec{d}|^2}$$
$$= \sqrt{(10\text{ ns})^2 - 0 + (30\text{ ns})^2} = 31.6\text{ ns} \qquad \text{(R3.11)}$$

R4 Proper Time

Chapter Overview

Introduction

In chapter R3, we derived the *metric equation*, which links the coordinate differences between two events measured in a given reference frame to the frame-independent spacetime interval between those events. In this chapter, we will learn how to use the metric equation to calculate the proper time along *any* worldline.

Section R4.1: A Curved Footpath

Suppose we know the function $x(y)$ that describes a certain path on a two-dimensional plane. We can calculate the pathlength $\Delta\ell_{AB}$ between any two points A and B along that path as follows. First, we divide the path into tiny segments whose endpoints are coordinate displacements of dx and dy apart. Second, we use the Pythagorean theorem $d\ell = (dx^2 + dy^2)^{1/2} = [(dx/dy)^2 + 1]^{1/2}\,dy$ to find each segment's length $d\ell$. Finally, we sum over all segments to find the total pathlength. In the limit that the segments become infinitesimally short, the sum becomes an integral.

Section R4.2: Curved Worldlines in Spacetime

We can perform an exactly analogous calculation to calculate the proper time between two events A and B as measured by a clock following an arbitrary worldline. Suppose we know the functions $x(t)$, $y(t)$, and $z(t)$ that describe the clock's position as a function of time as it moves along the worldline. (1) We divide the worldline into segments so short that the worldline is essentially straight between any segment's endpoints. (2) The proper time along each segment is then essentially equal to the spacetime interval along that segment: $d\tau \approx ds = (dt^2 - dx^2 - dy^2 - dz^2)^{1/2} = [1 - (dx/dt)^2 - (dy/dt)^2 - (dz/dt)^2]^{1/2}\,dt = (1 - |\vec{v}|^2)^{1/2}\,dt$. (3) The total proper time along the worldline is the sum of these infinitesimal proper times. In the limit that the segments become infinitesimally short, the sum becomes an integral:

$$\Delta\tau_{AB} = \int_{t_A}^{t_B} (1 - |\vec{v}|^2)^{1/2}\,dt \qquad \text{(R4.6)}$$

- **Purpose:** This equation describes how we can use measurements performed in an inertial frame to compute the proper time $\Delta\tau_{AB}$ measured by a clock traveling between any two events A and B along an arbitrary worldline, where t_A is the time of event A, t_B is the time of event B, dt is the coordinate time differential, and $|\vec{v}|$ is the clock's speed (as a function of time), all measured in a specific inertial reference frame.
- **Limitations:** This equation works only in an inertial reference frame.
- **Note:** *If* the clock's speed (not necessarily its velocity) is constant, then this equation reduces to

$$\Delta\tau_{AB} = (1 - |\vec{v}|^2)^{1/2}\,\Delta t_{AB} \qquad \text{only if } |\vec{v}| = \text{constant} \qquad \text{(R4.7)}$$

Section R4.3: The Binomial Approximation

The binomial approximation asserts that

$$(1 + x)^a \approx 1 + ax \quad \text{if } x \ll 1 \tag{R4.17}$$

- **Purpose:** This equation is a useful trick for simplifying calculations when we want to compute an arbitrary power a of a very small quantity x added to 1.
- **Limitations:** You will not get very good results if $|ax|$ is much larger than 0.1.
- **Note:** In this chapter, we will most often use this approximation to get

$$(1 - |\vec{v}|^2)^{1/2} = [1 + (-|\vec{v}|^2)]^{1/2} \approx 1 - \tfrac{1}{2}|\vec{v}|^2 \tag{R4.18}$$

This approximation usually works best when the problem is phrased (or the answer can be phrased) so that the 1 cancels out, but in the *worst* case, one can calculate $1 - ax$ by hand, something that is usually impossible for $(1 + x)^a$.

Section R4.4: Ranking the Three Kinds of Time

This section presents a proof of the following statement: The proper time measured by a clock traveling between two events A and B is *longest* if the clock follows a straight (constant-velocity) worldline.

Since the metric equation implies that $\Delta t^2 = \Delta s^2 + |\Delta\vec{d}|^2 \geq \Delta s^2$ in any arbitrary reference frame, we have in general

$$\Delta t \geq \Delta s \geq \Delta\tau \tag{R4.22}$$

- **Purpose:** This equation describes the hierarchical relationship between the coordinate time Δt between two events (measured in any arbitrary inertial frame), the spacetime interval Δs between those events, and the proper time $\Delta\tau$ between those events (measured along any arbitrary worldline going between them).
- **Limitation:** This equation applies only if $\Delta s^2 \geq 0$.
- **Note:** The equality $\Delta t = \Delta s$ applies if Δt is measured where the events occur at the same place; the equality $\Delta s = \Delta\tau$ applies if $\Delta\tau$ is measured along a straight worldline.

Section R4.5: Experimental Evidence

This section discusses two relatively recent experiments that have tested the validity of equation R4.6.

Section R4.6: The Twin Paradox

Consider a pair of twins. One leaves earth, travels to a distant star at a speed close to that of light, and then returns. A naive use of equation R4.6 might lead each twin to conclude that the other is younger (since each considers him- or herself to be at rest while the other is moving). This is the **twin paradox**. However, the twin's situations are not really symmetric: the traveling twin is *not* in an inertial reference frame, but the twin on earth is (at least approximately). Therefore, the earth-based twin can legally use equation R4.6 (at least approximately), whereas the traveling twin cannot. The traveling twin really ends up being younger than the earth-based twin.

R4.1 A Curved Footpath

The metric equation $\Delta s^2 = \Delta t^2 - \Delta x^2 - \Delta y^2 - \Delta z^2$ connects the spacetime coordinate differences between two events measured in some inertial reference frame to the spacetime interval Δs between the same events measured by an inertial clock present at both events. In this section and the next, we will use the metric equation to connect coordinate time in a given inertial frame to the proper time $\Delta\tau$ measured by any clock present at both events, inertial or not.

One might think that we must use the theory of *general* relativity to properly analyze the behavior of accelerating (noninertial) clocks. In fact, we can quite adequately analyze the behavior of such clocks with the metric equation alone if we remember the analogy between proper time and pathlength.

We compute the length of a curved path by breaking it up into tiny straight pieces

Consider a footpath around a small pond. Figure R4.1a shows a scale drawing of the path with a superimposed coordinate system. We could measure the path's length from point A to point B with a long, flexible tape measure. But once we have set up a coordinate system, we can also *compute* the path's length as follows. Suppose we divide up the path into a large number of infinitesimally small sections, as shown in figure R4.1b.* If we make these sections small enough, each will be approximately straight. In this limit, the pathlength $d\ell$ of a given segment as measured by a flexible tape measure will be almost equal to the straight length computed (using the Pythagorean theorem) from the coordinate differences of the segment's endpoints:

$$d\ell^2 \approx dx^2 + dy^2 \qquad \text{or} \qquad d\ell = \sqrt{dx^2 + dy^2} \tag{R4.1}$$

The path's total length $\Delta\ell_{AB}$ from A to B is the sum of all the segment lengths, which in the limit where the segments are truly infinitesimal becomes the integral

$$\Delta\ell_{AB} = \int_{\text{path}} d\ell = \int_{\text{path}} \sqrt{dx^2 + dy^2} \tag{R4.2}$$

Note that since each segment's length $d\ell$ is greater than its northward extension dy, the total pathlength between points A and B will be greater than the straight-line northward distance of 225 meters between A and B. We can say quite generally, therefore, that $\Delta\ell_{AB} \geq |\Delta\vec{d}_{AB}|$.

We can describe the path mathematically by using the function $x(y)$: this function specifies the path's x coordinate at each possible y coordinate. If we know this function, we can write the integral above as a single-variable integral over x by pulling a factor of dy out of the square root:

Formula for computing the length of a path $x(y)$

$$\Delta\ell_{AB} = \int_{y_A}^{y_B} \sqrt{1 + (dx/dy)^2}\, dy \tag{R4.3}$$

[Note that we are considering y to be the independent variable and x to be the dependent variable in equation R4.3. This reversal of convention is necessary because $y(x)$ is not well defined for the path shown in figure R4.1.]

As we have discussed before, although this equation uses the coordinates x and y measured in a given coordinate system, the pathlength itself is an invariant quantity: we'll get the same answer (the answer that a flexible tape measure would give) no matter *what* coordinate system we use.

---

*The analogy with pathlength presented here follows E. F. Taylor and J. A. Wheeler, *Spacetime Physics*, San Francisco; Freeman, 1963, pp. 32–34.

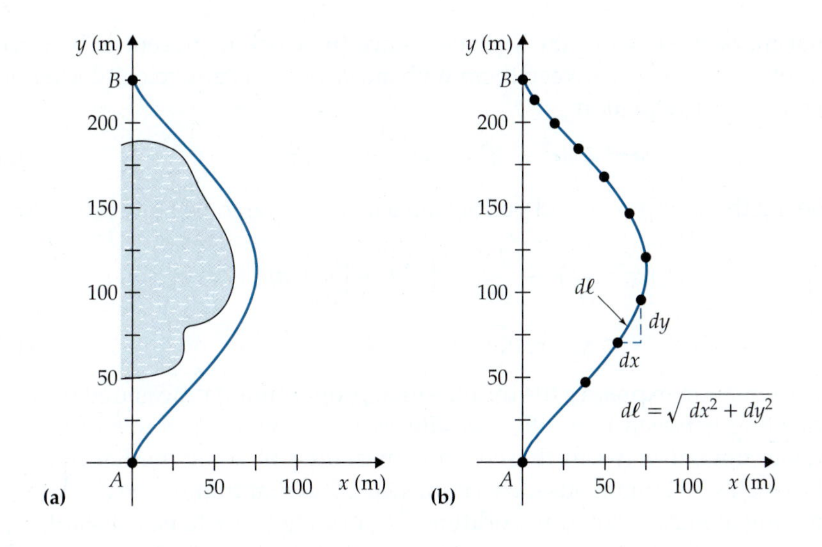

Figure R4.1
(a) Map of a footpath around a small pond, with a superimposed xy coordinate system. (b) We can calculate the path's length by subdividing it into many segments (each small enough to be nearly straight), calculating the length $d\ell$ of each segment, and summing to find the total length $\Delta\ell$.

R4.2 Curved Worldlines in Spacetime

The situation in spacetime is directly analogous

The analogy to events in spacetime is direct. Consider the worldline of a particle that in a certain inertial reference frame travels out from the origin a certain distance along the x axis and then returns. Such a worldline is shown on the spacetime diagram in figure R4.2a with the coordinate axes of that frame superimposed. Such a worldline describes an *accelerating* particle: we can see from the graph that the particle's x-velocity $v_x = dx/dt$ (which is the inverse slope of its worldline on the diagram) changes as time progresses.

A clock traveling with the particle measures the proper time $\Delta\tau_{AB}$ between events A and B along this worldline (by the definition of proper time). But once we have measured the particle's worldline in an inertial reference frame (any inertial frame), we can *calculate* what this clock will read between events A and B by using the metric equation in a manner analogous to our calculation of the pathlength between points A and B in section R4.1.

We compute $\Delta\tau$ along a worldline by dividing the worldline into many tiny straight segments

Suppose we divide the particle's worldline up into many infinitesimal segments, each of which is nearly a straight line on the spacetime diagram (figure R4.2b). We choose each segment to be short enough that the particle's velocity is approximately constant as it traverses that segment. If this is true, then the proper time $d\tau$ that a clock would measure along each segment will be almost equal to the spacetime interval ds between the events

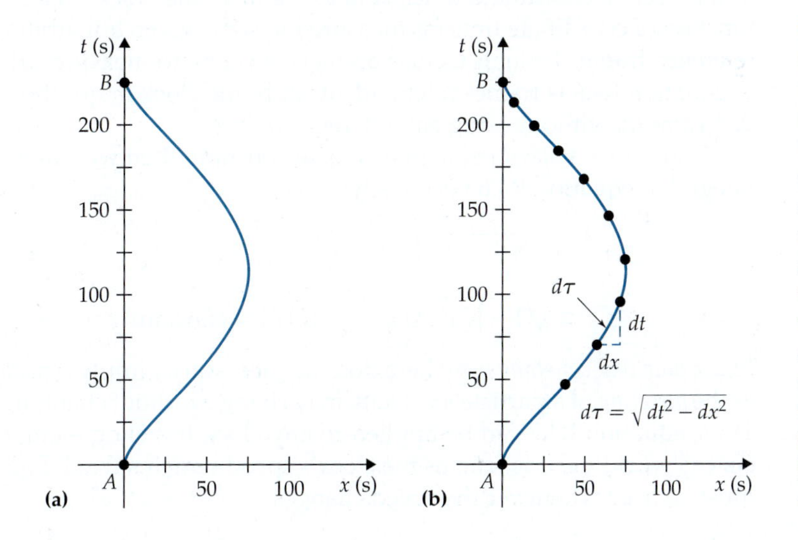

Figure R4.2
(a) A spacetime diagram of the motion of a particle's worldline based on measurements obtained in some inertial reference frame. (b) We can find the proper time along the worldline by subdividing the worldline into many segments (each small enough to be nearly straight), finding the proper time $d\tau$ along each segment, and then summing to find the total proper time. Note that because of the minus sign in the metric equation, $d\tau \le dt$ here, even though it *looks* like $d\tau \ge dt$ on the diagram.

that mark the ends of the segment, since the clock is present at both these events and travels between them with an *almost* constant velocity. Therefore, by the metric equation

$$d\tau^2 \approx ds^2 = dt^2 - dx^2 - dy^2 - dz^2 \tag{R4.4}$$

Taking the square root and pulling out a factor of the coordinate time dt yields

$$d\tau = \sqrt{1 - \left(\frac{dx}{dt}\right)^2 - \left(\frac{dy}{dt}\right)^2 - \left(\frac{dz}{dt}\right)^2}\, dt$$
$$= \sqrt{1 - v_x^2 - v_y^2 - v_z^2}\, dt = \sqrt{1 - |\vec{v}\,|^2}\, dt \tag{R4.5}$$

This equation expresses the infinitesimal proper time $d\tau$ measured by a clock traveling between two *infinitesimally separated* events in terms of the coordinate time dt between those events measured in some inertial frame and the clock's instantaneous *speed* $|\vec{v}\,|$ measured in that frame. The clock may be moving along *any* smooth worldline ($|\vec{v}\,|$ does not *have* to be constant).

To find the *total* proper time measured between events A and B by a clock traveling along the worldline, we sum the proper times measured for each nearly straight worldline segment, which in the limit of truly infinitesimal segments amounts to integrating equation R4.5:

The general formula for computing the proper time along a worldline

$$\Delta\tau_{AB} = \int_{t_A}^{t_B} (1 - |\vec{v}\,|^2)^{1/2}\, dt \tag{R4.6}$$

- **Purpose:** This equation tells us how we can use measurements performed in an inertial frame to compute the proper time $\Delta\tau_{AB}$ measured by a clock traveling between any two events A and B along an arbitrary worldline, where t_A is the time of event A, t_B is the time of event B, dt is the coordinate time differential, and $|\vec{v}(t)|$ is the clock's speed (as a function of time) all measured in a given inertial reference frame.
- **Limitations:** This equation works only in an inertial reference frame.

If we know the clock's speed $|\vec{v}(t)|$ as a function of time, then the integral is simply an ordinary one-variable integral with respect to t, which we can evaluate (at least in principle). This equation links the total proper time $\Delta\tau_{AB}$ between two events (measured by a clock traveling between events A and B) to the events' coordinate times t_A and t_B and to the clock's speed $|\vec{v}(t)|$ as a function of coordinate time (as measured in some given but arbitrary inertial reference frame. Though we use an inertial frame to measure $|\vec{v}\,|$, the *result* of equation R4.6 is frame-independent, since the clock in question measures $\Delta\tau_{AB}$ *directly* without using any reference frame.

If the clock's *speed* $|\vec{v}\,|$ happens to be constant, then we can evaluate the integral in equation R4.6 very easily:

The formula for the special case where the object's *speed* is constant

$$\Delta\tau_{AB} = \sqrt{1 - |\vec{v}\,|^2} \int_{t_A}^{t_B} dt = \sqrt{1 - |\vec{v}\,|^2}\,(t_B - t_A)$$
$$\equiv \sqrt{1 - |\vec{v}\,|^2}\,\Delta t_{AB} \qquad \text{if } |\vec{v}\,| = \text{constant} \tag{R4.7}$$

Please note that constant speed here does not necessarily imply *constant velocity,* as the *direction* of a particle's velocity may change without changing its speed. Thus, equation R4.7 can be applied to any clock traveling along straight or curved worldlines, as long as the clock's speed remains fixed. Equation R4.6 must be used whenever the speed changes.

Any proper time between two events is smaller than the coordinate time between the same events

Note that since $(1-|\vec{v}|^2)^{1/2} \leq 1$ always, the proper time that *any* clock present at two events measures between those events will be *smaller* than (or at most equal to) the coordinate time between those events measured in any inertial frame: $\Delta\tau_{AB} \leq \Delta t_{AB}$ always. However, if the clock's speed is small compared to that of light ($|\vec{v}| \ll 1$), that clock will register almost the same time between the events as measured in the inertial frame: $\Delta\tau_{AB} \approx \Delta t_{AB}$. (This is true whether $|\vec{v}|$ is constant or not.)

Two things to remember

When applying either equation R4.6 or R4.7, it is important to remember two things. First, the coordinate time Δt and the proper time $\Delta\tau$ represent the time interval between two events measured in two fundamentally different ways (just as the northward displacement Δy and the pathlength $\Delta\ell$ represent two fundamentally different ways of measuring the spatial separation of two points on the earth's surface). The coordinate time between events is measured with a pair of *synchronized* clocks in an inertial frame, while the proper time is measured by a *clock present at both events*. One cannot use these equations to link readings on just any old clocks.

Second, the quantities $\Delta\tau$ and Δt that appear on both sides of equation R4.7 always refer to the time between the same *pair* of events, measured in these two different ways. Perhaps the most common error made by beginners in applying that equation lies in implicitly using *different* pairs of events to delimit the time intervals $\Delta\tau$ and Δt. To avoid this, be sure to think *carefully* about the events involved!

The equations for proper time fail when $|\vec{v}| > 1$

Note also that equations R4.6 and R4.7 break down if $|\vec{v}| > 1$: in such a case, they predict that the time registered by the traveling clock is an *imaginary* number (which is even worse than being a negative number!). Remember that these equations are all based on the metric equation, whose derivation (see section R3.2) is only valid for pairs of events for which $\Delta t > |\Delta\vec{d}|$, that is, events between which we can send a clock traveling with $|\vec{v}| < 1$. Therefore, the equations presented so far do not specify what a clock traveling faster than the speed of light would read between two events. We will see in chapter R7 that the principle of relativity in fact implies that it is *impossible* for a clock to travel faster than the speed of light in any reference frame. The failure of these equations for the case where $|\vec{v}| > 1$ is our first indication of this.

Example R4.1

Problem: Suppose you are at rest in an inertial frame (the Home Frame) and are whirling a clock around your head at a constant rate on the end of a string 3.0 m long. A friend compares the reading of the whirling clock as it speeds by with readings from a stationary clock. Find out how long it takes the whirling clock to go once around its circular path if its reading for one cycle is 0.01 percent smaller than the period read by your friend's clock.

Solution The first step is to rephrase the problem in terms of *events*. Let event A be the whirling clock passing by the stationary clock. Let event B be the next such passage event. The whirling clock measures a proper time between these events. The stationary clock measures a *different* proper time between the events (because it is also present at both events). Since the stationary clock is at rest in the Home Frame, the time it registers is the same as the coordinate time between the events in the Home Frame. The whirling clock, on the other hand, is noninertial (its velocity is constantly changing direction as it goes around the circle). But since its *speed* is constant, we can use equation R4.7 to find the proper time it measures between events A and B.

We are given that the result is 99.99% of the time measured by the stationary clock, so

$$\Delta\tau_{AB,\text{whirl}} = \sqrt{1-|\vec{v}|^2}\,\Delta t_{AB} = 0.9999\Delta t_{AB} \tag{R4.8}$$

implying that $\sqrt{1-|\vec{v}|^2} = 0.9999$, or $1-|\vec{v}|^2 = 0.9999^2$, implying that

$$|\vec{v}|^2 = 1 - 0.9999^2 \quad \Rightarrow \quad |\vec{v}| = \sqrt{1-0.9999^2} \approx 0.014 \tag{R4.9}$$

The radius of the circle in seconds is $(3.0\text{ m})(1\text{ s}/3.00\times10^8\text{ m}) = 1.0\times10^{-8}\text{ s} =$ 10 ns. The coordinate time that a clock traveling at $|\vec{v}| = 0.014$ would take to go once around this circle is

$$\Delta t_{AB} = \frac{2\pi R}{|\vec{v}|} = \frac{2\pi(1.0\times10^{-8}\text{ s})}{0.014} \approx 4.45\times10^{-6}\text{ s} \tag{R4.10}$$

This implies a frequency of revolution of $1/\Delta t_{AB} \approx 225{,}000$ Hz. This answer makes it clear that the scenario presented in this problem is completely unrealistic. Yet this speed is what would be necessary to get even a 0.01% difference in the rate of the whirling clock relative to a stationary clock. It is no wonder that we think of time as being universal and absolute!

Example R4.2

Problem: Suppose the speed of a certain spaceship relative to an inertial frame fixed to the sun is given by $|\vec{v}(t)| = |\vec{a}|t$, where $|\vec{a}| = 10\text{ m/s}^2$. How long does it take the ship to accelerate from rest to a speed of 0.5 (in SR units) relative to the sun, as measured by clocks on the *ship*?

Solution Again, the first step is to rephrase the problem in terms of events. Let event A be the event of the ship starting to accelerate and event B the event of its passing $|\vec{v}_0| = 0.5$ (in SR units). Since the ship is present at both events, its clock measures the proper time between them. Since the speed of the ship is *not* constant in this case, we must use equation R4.6 to compute this proper time:

$$\Delta\tau_{AB} = \int_{t_A}^{t_B}\sqrt{1-|\vec{v}|^2}\,dt = \int_{t_A}^{t_B}\sqrt{1-|\vec{a}|^2t^2}\,dt \tag{R4.11}$$

In spite of the simple form of the equation $|\vec{v}| = |\vec{a}|t$, this is not a simple integral to evaluate. We can put it in a somewhat simpler form by doing the following. First, note that $t_A = 0$, because if $|\vec{v}| = |\vec{a}|t$, then $t = 0$ when the ship is at rest but beginning to accelerate. Second, note that $d|\vec{v}| = |\vec{a}|\,dt$ in this case, so we can change the variable in the integral from t to $|\vec{v}|$ as follows:

$$\Delta\tau_{AB} = \frac{1}{|\vec{a}|}\int_0^{t_B}\sqrt{1-|\vec{a}|^2t^2}\,|\vec{a}|\,dt = \frac{1}{|\vec{a}|}\int_0^{0.5}\sqrt{1-|\vec{v}|^2}\,d|\vec{v}| \tag{R4.12}$$

This integral now has a simple enough form that we can try to look it up in a table of integrals. My table of integrals says that

$$\int\sqrt{1-u^2}\,du = \frac{u\sqrt{1-u^2}}{2} + \frac{\sin^{-1}u}{2} \tag{R4.13}$$

Therefore,

$$\Delta\tau_{AB} = \frac{1}{2|\vec{a}|}(0.5\sqrt{1-0.5^2} + \sin^{-1}0.5 - 0\sqrt{1-0^2} - \sin^{-1}0)$$

$$= \frac{1}{2|\vec{a}|}(0.433 + 0.524 - 0 - 0) = \frac{0.478}{|\vec{a}|} \tag{R4.14}$$

To finish the calculation, we need to know what $|\vec{a}|$ is in SR units (since equation R4.6 and everything we have done presumes that we are working in

SR units). To change $|\vec{a}| = 10\text{ m/s}^2$ to SR units, we must convert the meters appearing in this expression to seconds:

$$|\vec{a}| = 10\frac{\cancel{\text{m}}}{\cancel{\text{s}^2}}\left(\frac{1\cancel{\text{s}}}{3.0\times10^8\cancel{\text{m}}}\right) = \frac{1}{3.0\times10^7\text{ s}} \quad \text{(R4.15)}$$

Substituting this into equation R4.14 yields

$$\Delta\tau_{AB} = 0.478(3.0\times10^7\text{s}) = 1.43\times10^7\cancel{\text{s}}\left(\frac{1\text{ y}}{3.16\times10^7\cancel{\text{s}}}\right) \approx 0.454\text{ y} \quad \text{(R4.16)}$$

Note that the units come out right, and the answer seems reasonable.

Because integrals in cases where $|\vec{v}| \neq$ constant are so difficult, we will generally stick to cases where $|\vec{v}| =$ constant in this course. I included this example to illustrate how one can do a calculation where $|\vec{v}| \neq$ constant, but I will rarely ask you to do anything nearly as difficult.

R4.3 The Binomial Approximation

Motivation for the binomial approximation

The square root that appears in equations R4.6 and R4.7 is rather difficult to evaluate for very small speeds ($|\vec{v}| << 1$). The speeds of objects we encounter on an everyday basis are on the order of $|\vec{v}| = 10^{-8}$ in SR units, meaning that $|\vec{v}|^2 \approx 10^{-16}$. When one tries to evaluate the square root $(1-|\vec{v}|^2)^{1/2}$ in such a case, one's calculator usually simply returns just 1.0, since few calculators keep track of enough decimal places to accurately register the subtraction of 10^{-16} from 1. Such an answer is not really helpful. In such cases, however, there is an approximation we can use to help us convert the square root to a more usable form.

One can show (see problems R4D.2, R4D.3, and R4A.1) that

General statement of the binomial approximation

$$(1+x)^a \approx 1 + ax \quad \text{if } x << 1 \quad \text{(R4.17)}$$

- **Purpose:** This equation is a useful trick for simplifying calculations when we are trying to find an arbitrary power a of a very small quantity x added to 1.
- **Limitations:** You will not get very good results if $|ax|$ is much larger than 0.1.

Equation R4.17, which we call the **binomial approximation**, has *many* applications in physics (memorize it!). In relativity, we will most often use it to calculate $(1-|\vec{v}|^2)^{1/2}$ for very small $|\vec{v}|$. To do this, we identify $x \equiv -|\vec{v}|^2$ and $a = \frac{1}{2}$, yielding

$$\sqrt{1-|\vec{v}|^2} \approx 1 + \tfrac{1}{2}(-|\vec{v}|^2) = 1 - \tfrac{1}{2}|\vec{v}|^2 \quad \text{(R4.18)}$$

Exercise R4X.1

Using your calculator, check the accuracy of the approximation in equation R4.18 for $|\vec{v}| = 0.1, 0.01$, and 0.001. In particular, what is the fractional error in $1-\frac{1}{2}|\vec{v}|^2$ as an approximation for $(1-|\vec{v}|^2)^{1/2}$ in each case?

Example R4.3

Problem: You and a friend stand at a street corner and synchronize your watches. You leave your friend (call this event A) and walk around the block, traveling at a constant speed of 2 m/s (about 4.5 mi/h). After a time $\Delta t_{AB} =$ 550 s as measured by your friend's watch, you return (call this event B). How much less than 550 s does your watch register between events A and B?

Solution Take your friend's frame to be inertial; your friend's watch thus measures the coordinate time between A and B in that frame (also the spacetime interval!). Since you are not moving at a constant *velocity*, you measure a proper time $\Delta\tau_{AB}$ between those events. In your friend's frame, you have a constant *speed* of $|\vec{v}| = 2$ m/s (or in SR units, $|\vec{v}| \approx 6.7 \times 10^{-9}$), so we can use equation R4.6 to calculate your proper time. Your speed is also extremely small compared to 1, so we can employ the binomial approximation to evaluate the square root. The proper time that you measure between events A and B is therefore

$$\Delta\tau_{AB} = \sqrt{1-|\vec{v}|^2}\,\Delta t_{AB} \approx \left(1 - \tfrac{1}{2}|\vec{v}|^2\right)\Delta t_{AB} \tag{R4.19}$$

Now, if we were to plug $\Delta t_{AB} = 550$ s into this equation, we would *still* find that $\Delta\tau_{AB} = 550$ s to the accuracy of a typical calculator, in spite of our use of the binomial approximation. The entire *point* of the binomial approximation, however, is that it makes it much easier to calculate the *difference* between your time $\Delta\tau_{AB}$ and your friend's time Δt_{AB}, which is what the problem requests. This difference is

$$\begin{aligned}\Delta\tau_{AB} - \Delta t_{AB} &\approx \left(1 - \tfrac{1}{2}|\vec{v}|^2\right)\Delta t_{AB} - \Delta t_{AB} = \left(1 - \tfrac{1}{2}|\vec{v}|^2 - 1\right)\Delta t_{AB} \\ &= -\tfrac{1}{2}|\vec{v}|^2\,\Delta t_{AB} = -\tfrac{1}{2}(6.7 \times 10^{-9})^2\,(550\text{ s}) \approx -1.2 \times 10^{-14}\text{ s}\end{aligned} \tag{R4.20}$$

meaning that the time is smaller than your friend's time by about 12 fs.

Note that if you *really* want to evaluate $\Delta\tau_{AB}$, you can subtract 1.2×10^{-14} s from 550 s by hand to get 549.999999999999988 s. However, since we probably don't know Δt_{AB} to 15 decimal places to begin with, the difference between $\Delta\tau_{AB}$ and Δt_{AB} is not even remotely measurable. Again, it is no wonder that we all intuitively have the idea that time is universal and absolute!

Example R4.3 illustrates that the binomial approximation is most useful when we want to calculate the *difference* between the coordinate time in some frame and a slowly moving clock's proper time. While we cannot easily calculate an expression like $\Delta\tau_{AB} - \Delta t_{AB} = \left[(1-|\vec{v}|^2)^{1/2} - 1\right]\Delta t_{AB}$ when $|\vec{v}|$ is small, we can quite easily calculate $\Delta\tau_{AB} - \Delta t_{AB} = \left(1 - \tfrac{1}{2}|\vec{v}|^2 - 1\right)\Delta t_{AB}$ because the 1s cancel. Try to make the 1s cancel similarly in any problems in which you use the binomial approximation.

If you really do need to calculate $\Delta\tau_{AB} = (1-|\vec{v}|^2)^{1/2}\Delta t_{AB}$ directly, though, the binomial approximation is still useful because it is much easier to calculate a simple difference by hand than it is to calculate a square root by hand (and a hand calculation is going to be necessary either way if your calculator yields one when you try to compute the square root).

R4.4 Ranking the Three Kinds of Time

Note that equation R4.6 implies that generally the proper time measured by a clock between two events will indeed depend on the worldline that the clock follows between the events: specifically, the proper time depends on the particular way that the clock's speed $|\vec{v}|$ varies with time. This is analogous to the way that the pathlength between two points on a plane depends on the curvature of the path along which it is measured.

$\Delta\tau$ along a *straight* worldline between two events is the *longest* possible proper time between those events

In Euclidean geometry, the straight-line distance between two points is the shortest possible pathlength between the two points. In this section, we will prove that an *inertial* clock that travels between two events (which thus measures the spacetime interval Δs between them) measures the *longest possible proper time* between those events, longer than any noninertial clock: $\Delta s \geq \Delta\tau$ *for all possible worldlines between the events.*

Proof of this theorem

How might we prove such a theorem? Consider an arbitrary pair of events, A and B. Suppose we measure the time between these events by using two clocks that are present at both events. Clock I follows an inertial worldline between the events, while clock N follows a noninertial worldline. Since clock I is inertial, its proper time $\Delta\tau_I$ will be equal (by definition) to the spacetime interval Δs_{AB} between the events. We will take advantage of the fact that we can calculate the proper time for *any* given worldline by using *any* inertial reference frame that we please, since the *result* is frame-independent. With this in mind, we can most conveniently evaluate the proper times for clocks I and N in that particular inertial frame where clock I is at *rest*. Since that clock is inertial, the frame in which it is at rest will also be inertial: let's call this the Home Frame.

Calculating the proper time along *any* worldline from event A to event B in this reference frame involves evaluating the integral $\int_{t_A}^{t_B}(1-|\vec{v}|^2)^{1/2}\,dt$, where the function $|\vec{v}(t)|$ and the endpoints t_B and t_A are all determined in the Home Frame. Now, for clock I, $(1-|\vec{v}|^2)^{1/2}=1$, since that clock is at rest in the Home Frame by construction. Since clock N travels along a *different* worldline, it must at least *sometimes* have $|\vec{v}|\neq 0$ in the Home Frame, so the integrand $(1-|\vec{v}|^2)^{1/2}$ must be *less* than 1 for at least *part* of the range of integration. So we must have

$$\Delta s_{AB} = \Delta\tau_I = \int_{t_A}^{t_B} 1\,dt > \int_{t_A}^{t_B}(1-|\vec{v}|^2)^{1/2}\,dt = \Delta\tau_N \tag{R4.21}$$

(note that both integrals have the same endpoints). Since the proper times $\Delta\tau_I$ and $\Delta\tau_N$ are frame-independent, this inequality must be true no matter *what* inertial frame we use to actually calculate $\Delta\tau_I$ and $\Delta\tau_N$. Q.E.D.

Note that in a spacetime diagram based on measurements made in an inertial reference frame, an inertial clock will have a straight worldline (since it moves with constant velocity with respect to any inertial frame) whereas a noninertial clock will have a curved worldline. The theorem we have just proved thus says that *a straight worldline between any two events on a spacetime diagram is the worldline of greatest proper time between the events.*

Analogy to a straight line being the shortest distance between two points

That the spacetime interval between two events in spacetime represents the *longest* proper time between those events while the distance between two points on a plane represents the *shortest* pathlength between those points is a direct consequence of the minus signs that appear in the metric equation where plus signs appear in the corresponding Pythagorean relation. This is another of the basic differences between spacetime geometry and the Euclidean geometry of points on a plane. Even so, a straight worldline (or path) in both cases leads to an *extreme* value for the proper time (or pathlength).

The metric equation $\Delta s = (\Delta t^2 - |\Delta\vec{d}|^2)^{1/2}$ implies that the coordinate time Δt between two events measured in any inertial reference frame is greater than (or at minimum equal to) the spacetime interval Δs between those events). The three kinds of time interval you can measure between two events must therefore stand in the strict relation:

An important inequality involving Δt, Δs, and $\Delta\tau$

$$\Delta t \geq \Delta s \geq \Delta\tau \qquad \text{(R4.22)}$$

where the first inequality $\Delta t \geq \Delta s$ becomes an equality if the events occur at the *same place* ($|\Delta\vec{d}| = 0$) in the inertial reference frame where Δt is measured (so that a single clock in that frame is present at both events) and the second inequality $\Delta s \geq \Delta\tau$ becomes an equality if the clock measuring the proper time follows the one and only *inertial path* that connects the events (so that the proper time it measures between the events *is* the spacetime interval between them as well).

R4.5 Experimental Evidence

Equation R4.6 implies that if two clocks are synchronized at event A and then travel to event B along different worldlines, the clocks will generally *not* be synchronized when they arrive at event B, since their speeds (as measured in some inertial frame) will not generally be the same as they follow their different worldlines. This prediction severely conflicts with our Newtonian intuition about time but is a testable consequence of the principle of relativity.

A muon decay experiment

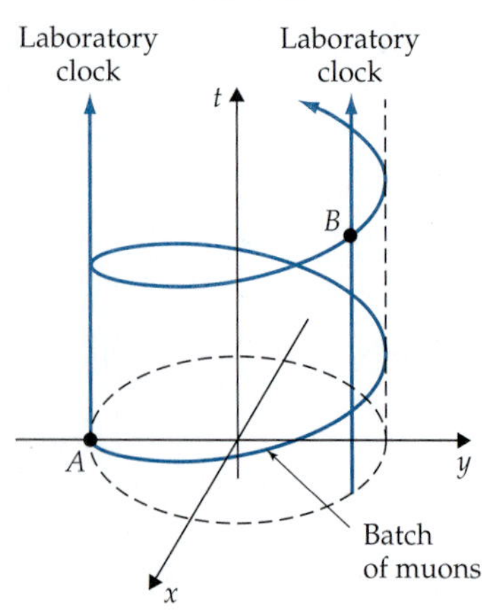

Figure R4.3
The worldlines of two laboratory clocks and a batch of orbiting muons, as drawn by an observer in the laboratory frame. A clock traveling with the muons would register less time between events A and B than the laboratory clock does. (In the experiment described, the muons actually circled the ring more than 300 times between events A and B.)

In one well-known experiment,* a particle accelerator generated muons which then traveled around a circular storage ring at a constant speed of about $|\vec{v}| = 0.99942$. Although the muons' *speed* was constant, their *velocity* was not, because they were traveling in uniform circular motion. In fact, the worldline of such a muon is quite curved and looks something like that shown in the spacetime diagram in figure R4.3. Figure R4.4 shows a muon storage ring almost exactly like the one used in this experiment.

Imagine that a pulse of muons is injected into the ring: call this event A. The muons subsequently travel around the ring. Let event B be the time and place where one-half of the bunch has decayed. The bunch of muons, since it is present at both events, measures a proper time $\Delta\tau_{AB}$ between events A and B. The laboratory clocks measure coordinate time Δt_{AB} between those events. The muons travel along a curved worldline of constant speed but with ever-changing velocity. Because the bunch's *speed* is constant, though, we can use equation R4.7 to relate the proper time to the coordinate time:

$$\Delta\tau_{AB} = \sqrt{1-|\vec{v}|^2}\,\Delta t_{AB} \qquad \text{(same as equation R4.7)} \qquad \text{(R4.23)}$$

As we have discussed before, the fact that muons decay with a certain fixed half-life makes them effectively little clocks, and measuring the decay rate of a batch of muons amounts to reading those clocks. Since the half-life of a muon at rest was known from other experiments to be 1.52 μs when muons are at rest, $\Delta\tau_{AB} = 1.52$ μs in this experiment. All that remains to be done is to measure the half-life Δt_{AB} of the muons in the storage ring according to the laboratory clock and their exact speed in that storage ring. In this particular experiment, muons were observed to have a half-life of (44.623 ± 0.18) μs in the laboratory frame (about 30 times longer than normal, in stark contrast to the Newtonian prediction), and their speed was found to be such that $1/(1-|\vec{v}|^2)^{1/2} = 29.327 \pm 0.004$. Equation R4.7 predicts that $\Delta\tau_{AB} - \Delta t_{AB}\,(1-|\vec{v}|^2)^{1/2} = 0$. This experiment showed that

$$\frac{\Delta\tau_{AB} - \Delta t_{AB}\,(1-|\vec{v}|^2)^{1/2}}{\Delta\tau_{AB}} = (4 \pm 18) \times 10^{-4} \qquad \text{(95\% confidence)}$$

meaning that this experiment shows that equation R4.7 is good to at least three significant figures. As a point of interest, these muons are accelerating

*J. Bailey et al., *Nature*, vol. 268, 1977.

Figure R4.4
The muon storage ring at Brookhaven National Laboratory before its 2013 move to Fermilab. This ring has almost exactly the same dimensions and features as the storage ring at the CERN laboratory in Switzerland that was used for the experiment described in this section. (Credit: © Brookhaven National Laboratory)

in the laboratory frame at a rate of about $10^{15}|\vec{g}|$! Therefore, equation R4.7 (at least) is seen to apply even in cases of extreme acceleration.

Exercise R4X.2

Check that the muon's laboratory-frame acceleration in this experiment really is about $10^{15}|\vec{g}|$ (the radius of the muons' circular path was 7.01 m).

An experiment involving a clock flown around the world

In 1971, J. C. Hafele and R. E. Keating* performed a test of the more general equation R4.6 on a more human scale. They synchronized a pair of very accurate atomic clocks, and then one was put on a jet plane and sent around the world while the other remained in the laboratory (see figure R4.5). Upon its return, the jet clock (which followed a noninertial worldline in its trip around the earth) was compared with the inertial clock that remained at rest in the frame of the earth. With suitable corrections for the effects of gravity (predicted by *general* relativity), the results (a difference of a few hundred nanoseconds) were found to be in complete agreement with equation R4.6.

The Global Positioning System also essentially performs a continuously running test of equation R4.6. Each orbiting GPS satellite carries an atomic clock whose rate is compared to atomic clocks on the earth's surface, and the rate of each orbiting clock is continually corrected for the effect of equation R4.6, as well as for gravitational effects predicted by general relativity. If these corrections were not continually applied (or if equation R4.6 did not supply the right correction), the GPS would deliver inconsistent position results. That the GPS works is therefore testimony to the validity of equation R4.6.

Many other experiments have been performed to check this prediction of the theory of relativity, and all have been in complete agreement with the predictions of equation R4.6. Outrageous as it may seem, the idea that two clocks present at the same two events do not register the same time between those events is a well-established experimental fact.

Figure R4.5
Hafele and Keating carry one of their atomic clocks out of an airplane in the process of performing their 1971 experiment. (Credit: © Israel Sun/AP Images)

*J. C. Hafele and R. E. Keating, *Science,* vol. 117, p. 168, July 14, 1972.

R4.6 The Twin Paradox

As a result of misinterpreting the meaning of equation R4.7, many people (including competent professors of physics) have been unnecessarily perplexed by apparent paradoxes in the theory of relativity. Physicists call one of the most famous of these the **twin paradox** (or sometimes the *clock paradox*). This problem generated reams of journal articles (as late as the 1960s) before the inadequacy of the language and concepts commonly used to describe relativity at that time became sufficiently well understood.

Here is a statement of the apparent paradox. Andrea and Bernard are twins. When they are both 25 years old, Andrea accepts a commission to be an exobiologist on an expedition to Sirius, which is about 8.6 light-years from earth. So she flies away on a ship that is capable of near-light speeds, leaving her brother Bernard on earth. The years roll by and the world waits. Finally, hurtling out of the emptiness of space, the spacecraft returns.

Bernard expects Andrea to be younger

As he waits for his sister to emerge from the newly arrived spacecraft, Bernard (now a distinguished man of 50) muses on the bit of relativity that he remembers from college. He recalls that "$\Delta\tau_{AB} = \int(1-|\vec{v}|^2)^{1/2}\,dt$." Since Andrea has been moving with a large speed $|\vec{v}|$ for much of the trip, Andrea's clocks should measure much less time for the trip than his clocks register. This includes biological as well as mechanical clocks, and so Bernard expects to see a substantially younger sister emerge from the hatch, still displaying their once common youthful vitality. Bernard chews his lip, wondering what it will be like to have a younger "twin" sister.

Andrea expects Bernard to be younger

Similar thoughts run through Andrea's mind as she prepares to disembark. In Andrea's frame, however, she and the spacecraft were motionless, and the earth (and thus Bernard) has moved backward 8 light-years and returned. Andrea (who had the same course in college) thinks that since it is Bernard whose speed has been nonzero, it will be *Bernard* who is younger.

Doesn't the principle of relativity imply that they should have the same age?

The paradox is clear: each expects the other to be younger from their partial recollection of relativity theory. To this confusion, we can add a third perspective. The principle of relativity states that the laws of physics are the same in every inertial frame. This means we cannot physically distinguish two inertial frames: if you perform identical experiments within each reference frame, you must get the same results. But isn't the aging process essentially a physical experiment that each person performs in his or her reference frame? If *either* twin is younger than the other, won't that distinguish between the frame of the earth and the frame of the spaceship, contrary to the principle? So shouldn't they have the same age?

Solution to the paradox

We have in fact *already* resolved this paradox in this chapter: we simply need to rephrase it in more appropriate language. The first task is to clarify what *events* we are talking about. Let us define event A to be the ship's departure from earth. Let its arrival back on earth be event B. The twins are present at both events A and B, so their clocks (including their biological clocks) measure (different) proper times between the events. The question is, which of these twins measures the longer proper time between these events? To find out, we need to sketch their worldlines as measured in some inertial reference frame.

For our master inertial reference frame, let us choose a frame at rest with respect to the sun. Since the sun is freely falling around a very distant galactic center of mass, this will be an excellent approximation to an inertial frame.

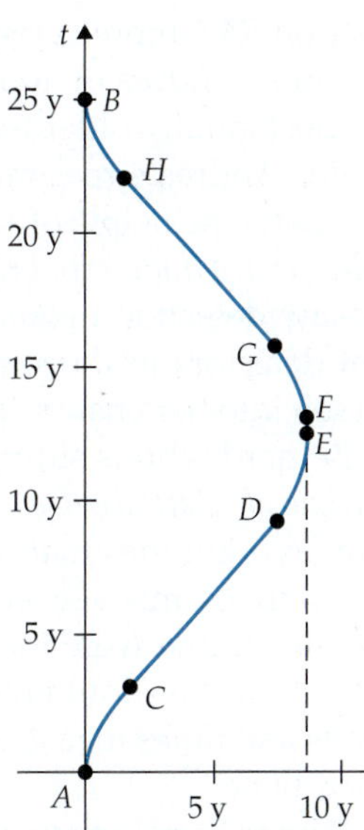

Figure R4.6
The worldline of Andrea's spaceship as drawn by an observer in the frame of reference of the sun.
A: Andrea's departure from earth
C: ship reaches cruising speed
D: ship begins decelerating
E: ship arrives at Sirius
F: ship departs from Sirius
G: ship reaches cruising speed
H: ship begins decelerating
B: ship arrives at earth

Now let us sketch the twins' worldlines as observed in this frame. Andrea takes off from the earth at event *A*, which we can take to be the origin event (i.e., the event that defines $t = x = 0$ in the sun frame). Her spacecraft travels slowly at first, but gradually picks up speed as it strains toward the speed of light. After a year or so, it reaches a cruising speed. But before it reaches Sirius, it must begin to slow down. Finally, it coasts into the star system and lands. The process of acceleration and deceleration repeats on the way home. Event *B* is the event of her return to earth. Figure R4.6 shows a spacetime diagram of the resulting worldline.

On this diagram, the earth's (and thus Bernard's) worldline is a tiny helix winding around the *t* axis, twisting around it about 25 times in the roughly 25-y duration of the flight. So Bernard does *not* measure coordinate time between events *A* and *B*, as his vague argument seems to suggest: rather, he measures a *proper time* between those events that is somewhat different from coordinate time. But since Bernard is never more than about 8 light-minutes ($\approx 1.5 \times 10^{-5}$ y) from the sun, the squiggles of his little helix are far too tiny to show up on the diagram. Thus, his worldline is essentially a straight line up the *t* axis. Moreover, since Bernard's speed in his worldline (roughly equal to the earth's orbital speed) is about 30 km/s $\approx 10^{-4}$ in SR units, the proper time measured by his clock only differs from that measured by coordinate clocks in the sun's frame by about $[1 - (1-|\vec{v}|^2)^{1/2}]\,\Delta t_{AB} \approx +\frac{1}{2}|\vec{v}|^2 \Delta t_{AB} = (5 \times 10^{-9})(25\text{ y}) \approx 4$ s over the time period between events *A* and *B*. We see for Bernard that the proper time he measures between events *A* and *B* is essentially the same as the time measured in the sun's frame between those events.

But for Andrea, the situation is different. In the sun's frame, Andrea spends quite a bit of time traveling at nearly the speed of light: her average speed in this frame is 17.2 ly/25 y ≈ 0.69. Therefore, the factor $(1-|\vec{v}|^2)^{1/2}$ that appears in the formula for her proper time will be quite a bit smaller than 1 for major portions of the trip. As a result, Andrea's clocks (including her body's biological clock) register much less time between *A* and *B* than Bernard's clocks do. Even though 25 y passes on earth, clocks traveling with Andrea will measure a proper time of (approximately) $(1 - 0.69^2)^{1/2}(25\text{ y}) \approx 18$ y between her departure and arrival (we would need to do the integral of equation R4.6 more carefully to get a more accurate answer). So it is indeed a younger Andrea of about age 43 that bounds out of the spacecraft to greet her substantially older twin, Bernard.

The core issue: Andrea's frame is *not inertial*

But Andrea's reasoning seems perfectly logical. Why is it wrong? And what of the principle of relativity? The answer to both questions is the same: *Andrea is in a noninertial reference frame*. Every time the spacecraft engines fire, first-law detectors in *Andrea's* frame register a violation of that law (equivalent detectors in the sun's frame would read nothing). It is *Andrea* who is pressed into her chair as the engines accelerate and decelerate the spacecraft, not Bernard who is sitting at home. Since Andrea's frame is not inertial, the *principle of relativity does not apply to her*. This exposes the error in the argument favoring the *equality* of the twins.

Andrea's mistake is to apply the proper time formula as if her own reference frame were inertial (and thus as if her clocks measure coordinate time Δt_{AB}). She should compute proper times between events A and B using speeds and times measured in a *real* inertial reference frame, which her frame is clearly not.

Bernard's *reasoning* is no better than Andrea's, since he is not in an inertial frame either! But since Bernard's frame is *nearly* inertial, when he applies the proper time formula to the times measured between events A and B, he gets an answer that is at least *approximately* correct. Andrea *really is* about 7 y younger than Bernard. There is no ambiguity and no paradox.

A footpath analogy

This situation should not seem any more paradoxical than the following (more familiar) situation. Imagine that Alex and Brin both set off from a given point A on the surface of the earth. Brin (analogous to Bernard here) takes an approximately straight path from point A to the destination point B, while Alex takes a curved path. Should they be shocked when they arrive at B and find that Alex has walked more miles than Brin? Hardly! Alex might try to claim it was Brin who departed and then returned and so must have taken the curved path, but this is misleading. The curvature of Alex's path is an absolute physical property of that path, a property that can be displayed in any fixed coordinate system. Alex's personal coordinate system, whose axes change direction every time Alex takes a new turn, is not an appropriate coordinate system for displaying a path's curvature. We don't have trouble accepting the nonequivalence of the distance measured along Alex's curved and Brin's straight footpaths in *this* case, so we should not have trouble accepting the nonequivalence of the proper times measured along Andrea's and Bernard's worldlines in the first case.

Part of the problem is people sometimes say that $\Delta\tau_{AB} = \int(1-|\vec{v}|^2)^{1/2}dt$ (equation R4.6) means that "moving clocks run slow." This is a completely misleading statement that can lead to all kinds of misconceptions and errors. You should not focus on which clocks are "moving," because that is frame-dependent. Instead focus on which clocks are measuring *what kind of time*. Coordinate times are always greater than or equal to the spacetime interval, which in turn is greater than or equal to any proper time. Focusing on the kind of time being measured, therefore, gives you certain and unambiguous information about who measures the greater or smaller time.

Exercise R4X.3

Why is the statement that Andrea's proper time is equal to $(1-0.69^2)^{1/2}$ (25 y) an approximation? Do you think it is likely to be a bit too high or a bit too low as an estimate?

TWO-MINUTE PROBLEMS

R4T.1 Consider the spacetime diagram below, which shows events A and B and the worldlines of clocks that one might use to measure the time between those events.

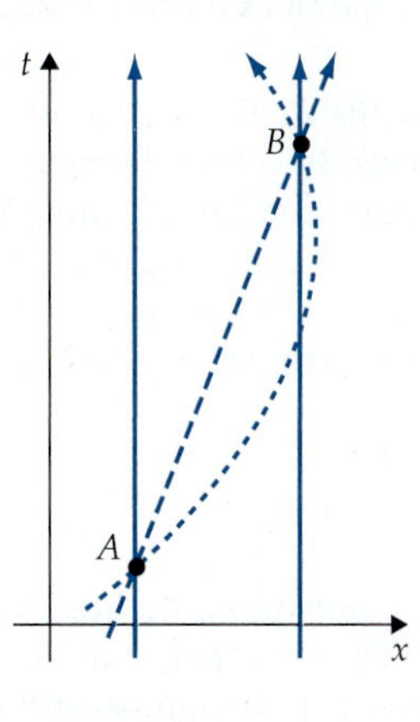

The clock(s) following which worldline(s) measure(s)
(a) the spacetime interval between events A and B?
(b) a proper time between events A and B?
(c) a coordinate time between events A and B?
(d) the shortest time interval between events A and B?
(e) the longest time interval between events A and B?
A. ______
B. — —
C. - - - - -
D. ______ and - - - - -
E. — — and - - - - -
F. ______ and — —
T. None of these choices

R4T.2 Suppose we carefully synchronize two identical atomic clocks initially standing next to each other (call them A and B). We put clock B on a jet plane, which then flies around the world at an essentially constant speed of 300 m/s, returning 134,000 s (37.1 h) later. We then again compare the two clocks. Assume the earth's surface defines an inertial reference frame, and ignore the possible effects of gravity.
(a) Which clock measures the spacetime interval between the synchronization and comparison events?
(b) Which clock measures a coordinate time between the synchronization and comparison events?
(c) Which clock measures the shorter time interval between the synchronization and comparison events (or do both measure the same time)?
A. Clock A
B. Clock B
C. Both
D. Neither

R4T.3 In the round-the-world experiment described in problem R4T.2, what is the minimum accuracy over the experiment's duration that the clocks must have to clearly display the relativistic effect?
A. Both clocks must be accurate to the nearest 10 ms.
B. Both clocks must be accurate to the nearest 10 μs.
C. Both clocks must be accurate to the nearest 10 ns.
D. Both clocks must be accurate to the nearest 10 ps.

R4T.4 Jennifer bungee-jumps from a bridge (event A). Jennifer's bungee cord is perfectly elastic, so she bounces exactly back up to the bridge and lands on her feet (event B). The time between these events is measured by Jennifer's watch, a stopwatch held by Jennifer's friend, Rob, who is standing on the bridge, and by two passengers (one present at event A and one present at event B) who are riding on a train traveling at a constant velocity across the bridge at the time (the passengers have synchronized watches and compare readings later). Assuming the earth's frame is inertial, who measures
(a) the longest time interval between these events?
(b) the shortest time interval between these events?
(c) the spacetime interval between the events?
A. Jennifer
B. Rob
C. The train passengers

R4T.5 In the situation described in problem R4T.4, the train passengers are moving, but Rob is at rest. Therefore, the train passengers measure less time between the events than Rob does. T or F?

R4T.6 Suppose we synchronize two atomic clocks at a point at 45° south latitude, and then move one clock directly north to the earth's equator and the other directly south to the south pole, where they remain for some years in climate-controlled enclosures that keep them at the same temperature and humidity. We then reunite the clocks at the origin point and compare them again. Which (if either) has registered a shorter time between the synchronization and comparison events?
A. The clock at the equator
B. The clock at the south pole
C. Both clocks read the same time.

R4T.7 GPS satellites go around the earth in orbits that have a common radius of 26,600 km and a period of 12 h. Roughly how much less time would an atomic clock on a GPS satellite register between two events separated by exactly 24 h than clocks in the reference frame of the earth (ignoring gravitational effects on the satellite's clock rate)?
A. About 10 ms
B. About 1 ms
C. About 100 μs
D. About 10 μs
E. About 1 μs
F. About 10 ns

R4T.8 The coordinate time between two given events is shortest in the inertial frame where their spatial separation is the smallest. T or F?

HOMEWORK PROBLEMS

R4B.1 A spaceship leaves earth (event A), travels to Pluto (which is 5.0 h of distance away at the time), and then returns (event B) exactly 11.0 h later. If the spaceship's acceleration time is very short, so that it spends virtually all its time traveling at a constant speed, estimate the time measured between events A and B by the ship's clock.

R4B.2 A spaceship leaves earth (event A), travels to Alpha Centauri (which is 4.3 y of distance away), and then returns (event B) exactly 9.0 y later. If the spaceship's acceleration time is very short, so that it spends virtually all its time traveling at a constant speed, estimate the time measured between events A and B by the ship's clock.

R4B.3 The designers of particle accelerators use electromagnetic fields to boost particles to relativistic speeds while at the same time constraining them to move in a circular path inside a donut-shaped evacuated cavity. Imagine a particle traveling in such an accelerator in a circular path of radius 7.01 m at a constant speed of 0.9994 (as measured by laboratory observers). Let events A be the particle passing a certain point on its circular path, and let event B be the particle passing the point of the circle directly opposite that point, as shown below.

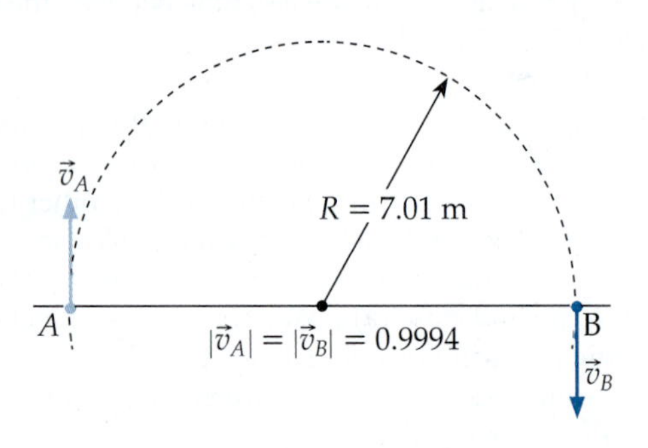

(a) What are the coordinate time Δt and the distance $|\Delta \vec{d}|$ between these events in the laboratory frame? [*Hint:* $|\Delta \vec{d}| \neq \pi\,(7.01\text{ m})$! Think about it more carefully!]
(b) What is the spacetime interval Δs between the events?
(c) What is the proper time $\Delta \tau$ between the events, as measured by a clock traveling with the particle? About how many times greater than $\Delta \tau$ is Δt?

R4B.4 Suppose a new bullet train can go all the way around the world in 6.235 s as measured by clocks at the station. Assuming the train cruises at a constant speed, how long do the passengers' watches register for a complete circumnavigation of the globe? The radius of the earth is about 6380 km.

R4B.5 We synchronize two atomic clocks, and then put one in a high-speed train car that subsequently goes 50 times around a circular track (radius 10.0 km) at a constant speed of 300 m/s. We then again compare the two clocks. By how much do the clocks now differ?

R4B.6 We synchronize two atomic clocks, and then put one in a race car that subsequently goes 50 times around a circular track (radius 5.0 km) at a constant speed of 60 m/s. By how much do the clocks differ afterward?

R4B.7 A jogger runs 22 times around a 0.50-km track in 48 min, as measured by a friend sitting at rest on the side.
(a) If the jogger and friend synchronize watches before the run, by how much do they differ afterward?
(b) Is this the reason why many people expect joggers to live longer than people who don't jog? Explain.

Modeling

R4M.1 Suppose a person commutes 50 mi each way to work and back 250 days per year for 35 y. During the commute, the person drives almost entirely on the freeway at an approximately constant speed of 65 mi/h.
(a) How much less time has this person's watch registered in 35 y than that of someone who has stayed at home?
(b) Is it true that commuters live significantly longer, or does it just seem longer?

R4M.2 Allison, Brad, Chris, and Dylan are enjoying a trip to the amusement park. Allison goes to ride the gigantic Ferris wheel (which is 47.75 m in diameter), but Brad (who doesn't like heights) elects simply to watch. Chris and Dylan are riding the monorail that passes along a straight track just below the Ferris wheel's bottom. The monorail moves at a constant speed of 9 m/s. The first time Allison passes Brad (who is standing right next to Allison but on the ground), Allison and Brad look at their watches, and Chris (who happens to be passing under Allison's seat on the Ferris wheel at that instant) looks at the clock on her seat's fancy electronic display. Call this event A. When Allison passes Brad the next time (call this event B), Allison and Brad again look at their watches, and Dylan, who happens to be passing under Allison's position at just that instant, looks at his display (which we will assume is synchronized with Chris's display). Everyone determines the time interval that they measure between these momentous events (Chris and Dylan by subtracting Chris's value for event A from Dylan's value for event B). Brad measures exactly 50 s between events A and B.
(a) Who measures the shortest time between these events? Who measures the longest? Explain. [*Hint:* What kind or kinds of time does each person measure?]
(b) How much larger or smaller is the time that Allison measures than the time that Brad measures?
(c) How much larger or smaller is the time that Chris and Dylan measure than Brad's time? Explain. (*Hint:* Note that their time will be *very close* to Brad's time.)
(d) Chris and Dylan are moving in the ground frame. Shouldn't they therefore measure *less* time between the events than Bob? Explain why the "moving clocks run slow" idea is very misleading here.

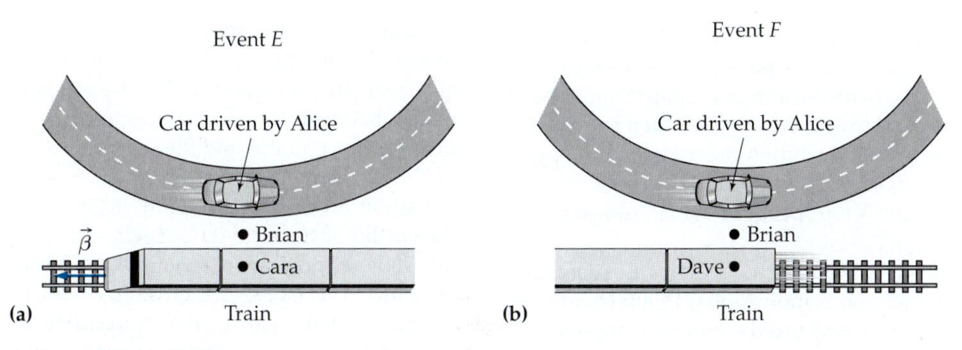

Figure R4.7
(a) Event E and (b) event F in the situation described in problem R4M.3.

R4M.3 Alice is driving a race car around an essentially circular track at a constant speed of 60 m/s. Brian, who is sitting at a fixed position at the edge of the track, measures the time that Alice takes to complete a lap by starting his watch when Alice passes by his position (call this event E) and stopping it when Alice passes his position again (call this event F). This situation is also observed by Cara and Dave, who are passengers in a train that passes very close to Brian. Cara happens to be passing Brian just as Alice passes Brian the first time, and Dave happens to pass Brian just as Alice passes Brian the second time (see figure R4.7). Assume the clocks used by Alice, Brian, and Cara are close enough together that we can consider them all to be "present" at event E; similarly, that those used by Alice, Brian and Dave are "present" at event F. Assume the ground frame is an inertial reference frame.

(a) Who measures the shortest time between these events? Who measures the longest? Explain.

(b) If Brian measures 100 s between the events, how much less time does Alice measure between the events?

(c) If Cara's and Dave's train moves at a speed of 30 m/s, how much larger or smaller is the time that they measure compared to Brian's time? Explain carefully.

(d) Chris and Dylan are moving in the ground frame. Shouldn't they therefore measure *less* time between the events than Bob? Explain why the "moving clocks run slow idea" is very misleading here.

R4M.4 The half-life of a muon at rest is 1.52 μs. One can store muons for a much longer time (as measured in the laboratory) by accelerating them to a speed very close to that of light and then keeping them circulating at that speed in an evacuated ring. Assume you want to design a ring that can keep muons moving so fast that they have a laboratory half-life of 0.25 s (about an eye-blink).

(a) How fast must the muons be moving? (*Hint:* Define $u \equiv \Delta\tau/\Delta t$, write an equation that links u to $|\vec{v}|$, then solve for $|\vec{v}|$ and use the binomial approximation. You may need to do the final calculation of $|\vec{v}|$ by hand.)

(b) If the ring is 7.01 m in diameter, how long will it take a muon to go once around the ring in the lab frame?

R4M.5 Suppose some astronauts travel in a near-earth orbit at an altitude of 200 km for 225 orbits.

(a) In unit N, we saw that the speed of an object in a circular orbit of radius R around an object with mass M is $|\vec{v}| = \sqrt{GM/R}$, where G is the universal gravitational constant. Argue that in SR units, $G = G_{[SI]}/c^3 = 2.475 \times 10^{-36}$ s/kg.

(b) About how much less time passed between the departure and arrival of the spaceship according to the astronauts' clocks than passed on the ground? Assume for the sake of simplicity that the surface of the earth defines an inertial reference frame.

R4M.6 The satellites used in the Global Positioning System go around the earth in circular orbits whose radius is 26,600 km and period is 12 h exactly. Assume for the sake of simplicity that the earth is not rotating, so that a clock on its surface is in an inertial frame.

(a) In unit N, we saw that the speed of an object in a circular orbit of radius R around an object with mass M is $|\vec{v}| = \sqrt{GM/R}$, where G is the universal gravitational constant. Argue that in SR units, $G = G_{[SI]}/c^3 = 2.475 \times 10^{-36}$ s/kg.

(b) Let event A be a certain GPS satellite passing a given position in space and event B be it passing that point again after one complete orbit. At each event, this satellite sends a radio signal to a clock directly below it on the (nonrotating) earth, which receives the signals at events C and D, respectively. What is the difference between the time an atomic clock on board the satellite registers between events A and B and the time a clock on the earth's surface registers between events C and D? Express your result *symbolically* in terms of G, M, and R (don't crunch numbers yet), though you can assume that $GM/R \ll 1$. (*Hint:* Argue that the signal's travel time is the same in both cases.)

(c) Now calculate numerically how much less time a clock on the GPS satellite measures for a complete orbit than the clock on the ground does.

(See problem R4A.3 for a discussion of how the earth's gravity affects GPS satellite clock rates.)

R4M.7 Integrating equation R4.6 is a lot less tricky if $|\vec{v}|$ is always small enough that we can use the binomial approximation. Suppose a spaceship starts from rest from Space Station Alpha floating in deep space and accelerates at a

R4M.7 (continued) constant rate of $|\vec{a}| = 10$ m/s^2 (relative to the station) for 1.0 Ms ($\approx$12 days), decelerates for the same amount of time to arrive at rest at Space Station Beta (which floats at rest relative to Alpha), and then repeats the same acceleration and deceleration processes in the opposite direction to get back to space station Alpha.

(a) Find the spaceship's top speed. Is it small compared to the speed of light?

(b) About how much less time has elapsed on the spaceship compared to clocks on Station Alpha? (*Hint:* Divide the trip into four pieces, and use the binomial approximation to convert the integral in equation R4.6 to two simpler integrals for each piece.)

Derivations

R4D.1 Compare $(1-|\vec{v}|^2)^{1/2}$ and $1-\frac{1}{2}|\vec{v}|^2$ for the following values of $|\vec{v}|$: 0.5, 0.2, 0.1, 0.05, 0.01, 0.002. Which is the largest of these values for which the difference between the two expressions is smaller than 1%? Smaller than 0.01%?

R4D.2 Here is a simple way to understand the binomial approximation. Consider $(1 + x)^2 = (1 + x)(1 + x)$. If you multiply this out, you get $(1 + x)^2 = 1 + 2x + x^2$. Now, if $|x| \ll 1$, then $x^2 \ll x$, so we have $(1 + x)^2 \approx 1 + 2x$, just as the binomial approximation states. Apply the same kind of reasoning to $(1 + x)^3$ and $(1 + x)^4$, and show that the binomial approximation works in these cases, too, when x is small enough that we can ignore x^2 (and higher powers of x) compared to x. (While this is not a proof, it may help you understand the basic issues involved.)

R4D.3 One can *prove* the binomial approximation more generally as follows. According to the definition of the derivative, we have for any function $f(x)$

$$\frac{f(x)-f(0)}{x} \approx \left[\frac{df}{dx}\right]_{x=0} \quad \text{if } x \text{ is very small} \qquad \text{(R4.24)}$$

where $[df/dx]_{x=0}$ tells us to evaluate the derivative at $x = 0$. Now consider the function $f(x) = (1 + x)^a$. Apply equation R4.24 and the chain rule to this function to get $f(x) \approx 1 + ax$.

R4D.4 In section R4.4, we proved that the coordinate time, the spacetime interval, and the proper time between a given pair of events stand in the relationship $\Delta t \geq \Delta s \geq \Delta\tau$, no matter what inertial frame is used to measure Δt and no matter what worldline is followed by the clock measuring $\Delta\tau$. It was asserted there that $\Delta t = \Delta s$ if and only if Δt is measured in a frame where the distance $|\Delta\vec{d}|$ between these events is zero. It was also *asserted* that $\Delta s = \Delta\tau$ if and only if the clock measuring $\Delta\tau$ is inertial. Write a short argument supporting each of these statements (for both directions of the "if and only if").

Rich-Context

R4R.1 Consider the Hafele–Keating experiment discussed in section R4.5. In this experiment, two atomic clocks were synchronized (event A), one was put on a jet and flown around the world, and then the clocks were compared (event B). Our task in this problem is to make a reasonably realistic prediction of the discrepancy that we would expect between the clocks. Suppose the plane starts at a point on the equator and flies around the world at a speed of 290 m/s relative to the ground. Assume the plane cruises at an altitude of about 35,000 ft $\approx$ 10.7 km above the earth's surface.

(a) Make a prediction (accurate to about $\pm$1%) of how much the clocks will disagree when they are compared at the end of the experiment if the plane flies east around the equator (do not ignore the earth's rotation). Describe any assumptions or approximations you are compelled to make. (*Hint:* Do your analysis in an inertial reference frame fixed to the earth's center but which does *not* rotate with the earth. You can use the *coordinate* time Δt measured in this frame between the initial synchronization and the final comparison events as a reference to compare earth clocks and plane clocks. It is tricky to calculate an *exact* number for this coordinate time, but in your final calculation, note that this time will be equal to the time measured on the earth's surface to *much* better than $\pm$1%.)

(b) Repeat your calculation, assuming the plane flies west. Why is your answer different from the one you found for part (a)?

(c) *General* relativity predicts that a clock that is a distance h higher in a gravitational field than a second clock will run *faster* than the lower clock by the factor $1+|\vec{g}|h/c^2$ (in SI units), where $|\vec{g}|$ is the earth's gravitational field strength (9.8 m/s^2). How does including this information change your answer to part (a)?

R4R.2 Because the earth is freely falling around the sun, its center defines a pretty good inertial reference frame. With this in mind, consider the fates of two identical twins, one living since infancy at the North Pole and the other living since infancy on the equator. Imagine that both twins die after exactly the same biological time has passed (as determined by their own bodies). If this is so, about how much longer will the equatorial twin live than the northern twin if both live for normal time spans? Describe any approximations or assumptions you have to make.

R4R.3 The Spacer's Challenge is a yearly spaceship race between Starbase Delta and Starbase Epsilon, which float at rest with respect to each other a distance D apart. Participants may travel between the two along any worldline they like, as long as they arrive at Starbase Epsilon at a time T within $\pm$0.1% of 1.60D after departing from Starbase Delta (as measured by official synchronized clocks on the two starbases). The winner is the spaceship having the smallest value of the distance (in SR units) traveled along its worldline plus the ship's proper time measured along that worldline. (These quantities have the same SR units.)

Now, traveling faster between the arrival and departure events will reduce your proper time, but will also require that you travel a greater distance along a curved worldline to ensure your total trip time remains the same in the starbase frame, so one must consider the tradeoffs. The simplest curved worldline between the starbases is a circle in space with radius $\frac{1}{2}D$. If you follow such a worldline, will you beat a spaceship that simply travels directly at a constant velocity between the starbases?

Advanced

R4A.1 If you know about Taylor series, you can prove the binomial approximation quite generally. Any continuous and differentiable function $f(x)$ can be expressed in terms of a Taylor series expansion as follows:

$$f(x) = f(0) + x\left[\frac{df}{dx}\right]_{x=0} + \frac{x^2}{2!}\left[\frac{d^2f}{dx^2}\right]_{x=0} + \frac{x^3}{3!}\left[\frac{d^3f}{dx^3}\right]_{x=0} + \cdots \quad \text{(R4.25)}$$

Apply this to the function $f(x) = (1 + x)^a$ and show that if you drop terms in this power series involving x^2 or higher, you end up with the binomial approximation. Also show how you would write the approximation if you were to keep terms involving x^2 but drop higher-order terms.

R4A.2 Consider an inertial frame at rest with respect to the earth. We observe an alien spaceship to move along the x axis of this frame in such a way that

$$x(t) = \frac{1}{\omega}[\sin(\omega t + \tfrac{1}{4}\pi) - b] \quad \text{(R4.26)}$$

where both x and t are measured in the inertial reference frame, $\omega = \pi/2$ rad/h, and $b = \sin(\pi/4)$. Assume also that the earth is located at the origin ($x = 0$) in this frame.

(a) Argue that the ship passes the earth at $t = 0$ and again at $t = 1.0$ h. (*Hint:* The value of ωt is $\pi/2$ at this time.)

(b) Draw a quantitatively accurate spacetime diagram of the spaceship's worldline, labeling the events where and when it passes the earth as events A and B.

(c) Show that the ship's x-velocity is $v_x = \cos(\omega t + \pi/4)$ as measured in the inertial frame attached to the earth. (*Hint:* You don't need to use any relativity!)

(d) Find the proper time measured by clocks on the alien ship between the events where it passes earth the first and second times. (*Hint:* $1 - \cos^2 x = \sin^2 x$.)

R4A.3 Consider the Global Positioning System satellites described in problem R4M.6. Again, for simplicity's sake, suppose the earth is not rotating. Let Δt_0 be the time between two events that bracket one complete satellite orbit as measured by a clock at rest with respect to the earth but so far away that the effect of the earth's gravity on its rate is negligible. (We'll call this "the clock at infinity.")

(a) The satellite's speed is $|\vec{v}| = \sqrt{GM/R}$, where G is the universal gravitational constant (2.475×10^{-36} s/kg in SR units), M is the earth's mass, and R = 26,600 km is the orbit's radius. Assuming that $GM/R \ll 1$, use the binomial approximation to find an expression (in terms of G, M, and R) for the discrepancy $\delta t_v = \Delta t_0 - \Delta\tau$ between what the clock at infinity and the satellite's clock measure for a full orbit due to the satellite's motion.

(b) General relativity states that a clock at rest a distance R from the center of a planet of mass M runs more slowly than the clock at infinity by the factor $\sqrt{1 - 2GM/R}$ (in SR units) due to the planet's gravitational field. Find a symbolic expression for the gravity-induced discrepancy δt_g between what the clock at infinity and the stationary clock at R measure for a full orbit.

(c) Similarly find the gravity-induced discrepancy δt_e between what the clock at infinity and a clock on the earth's surface at radius R_e measure for a full orbit.

(d) Find a symbolic expression for the *total* discrepancy δt between what the satellite's clock and the clock on the earth's surface measure for a full orbit, taking into account all of these effects.

(e) Evaluate δt numerically and interpret its sign.

ANSWERS TO EXERCISES

R4X.1 See the table below. All results are rounded to nine decimal places.

$\lvert\vec{v}\rvert$	$\sqrt{1-\lvert\vec{v}\rvert^2}$	$1-\frac{1}{2}\lvert\vec{v}\rvert^2$
0.1	0.994987437	0.995000000
0.01	0.999949999	0.999950000
0.001	0.999999500	0.999999500

We see that the approximation is accurate to four decimal places even when $|\vec{v}|$ is as large as 0.1.

R4X.2 The ratio of a muon's acceleration $|\vec{a}|$ in the laboratory frame to $|\vec{g}|$ is

$$\frac{|\vec{a}|}{|\vec{g}|} = \frac{|\vec{v}|^2}{|\vec{g}|R} = \frac{0.99942^2}{(9.8\,\text{m}/\text{s}^2)(7.01\,\text{m})}\left(\frac{3.0\times10^8\,\text{m}}{1\,\text{s}}\right)^2 \approx 1.3\times10^{15} \quad \text{(R4.27)}$$

R4X.3 Andrea is not moving at a constant speed, so the use of the constant-speed formula for proper time is not appropriate, even when one uses the average speed. On the other hand, she spends *most* of her time traveling at a constant speed, so this will be a reasonable approximation. To see whether this is likely to yield an answer too high or too low, let us consider a specific and fairly extreme case. Suppose Andrea travels at a speed of 0.86 for four-fifths of the time and is at rest for the remaining time, as measured in the sun-based frame: this yields the desired average of $\frac{4}{5}(0.86) = 0.69$. In this case, though, we can compute the proper time more accurately by breaking each half of the path into two segments, one where she is moving at a constant speed of 0.86 and one where she is at rest. Andrea's total elapsed proper time in this particular case is

$$(1 - 0.86^2)^{1/2}\,\tfrac{4}{5}\,(25\text{ y}) + (1)^{1/2}\,\tfrac{1}{5}\,(25\text{ y}) = 15\text{ y} \quad \text{(R4.28)}$$

instead of 18 y. If this example is any indication, using the average speed yields an estimate that is too high.

R5 Coordinate Transformations

Chapter Overview

Introduction

To delve further into the implications of the principle of relativity, we need to find a way of linking an event's coordinates t, x, y, and z in one inertial frame with its coordinates t', x', y', and z' in another. This chapter develops tools for doing this, and chapters R6 and R7 use these tools to explore applications.

Section R5.1: Overview of Two-Observer Diagrams

A **two-observer spacetime diagram** is a spacetime diagram where the coordinate axes of two different observers are superimposed on events in spacetime.

Section R5.2: Conventions

We will assume in this and future chapters that the two observers are using inertial reference frames in *standard orientation* (that is, whose frame axes point in the same directions in space) with the Other Frame moving along the x axis relative to the Home Frame, and that spatial origins of both frames coincide at $t = t' = 0$.

Section R5.3: Drawing the Diagram t' Axis

The t' axis is the set of all events that happen at $x' = 0$ (the Other Frame's origin). Since that origin moves along the x axis with x-velocity β relative to the Home Frame, the t' axis is simply a worldline with a slope of $1/\beta$. The section argues that we should calibrate this axis either by using hyperbolas or by drawing marks whose *vertical* separation is $\gamma \equiv (1 - \beta^2)^{1/2}$ larger than the marks on the t axis.

Section R5.4: Drawing the Diagram x' Axis

The **diagram x' axis** is the set of all events that happen at $t' = 0$ (this is to be distinguished from the Other Frame's **spatial x' axis**, which is one of the frame's three coordinate axes *in space*). The radar method implies that the diagram x' axis slopes *upward* with a slope of β. We calibrate this axis just as we did the t' axis.

Section R5.5: Reading the Two-Observer Diagram

Since all events that lie on a line parallel to the t' axis occur at the same *place* in the Other Frame, and all events that lie on a line parallel to the x' axis occur at the same *time*, to find the Other Frame coordinates of a given event, we draw lines *parallel* to the t' and x' axes from this event until each intersects the other axis.

Figure R5.1 summarizes the process of constructing a two-observer diagram.

Section R5.6: The Lorentz Transformation

Using a two-observer spacetime diagram, we can argue that the coordinate separations between any two events in the Home Frame and Other Frame are related by

$$\Delta t' = \gamma(\Delta t - \beta\Delta x) \quad \text{(R5.11a)} \qquad \Delta t = \gamma(\Delta t' + \beta\Delta x') \quad \text{(R5.12a)}$$

$$\Delta x' = \gamma(-\beta\Delta t + \Delta x) \quad \text{(R5.11b)} \qquad \Delta x = \gamma(+\beta\Delta t' + \Delta x') \quad \text{(R5.12b)}$$

$$\Delta y' = \Delta y \quad \text{(R5.11c)} \qquad \Delta y = \Delta y' \quad \text{(R5.12c)}$$

$$\Delta z' = \Delta z \quad \text{(R5.11d)} \qquad \Delta z = \Delta z' \quad \text{(R5.12d)}$$

- **Purpose:** These equations allow us to calculate the coordinate differences $\Delta t'$, $\Delta x'$, $\Delta y'$, and $\Delta z'$ in the Other Frame from the corresponding differences Δt, Δx, Δy, and Δz in the Home Frame or vice versa, where β is the Other Frame's x-velocity relative to the Home Frame and $\gamma \equiv (1 - \beta^2)^{-1/2}$.
- **Limitations:** The two frames must be inertial and in standard orientation.
- **Notes:** Equations R5.11 are called the **Lorentz transformation equations**, and equations R5.12 the **inverse Lorentz transformation equations**.

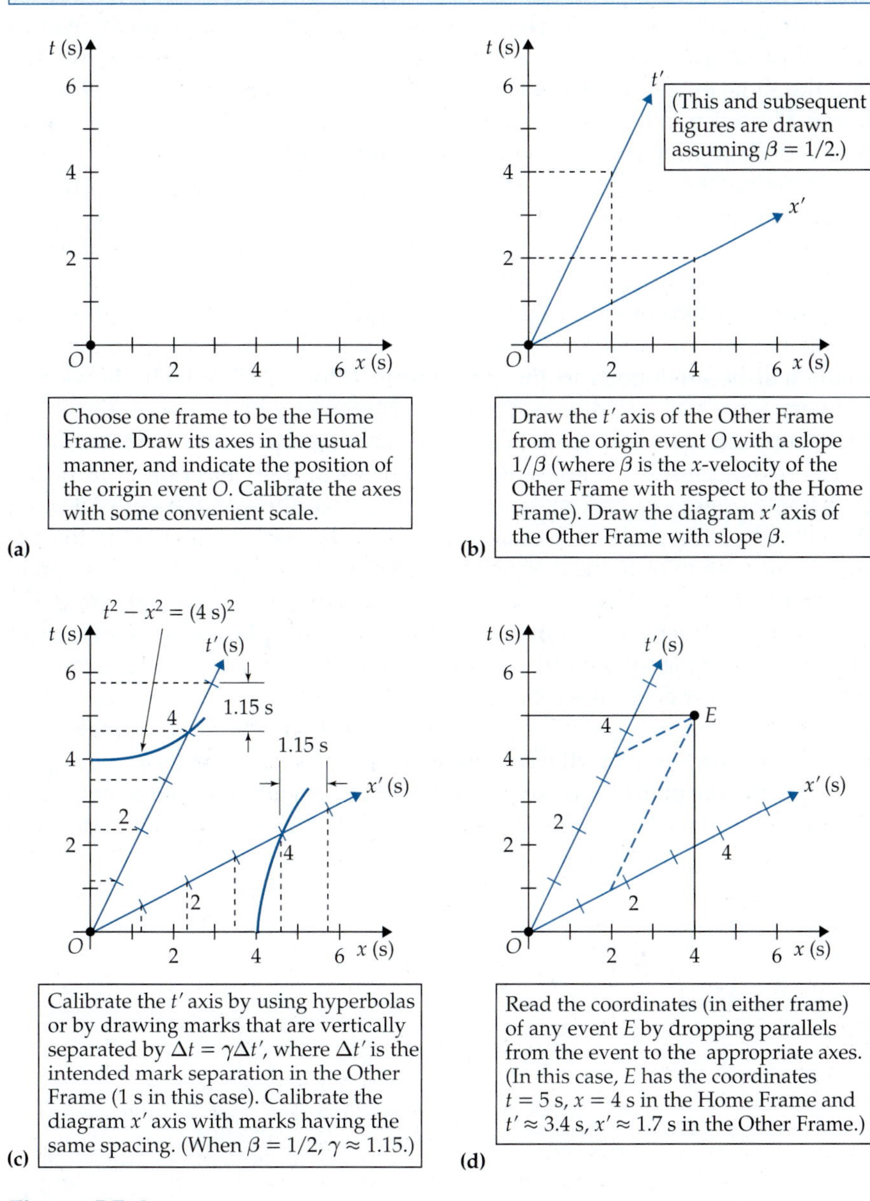

Figure R5.1
A summary of the steps in drawing and interpreting a two-observer diagram.

R5.1 Overview of Two-Observer Diagrams

In chapters R3 and R4, we learned how to use the metric equation to calculate the frame-independent spacetime interval Δs and the frame-independent proper time $\Delta\tau$ along a worldline, using frame-dependent coordinate measurements performed in some inertial frame. To delve further into the implications of the principle of relativity, we need to go a step further. We need to know how to link the coordinates t, x, y, and z in one inertial reference frame to the same event's coordinates t', x', y', and z' in another inertial reference frame. We need to understand such coordinate transformations to understand length contraction (see chapter R6) and generalize the Galilean velocity transformation equations (see chapter R7). This chapter therefore opens a three-chapter subdivision on such coordinate transformations and their implications.

Our specific task in this chapter is to understand how, given an event's spacetime coordinates t, x, y, and z in one inertial reference frame, we can find the same event's coordinates t', x', y', and z' in another inertial reference frame. Physicists call the equations that mathematically describe this coordinate transformation process the **Lorentz transformation equations**.

What is a two-observer spacetime diagram?

Deriving the Lorentz transformation equations is straightforward but somewhat abstract, and that abstraction can blunt one's intuition about what is really going on. Therefore, in this text we will address the same task by using a more visual and intuitive tool called **a two-observer spacetime diagram**. In a two-observer spacetime diagram, we superimpose the coordinate axes for two different observers on the same spacetime diagram. The result will be analogous to the drawing in figure R5.2, which shows two ordinary Cartesian coordinate systems (one rotated with respect to the other) superimposed upon the plane. Once we have set up such a Cartesian two-observer diagram, if we know the coordinates of a given point A in the xy coordinate system ($x_A = 7$ m, $y_A = 4$ m in the case shown), we can plot point A relative to point O. Then, we can simply *read* the coordinates of A in the $x'y'$ coordinate system from the diagram (the coordinates are $x'_A = 8$ m, $y'_A = 1$ m in this case). We do not have to use any equations or do any calculations at all!

Setting up such a diagram is straightforward for plane Cartesian coordinate systems: it merely involves drawing two sets of perpendicular axes (one rotated with respect to the other), scaling the axes, and drawing coordinate grid lines for each set of axes. Setting up the two sets of coordinate axes representing different inertial frames on a spacetime diagram is a *similar* process, but the peculiarities of spacetime geometry relative to plane geometry lead to some surprising dissimilarities as well. Therefore, we need to develop

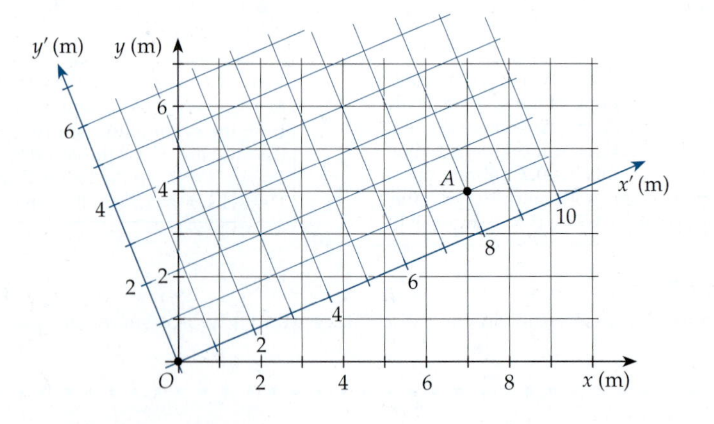

Figure R5.2
A drawing showing two sets of Cartesian coordinate axes superimposed on the same plane. We can easily read the coordinates of point A in both coordinate systems from such a diagram.

the procedure in a careful, step-by-step manner so we are sure to catch all these dissimilarities.

This is well worth the effort, though, because two-observer diagrams vividly display how different observers view the same events differently in a way that the abstract Lorentz transformation equations cannot. Those equations help us *calculate* how events in one frame are viewed in another, but two-observer diagrams help us really *understand* what is going on.

A two-observer diagrams is a powerful tool for displaying how different observers view events

R5.2 Conventions

To make the task of constructing two-observer spacetime diagrams easier, it is convenient to make several assumptions. First, *we assume that the two inertial reference frames are in standard orientation* with respect to each other, as defined in section R1.7. That is, we assume that our frames' corresponding spatial axes point in the same directions in space and that the frames' relative motion is directed along the common x direction. Since we can choose whatever orientation we desire for the axes of a spatial coordinate system, we do not really lose anything by choosing the frames to have this orientation, and we gain much in simplicity.

Standard orientation

Second, *we will work with only those events that occur along the common x and axes of these frames* (that is, those having coordinates $y = y' = z = z' = 0$). This is a substantial concession to convenience. We would really like to be able to handle any event, but plotting an event with arbitrary coordinates on a spacetime diagram would require that the diagram have four dimensions, which is impossible to represent on a sheet of paper. We choose therefore to limit our attention to events that can be easily plotted on a two-dimensional spacetime diagram. We will see that many interesting problems can still be treated within this restriction. So until you are told otherwise, you should assume that $y' = y = z' = z = 0$ for all events under discussion.

In two-observer diagrams, we can only display events that occur on the x axis

The first step in actually drawing a two-observer spacetime diagram is to (arbitrarily) pick one of the two frames to be the **Home Frame** and to call the other frame the **Other Frame**. Remember that I am capitalizing the terms *Home Frame* and *Other Frame* in this text to emphasize that these phrases are actually *names* for inertial frames. By convention, spacetime coordinates in the Home Frame t, x, y, z and coordinates in the Other Frame are t', x', y', z', as shown in figure R5.3. The other crucial distinction is that we *always* choose $\vec{\beta}$ to be the velocity of the Other Frame relative to the Home Frame (the Home Frame's velocity relative to the Other Frame is therefore $-\vec{\beta}$). The symbol β without the arrow refers to the x component of the Other Frame's velocity relative to the Home Frame, so it is negative if the Other Frame moves in the $-x$ direction relative to the Home Frame.

The distinctions between the Home Frame and the Other Frame

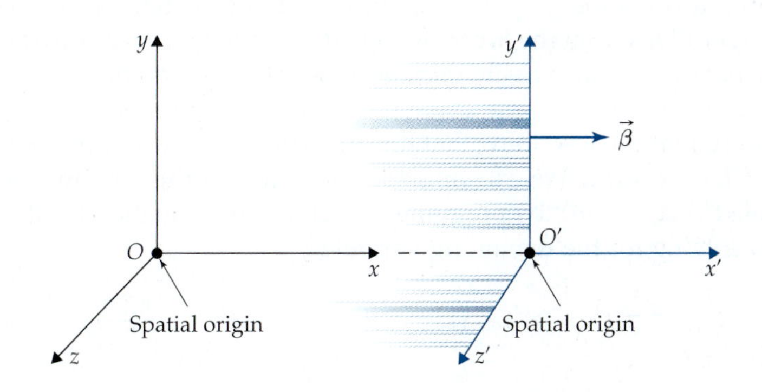

Figure R5.3
Two inertial reference frames in standard orientation. The frames are represented schematically here by bare orthogonal axes. Note that $\vec{\beta}$ *always* represents the Other Frame's velocity relative to the Home Frame, and β without the arrow represents that velocity's x component.

Now, our choice of a Home Frame does not *necessarily* mean we are considering that frame to be at rest and the Other Frame to be moving: we still want to reserve the freedom to consider either frame as being at rest. What this choice *does* imply is simply that we will represent the Home Frame t and x axes in the *usual* manner in a spacetime diagram (that is, we will draw its t axis as a vertical line and its x axis as a horizontal line).

R5.3 Drawing the Diagram t' Axis

The **time axis** for any frame on a spacetime diagram is by definition the line connecting all events having x coordinate $= 0$ in that frame. This means that the time axis is the worldline of the clock at the reference frame's spatial origin (all events happening at a frame's spatial origin have spatial coordinate $x = 0$ by definition). We draw the Home Frame's t axis as a vertical line by convention. How should we draw the t' axis of the Other Frame?

The slope of the t' axis

The Other Frame moves with x-velocity β along the x axis with respect to the Home Frame by hypothesis. This means (if we assume for the sake of argument that β is positive) that the Other Frame's spatial origin moves β units in the $+x$ direction relative to the Home Frame every unit of time. The worldline of the Other Frame's origin as plotted on the spacetime diagram is thus a straight line of slope $1/\beta$, as shown in figure R5.4a. Note that this line goes through the origin event O since the spatial origins of both frames coincide at $t = t' = 0$ if the frames are in standard orientation.

Drawing an appropriate scale on the t' axis

The next step is to put an appropriate scale on the Other Frame's t' axis. It is (unfortunately) *not* correct to simply mark this axis using the same scale as used for the t and x axes. How can we *correctly* scale the t' axis?

Assume that the marks on the t and t' axes are to be separated by some specific time difference δ in their respective frames (in figure R5.4b, for example, this difference is $\delta = 1$ s). The nth mark on the t' axis is actually an event that occurs at $t' = n\delta$ and $x' = 0$. The spacetime interval between this mark event and the origin event O (where $t = x = t' = x' = 0$) is

$$\Delta s_n^2 = (t')^2 - (x')^2 = (n\delta)^2 - 0^2 = (n\delta)^2 \quad \Rightarrow \quad \Delta s_n = n\delta \qquad \text{(R5.1)}$$

The nth mark on the Home Frame t axis is at $t = n\delta$ and $x = 0$, so this mark is *also* separated from the origin event by a spacetime interval of $\Delta s_n = n\delta$. Now, we saw in chapter R3 that all events on a spacetime diagram that are the same spacetime interval Δs from the origin event lie on the hyperbola $t^2 - x^2 = \Delta s^2$. So, to locate the nth mark on the t' axis, we can simply draw the hyperbola $t^2 - x^2 = \Delta s_n^2 = (n\delta)^2$ and draw the mark where the t' axis intersects that hyperbola. (This hyperbola also goes through the corresponding nth mark on the t axis.) We can repeat this for different values of n to draw as many marks as we need. This process is illustrated in figure R5.4b.

Hyperbola graph paper with pre-drawn hyperbolas makes this process easy. You can xerox the graph paper at the end of this chapter (or download graph-paper PDFs from the *Six Ideas* website). However, even if we lack such graph paper, we can also easily *calculate* where the marks should be.

Because the t' axis is a line going through the origin event O with a slope of $1/\beta$, the equation describing that line in terms of Home Frame coordinates is $t = x/\beta$, or $x = \beta t$. We can find the place where this line intersects the hyperbola $t^2 - x^2 = (n\delta)^2$ by plugging $x = \beta t$ into the equation for the hyperbola and solving for the remaining variable t:

$$(n\delta)^2 = t^2 - x^2 = t^2 - (\beta t^2) = (1 - \beta^2)t^2 \quad \Rightarrow \quad t = \frac{n\delta}{\sqrt{1-\beta^2}} \qquad \text{(R5.2)}$$

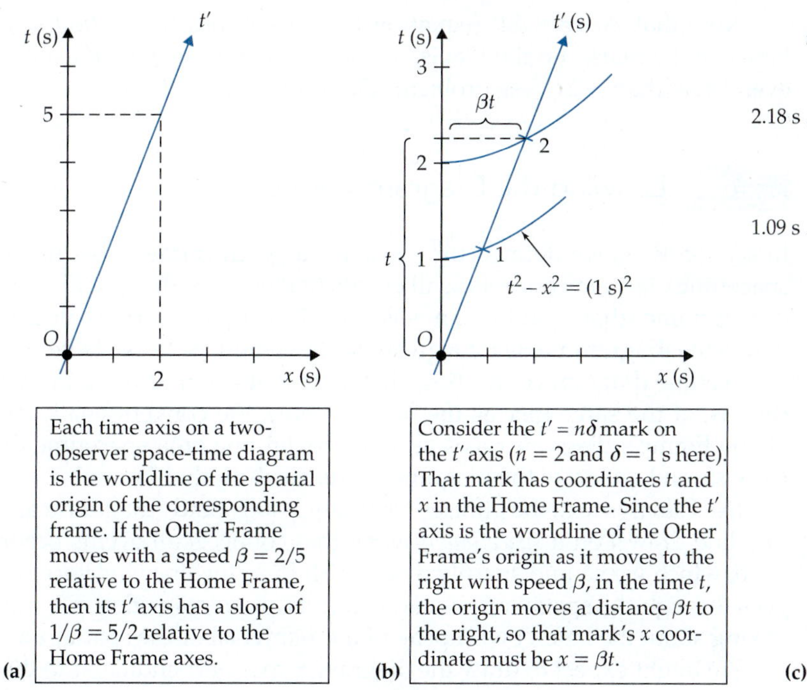

Figure R5.4
Steps in constructing the t' axis on a two-observer spacetime diagram.

Note that this specifies the Home Frame t coordinate (vertical coordinate) of the nth mark on the t' axis. Now, the quantity $1/\sqrt{1-\beta^2}$ occurs so often in relativity theory that it is given its own special symbol:

$$\gamma \equiv \frac{1}{\sqrt{1-\beta^2}} \tag{R5.3}$$

where γ is the Greek letter "gamma." (Note that γ is a number that is always larger than 1.) Using this symbol, equation R5.2 becomes

$$t = n(\gamma\delta) = \text{the } t \text{ coordinate of the } n\text{th mark on the } t' \text{ axis} \tag{R5.4}$$

This completely and correctly locates all the marks on the t' axis. Equivalently, we could say that since adjacent marks correspond to $\Delta n = 1$, the vertical coordinate *difference* Δt between adjacent marks on the t' axis as projected on the Home Frame t axis is

$$\Delta t = \Delta n(\gamma\delta) = \gamma\delta = \gamma\,\Delta t' \qquad \text{for events along the } t' \text{ axis} \tag{R5.5}$$

where $\Delta t' = \delta$ is the time between marks in the Other Frame.

To give a concrete example, suppose our Other Frame moves at an x-velocity of $\beta = \frac{2}{5}$ relative to the Home Frame, and we want to draw marks on the Other Frame t' axis that correspond to events that are $\delta = \Delta t' = 1$ s apart as measured in the Other Frame. In this case, $\gamma = \left[1 - \left(\frac{2}{5}\right)^2\right]^{-1/2} = 1.09$, so we should draw these marks so the vertical coordinate of the nth mark as projected onto the Home Frame t axis is $t = n\gamma\,\Delta t' = n(1.09)(1\text{ s}) = n(1.09\text{ s})$. Equivalently, we could draw the marks so the vertical separation (as projected onto the Home Frame t axis) between adjacent marks is

$$\Delta t = \gamma\,\Delta t' = (1.09)(1\text{ s}) = 1.09\text{ s} \tag{R5.6}$$

Figure R5.4c illustrates the process. This alternative process will prove very useful to us in section R5.6.

Note that $\Delta t = \gamma\,\Delta t'$ (equation R5.5) describes the *vertical* separation between the marks on the t' axis, *not* their spacing *along* the t' axis, which is even *larger* than $\gamma\,\Delta t'$ (see problem R5D.1 if you are curious).

R5.4 Drawing the Diagram x′ Axis

Definition of the diagram x' axis

In section R5.3, we defined the t axis for a given frame to be the line on a spacetime diagram connecting all events that occur at the spatial origin $x = 0$ of that frame (that is, at the same *place* as the origin event). Analogously, we define the **diagram x axis** for a given inertial reference frame to be the line on a spacetime diagram connecting all events that occur at $t = 0$ in that frame (that is, at the same *time* as the origin event). We conventionally draw the Home Frame's diagram x axis as a horizontal line on a spacetime diagram. How should we draw the diagram x' axis for the Other Frame?

The distinction between the diagram x axis and the spatial x axis

(*Note:* In this text, the phrases "diagram x' axis" and "diagram x axis" refer to a line drawn on a *spacetime diagram* that connects all events occurring at $t = 0$, which we should sharply distinguish from the line in *physical space* that goes through the spatial origin and connects all points with $y = z = 0$. When talking about the latter, I will speak of the *x direction* or the **spatial x axis**.)

We might expect to draw the diagram x' axis perpendicular to the t' axis. Unfortunately, drawing the x' axis in this way is *not* consistent with the definition of the diagram x' axis just stated. To figure out the *right* way to draw this axis, we have to consider carefully the implications of the idea that *the diagram x' axis connects events that are simultaneous in the Other Frame.*

The diagram x' axis must tilt upward

We begin by considering a set of events that illustrates the use of the radar method to determine coordinates in the Other Frame. Suppose that at some time $t' = -T$ (where T is some arbitrary number) we send a light flash from the master clock (located at the spatial origin of the Other Frame) in the $+x$ direction (let's call the emission of the flash event A). This flash reflects from some event B and returns to the Other Frame's master clock at time $t' = +T$ (let's call the reception of the flash by the master clock event C). Since the light flash must take the same time to return to the origin from event B as it took to get there, observers in the Other Frame will conclude that event B *must* have happened at a time halfway between $t'_A = -T$ and $t'_C = +T$ and so at time $t'_B = 0$. This implies that event B is *simultaneous* with the origin event O.

Now let's draw this set of events on a spacetime diagram based on the Home Frame (see figure R5.5a). Events A, O, and C all occur at $x' = 0$, so they all lie on the t' axis of the spacetime diagram. Since events A and C occur at $t' = -T$ and $t' = +T$, they must be symmetrically spaced along the t' axis on opposite sides of the origin event, as shown. Note that since the speed of light is 1 in every reference frame, these light-flash worldlines must have a slope of ± 1 on this diagram.

Now, the definition of clock synchronization in the Other Frame requires that events O and B both occur at time $t' = 0$. If we define the diagram x' axis to be the line connecting all events that occur at $t' = 0$, then it must connect events O and B. Thus, *the diagram x' axis must angle upward*, as shown.

The angle between the diagram x and x' axes is equal to that between the t and t' axes

In fact, by considering the geometric relationships implicit in figure R5.5a, we can see that *the diagram x' axis makes the same angle with the diagram x axis that the t' axis makes with the t axis.* The argument is easier if we imagine that the master clock also emits a right-going light flash at the origin event O: let event E be this flash meeting the incoming reflected flash. This new light flash is shown in figure R5.5b. (This new light flash has no physical importance: it just makes the following argument simpler.) Since light flash worldlines always have a slope of ± 1, they all make a 45° angle with respect

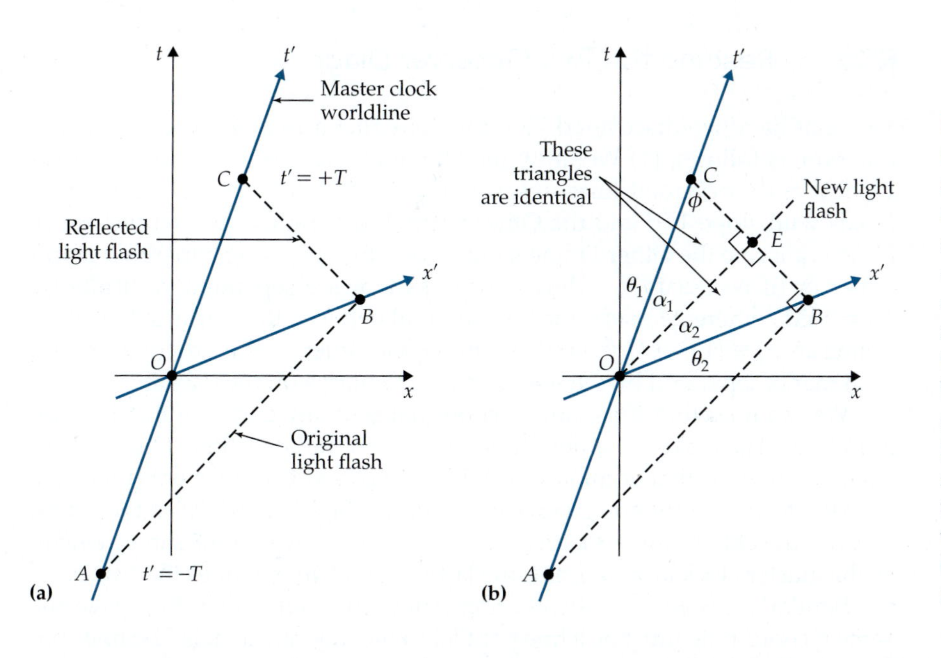

Figure R5.5

(a) This diagram shows that the x' axis on a two-observer diagram must tilt upward at an angle. If a flash emitted from $x' = 0$ at a certain time T *before* the origin event O is reflected from event B and returns to $x' = 0$ at the same time T *after* O, then event B must have occurred at $t' = 0$, and thus should lie on the x' axis, as shown. (b) This diagram shows that the x' axis must make the same angle with the x axis as the t' axis makes with the t axis. Triangles OEC and OEB are identical, so $\alpha_1 = \alpha_2 \Rightarrow \theta_1 = \theta_2$.

to the vertical or horizontal directions. This means that if light-flash worldlines cross at all, they always cross at right angles.

Now, I claim that triangles ABC and OEC in figure R5.5b are *similar* triangles: they are right triangles that share the common angle ϕ. Moreover, the hypotenuse of ABC is *twice* as long as that of OEC, since A and C are symmetrically placed about event O. This means that the sides of triangle ABC must be exactly twice as large as those of the triangle OEC, implying that line BC must also be twice as large as line EC (remember that if two triangles are similar, the lengths of their corresponding sides are proportional).

But if line BC is twice as large as EC, then lines BE and EC have equal lengths. This means that triangles OEC and OEB are *identical*, since they are both right triangles and their corresponding legs are equal in length. Therefore, $\alpha_1 = \alpha_2$, which in turn means that $\theta_1 \equiv 45° - \alpha_1$ is equal to $\theta_2 \equiv 45° - \alpha_2$. Thus, *the diagram x' axis makes the same angle with the diagram x axis that the t' axis makes with the t axis*, as previously asserted.

Another important consequence is that the length of line OC (which represents the coordinate time interval $\Delta t' = +T$) is the same as that of line OB (which represents the coordinate displacement $\Delta x' = T$, the distance in the Other Frame that the light signal had to travel to get to event B). This means that *the scale of both axes must be the same*; that is, the spacing of marks on the diagram x' axis is exactly the same as the spacing of marks on the t' axis!

Exercise R5X.1

Explain *why* $\Delta x' = T$ between events O and B.

The slope of the diagram x' axis

Note that $\tan\theta_1$ = run/rise for the t' axis = 1/(slope of t' axis) = β. Note also that $\tan\theta_2$ = rise/run for the diagram x' axis = slope of diagram x' axis. Since we have just seen that $\tan\theta_1 = \tan\theta_2$, we have

$$\text{Slope of } x' \text{ axis} = \beta \tag{R5.7}$$

So to be consistent with the principle of relativity, we *must* draw the Other Frame diagram t' and x' axes with slopes $1/\beta$ and β, respectively.

R5.5 Reading the Two-Observer Diagram

Summary of how to construct a two-observer spacetime diagram

So, what have we discovered? We can construct a two-observer spacetime diagram as follows: (1) We draw the Home Frame t axis diagram x axis as vertical and horizontal lines, respectively. (2) We draw the Other Frame's t' axis with slope $1/\beta$ and the Other Frame's diagram x' axis with slope β. (3) We calibrate the Other Frame's time axis either by using hyperbola graph paper or (if necessary) by drawing marks that are separated vertically by $\Delta t = \gamma\,\Delta t'$, where $\Delta t' = \delta$ is the time interval between the marks in the Other Frame and $\gamma \equiv (1 - \beta^2)^{-1/2}$. (4) We calibrate the Other Frame's diagram x' axis with marks separated by the *same* distance as marks on the t' axis.

Why we must draw lines *parallel* to the axes to find an event's coordinates?

We can now find the t' and x' coordinates of any event on the diagram as follows. The t' axis is by definition the line on the spacetime diagram connecting all events that occur at $x' = 0$. The line connecting all events that have coordinate $x' = 1$ s (or any given value $\neq 0$) will be a line *parallel* to the t' axis, because the Other Frame's lattice clock at $x' = 1$ s moves at the same velocity as the master clock at $x' = 0$, and the latter's worldline defines the t' axis.

Similarly, the line on the diagram connecting all events that have the same t' coordinate must be a line parallel to the diagram x' axis (the line connecting all events having $t' = 0$). Suppose, for example, that the line connecting all events that occur at $t' = 1$ s were *not* parallel to the line connecting all events that occur at $t' = 0$. Then, these lines would intersect at some point on the diagram. An event located at the point of intersection would thus occur at both $t' = 1$ s and $t' = 0$, which is absurd. A line connecting events having the same t' coordinate therefore *must* be parallel to any other such line.

So if the line connecting all events occurring at the same *time* in the Other Frame is parallel to the diagram x' axis, and the line connecting all events occurring at the same *place* in that frame is parallel to the t' axis, we find the coordinates of an event in the Other Frame by drawing lines through the event that are *parallel* to the diagram t' and x' axes (and *not* perpendicular to them). The places where these lines of constant x' and t' cross the coordinate axes indicate the coordinates of the event in the Other Frame (see figure R5.6).

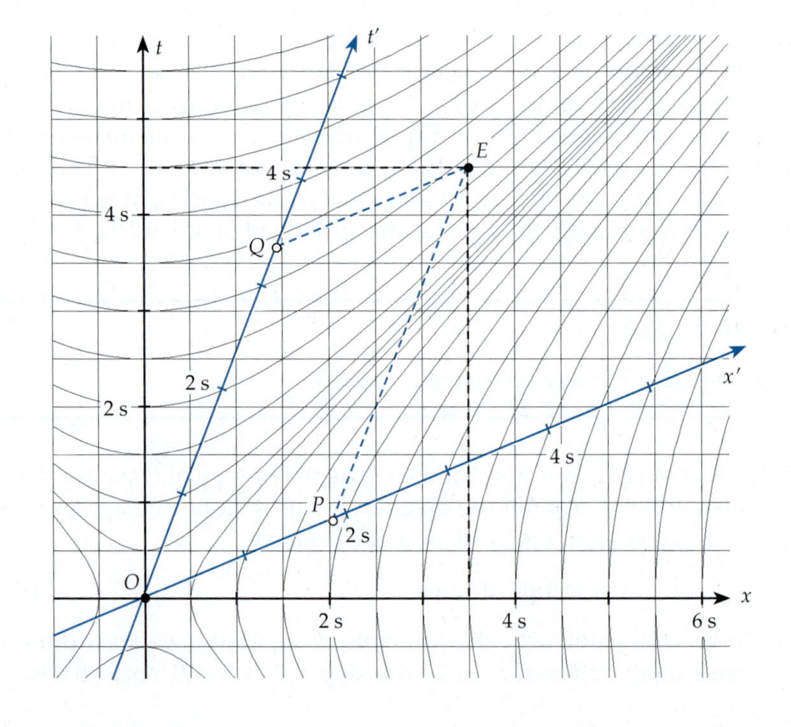

Figure R5.6
An event E occurs at $t = 4.5$ s and $x = 3.5$ s in the Home Frame. Events P and E occur at the same place in the Other Frame, since all events that occur at the same place in that frame lie along a line parallel to the t' axis. Similarly, events Q and E occur at the same time in the Other Frame. Therefore, the time coordinate of E is the same as that of Q in the Other Frame (that is, $t'_E = t'_Q \approx 3.4$ s in this case), and the position of E is the same as that of P (that is, $x'_E = x'_P \approx 1.8$ s in this case). This diagram also illustrates how to use hyperbola graph paper to calibrate the Other Frame axes. The frames' relative speed is $\beta \approx 2/5$.

Finding the coordinate values by dropping *parallels* instead of perpendiculars may seem strange to you, and will probably take some getting used to. Nonetheless, I hope you see from the argument earlier that dropping parallels is the only way to read the coordinates that makes any sense in this case.

R5.6 The Lorentz Transformation

A two-observer spacetime diagram provides a very powerful visual and intuitive tool for linking the coordinates of an event measured in one inertial reference frame with the coordinates of the same event measured in another inertial frame. Because it is visual in nature, it is much more immediate and less abstract than working with equations. But this tool does not give us the quantitative precision that only equations can provide.

In this section, we will develop a set of *equations* that link the coordinates of an event measured in the Home Frame with the coordinates of the same event measured in the Other Frame. These equations do *mathematically* exactly what the two-observer diagram does *visually*. Together, these two tools will enable us to discuss relativity problems with both clarity and precision.

Derivation of the Lorentz transformation equations

Consider an arbitrary event E, as illustrated in figure R5.7. Suppose we know the coordinates t'_E and x'_E of this event in the Other Frame. This means we can locate an event P which occurs at $t' = 0$ (that is, on the diagram x' axis) and at the same *place* as E in the Other Frame (that is, $x'_E = x'_P$). Let the time coordinate separation between events P and E be Δt_{PE} and the spatial coordinate separation between O and P be Δx_{OP} in the Home Frame. Proper calibration of the Other Frame axes requires that $\Delta t_{PE} = \gamma t'_E$ and $\Delta x_{OP} = \gamma x'_E$. Also, since the line connecting events P and E is parallel to the t' axis, its slope must be $1/\beta$, implying that the bottom leg of the triangle involving points P and E must have length $\beta \Delta t_{PE}$. Similarly, the slope of the diagram x' axis is β, so the vertical leg of the triangle involving points O and P must have length $\beta\,\Delta x_{OP}$. Figure R5.7 illustrates these relationships.

Now, you can see from the diagram that

$$t_E = \Delta t_{PE} + \beta \Delta x_{OP} = \gamma t'_E + \gamma \beta x'_E \quad \text{(R5.8}a\text{)}$$

$$x_E = \Delta x_{OP} + \beta \Delta t_{PE} = \gamma x'_E + \gamma \beta t'_E \quad \text{(R5.8}b\text{)}$$

Make sure you understand this derivation before you go on.

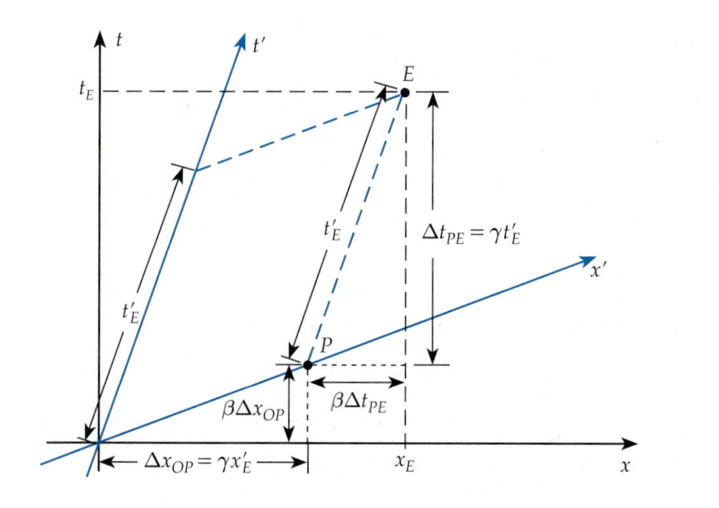

Figure R5.7
Pick an arbitrary event E. Then, choose event P to occur at $t' = 0$ (that is, on the x' axis) and at the same place as E in the Other Frame. Note that since the line connecting events P and E is parallel to the t' axis, its slope must be equal to $1/\beta$. Note also that the x' axis has a slope of β. These slopes, along with the procedure for calibrating axes that we developed in section R5.2, imply the relationships displayed on the diagram.

Since the event E is arbitrary, we can drop the subscript and simply say that the Home Frame coordinates t and x of *any* event can be expressed in terms of the Other Frame coordinates t' and x' of the *same* event as follows:

The inverse Lorentz transformation equations

$$t = \gamma(t' + \beta x') \tag{R5.9a}$$

$$x = \gamma(\beta t' + x') \tag{R5.9b}$$

We call these equations the **inverse Lorentz transformation equations**.

The "plain" **Lorentz transformation equations** (or LTEs for short), express an event's Other Frame coordinates in terms of its Home Frame coordinates. You can find them by solving equations R5.9 for t' and x', which yields

The (direct) Lorentz transformation equations

$$t' = \gamma(t - \beta x) \tag{R5.10a}$$

$$x' = \gamma(-\beta t + x) \tag{R5.10b}$$

Exercise R5X.2

Verify equations R5.10*a* and *b*.

Note that these equations are the same as those in equation R5.9 except that the sign of β is changed. This makes complete sense if you recall that the only difference between the Other Frame and the Home Frame is that β is the x component of the Other Frame's velocity relative to the Home Frame. Therefore, the Home Frame's x-velocity relative to the Other Frame is $-\beta$. Thus, equations R5.9 and R5.10 are the *same* once we account for this sign difference. This is indeed required by the principle of relativity the transformation is a law of physics that should apply to both inertial reference frames. Indeed, we might have *derived* the Lorentz transformation equations from the principle of relativity by insisting that the transformation equations be reflexive this way (see problem R5D.4).

Transformations of the y and z coordinates

We can easily generalize these equations to handle events having nonzero coordinates y and z. We saw in section R3.3 that if two inertial frames are in relative motion along a given line, a displacement perpendicular to that line has the same length in both frames. Since frames in standard orientation move relative to each other along their common x axis, this means that observers in the Home and Other frames will agree on an event's y and z coordinates, which are perpendicular to this line of relative motion.

$$y' = y \tag{R5.10c}$$

$$z' = z \tag{R5.10d}$$

Together, all four equations R5.10 represent the relativistic generalization of the Galilean transformation equations R1.2.

Often, we are not so much interested in an event's raw coordinates as we are in the coordinate *differences* between two events. Consider a pair of events A and B separated by coordinate differences $\Delta t \equiv t_B - t_A$ and $\Delta x \equiv x_B - x_A$ in the Home Frame. What are the corresponding coordinate differences $\Delta t' \equiv t'_B - t'_A$ and $\Delta x' \equiv x'_B - x'_A$ measured in the Other Frame? Applying equation R5.10*a* to t'_A and t'_B separately, we get

The Lorentz transformation equations for coordinate *differences*

$$\Delta t' \equiv t'_B - t'_A = \gamma(t_B - \beta x_B) - \gamma(t_A - \beta x_A)$$

$$= \gamma(t_B - \beta x_B - t_A + \beta x_A) = \gamma[(t_B - t_A) - \beta(x_B - x_A)]$$

$$\Rightarrow \quad \Delta t' \equiv \gamma(\Delta t - \beta \Delta x) \tag{R5.11a}$$

Similarly, you can show that

$$\Delta x' \equiv \gamma(-\beta\Delta t + \Delta x) \qquad \text{(R5.11b)}$$

$$\Delta y' = \Delta y \qquad \text{(R5.11c)}$$

$$\Delta z' = \Delta z \qquad \text{(R5.11d)}$$

These are the Lorentz transformation equations for coordinate differences. Note that they have the same form as the ordinary Lorentz transformation equations (equations R5.10): one simply replaces the coordinate quantities with the corresponding coordinate differences. The inverse Lorentz transformation equations for coordinate differences are completely analogous.

To summarize, then, the transformation equations (LTEs) that allow us to compute coordinate differences in one inertial frame from the corresponding differences in the other are

$$\Delta t' = \gamma(\Delta t - \beta\Delta x) \quad \text{(R5.11a)} \qquad \Delta t = \gamma(\Delta t' + \beta\Delta x') \quad \text{(R5.12a)}$$

$$\Delta x' \equiv \gamma(-\beta\Delta t + \Delta x) \quad \text{(R5.11b)} \qquad \Delta x = \gamma(+\beta\Delta t' + \Delta x') \quad \text{(R5.12b)}$$

$$\Delta y' = \Delta y \quad \text{(R5.11c)} \qquad \Delta y = \Delta y' \quad \text{(R5.12c)}$$

$$\Delta z' = \Delta z \quad \text{(R5.11d)} \qquad \Delta z = \Delta z' \quad \text{(R5.12d)}$$

- **Purpose:** These equations allow us to calculate the coordinate separations $\Delta t'$, $\Delta x'$, $\Delta y'$, and $\Delta z'$ in the Other Frame from the corresponding differences Δt, Δx, Δy, and Δz in the Home Frame, or vice versa, where β is the Other Frame's x-velocity relative to the Home Frame and $\gamma \equiv (1 - \beta^2)^{-1/2}$.
- **Limitations:** These equations apply only to inertial frames in standard orientation.
- **Notes:** Equations R5.11 are called the **Lorentz transformation equations (LTEs)**, and equations R5.12 the **inverse Lorentz transformation equations** (for coordinate differences in both cases).

The Lorentz transformation equations and two-observer diagrams are equivalant

I hope you can see that the derivation outlined in this section means that the inverse Lorentz transformation equations (or the "plain" Lorentz transformation equations) simply quantify more precisely what you could read from a two-observer spacetime diagram. I chose this particular method of deriving the Lorentz transformation equations deliberately to drive home the point that the equations simply express *mathematically* exactly the same thing that a two-observer spacetime diagram expresses *visually*. Two-observer spacetime diagrams are most useful for visually displaying and thinking about how different observers will interpret events; the Lorentz transformation equations are best for doing accurate calculations once the thinking is done. Example R5.1 illustrates how these tools complement each other.

Example R5.1

Problem: Consider an event that is observed in the Home Frame to occur at time $t = 6.0$ s and position $x = 10$ s. When and where does this event occur according to observers in an Other Frame that is moving with speed $\beta = \frac{2}{5}$ in the $+x$ direction with respect to the Home Frame? Determine this both graphically (using a spacetime diagram) and analytically (using the Lorentz transformation equations).

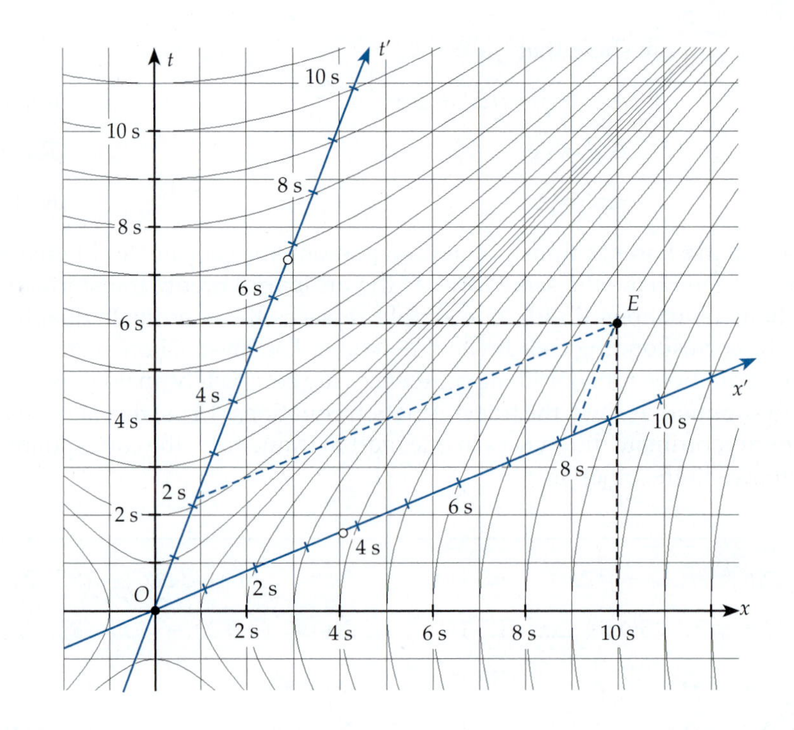

Figure R5.8
The two-observer diagram solving the question posed in this example. Event E has coordinates $t = 6.0$ s and $x = 10$ s in the Home Frame, and about $t' = 1.5$ s and $x' = 5.7$ s in the Other Frame.

Solution Figure R5.8 displays the graphical solution. I first constructed and scaled the Home Frame axes. I then plotted the event (here marked with an E) so that $t = 6.0$ s and $x = 10.0$ s. I then drew the Other Frame's t' axis with slope $\frac{5}{2}$ and the diagram x' axis with slope $\frac{2}{5}$, and calibrated those axes using the hyperbola graph paper. Finally, I dropped parallels from event E to the t' axis and the diagram x' axis, and read E's Other Frame coordinates from their intersection with those axes. I find that $t' \approx 2.2$ s and $x' \approx 8.3$ s.

Now let's *calculate* these coordinates, using the Lorentz transformation equations. If $\beta = \frac{2}{5}$, then $\gamma = (1 - \beta^2)^{1/2} = 1/\sqrt{1 - \frac{4}{25}} \approx 1.09$, so

$$t' = \gamma(t - \beta x) \approx 1.09[6.0\text{ s} - \tfrac{2}{5}(10.0\text{ s})] \approx 2.18\text{ s} \qquad \text{(R5.13}a\text{)}$$

$$x' = \gamma(-\beta t + x) \approx 1.09[-\tfrac{2}{5}(6.0\text{ s}) + 10\text{ s}] \approx 8.29\text{ s} \qquad \text{(R5.13}b\text{)}$$

which are in substantial agreement with the results read from the diagram. (One should expect an uncertainty of about 3% to 5% in results read from even the most carefully constructed spacetime diagram.)

Note that we use the hyperbolas *only* for calibrating the Other Frame axes

Important! Note that we use the hyperbolas *only* to calibrate the t' and x' axes: after that, we completely ignore them. In particular, we do *not* use them to read an event's coordinates in either frame.

Exercise R5X.3

(a) Determine (both graphically and analytically) the Other Frame coordinates of an event with Home Frame coordinates $t = 3.0$ s, $x = 10.0$ s.
(b) Determine (both graphically and analytically) the Home Frame coordinates of an event with Other Frame coordinates $t' = 8.0$ s and $x' = 4.0$ s.
Assume in both cases that the frames are as described in example R5.1. You may use figure R5.8 to do the graphical solution.

TWO-MINUTE PROBLEMS

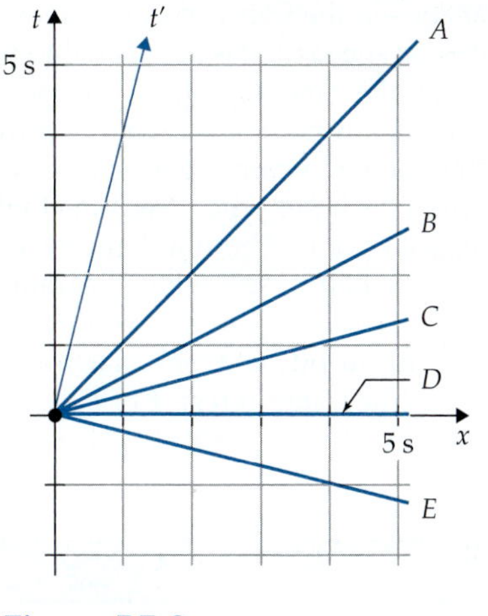

Figure R5.9

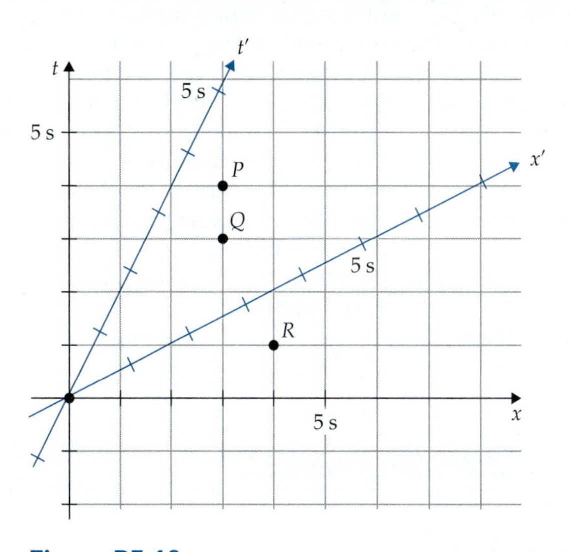

Figure R5.10

R5T.1 The Other Frame is moving in the $+x$ direction with x-velocity $\beta = 0.25$ with respect to the Home Frame. The two-observer spacetime diagram in figure R5.9 shows the diagram t and x axes of the Home Frame and the diagram t' axis of the Other Frame. Which of the choices in that figure best corresponds to the diagram x' axis?

R5T.2 The Other Frame is moving in the $+x$ direction with x-velocity $\beta = 0.25$ with respect to the Home Frame. The two-observer spacetime diagram in figure R5.9 shows the diagram t and x axes of the Home Frame and the diagram t' axis of the Other Frame. Which of the choices in that figure would best correspond to the diagram x' axis if the Newtonian concept of time were true?

R5T.3 Suppose the marks on the Home Frame t axis in figure R5.9 are 1.0 cm apart. What should be the *vertical* separation of the corresponding marks on the t' axis?

A. 0.94 cm
B. 0.97 cm
C. 1.0 cm
D. 1.03 cm
E. 1.07 cm
F. Other

R5T.4 Figure R5.10 shows a two-observer spacetime diagram for an Other Frame that moves at a speed of 0.5 relative to the Home Frame. What are the coordinates of event P in the Other Frame?

A. $t' = 3.4$ s, $x' = 2.6$ s
B. $t' = 5.2$ s, $x' = 2.6$ s
C. $t' = 2.9$ s, $x' = 1.2$ s
D. $t' = 3.7$ s, $x' = 3.4$ s
E. Other (specify)

R5T.5 Figure R5.10 shows a two-observer spacetime diagram for an Other Frame that moves at a speed of 0.5 relative to the Home Frame. What are the coordinates of event Q in the Other Frame?

A. $t' = x' = 5.2$ s
B. $t' = x' = 3.2$ s
C. $t' = x' = 2.6$ s
D. $t' = x' = 1.7$ s
E. Other (specify)

R5T.6 Figure R5.10 shows a two-observer spacetime diagram for an Other Frame that moves at a speed of 0.5 relative to the Home Frame. What are the coordinates of event R in the Other Frame?

A. $t' = -1.15$ s, $x' = 4.0$ s
B. $t' = 0.9$ s, $x' = 3.4$ s
C. $t' = 6.5$ s, $x' = 1.7$ s
D. $t' = 2.2$ s, $x' = 3.2$ s
E. Other (specify)

R5T.7 Consider two blinking warning lights 3000 m apart along a railroad track. These lights flash simultaneously in the ground frame. Let W be the event of the west light blinking, and let E be the event of the east light blinking. A train moves eastward along the track at a relativistic speed β. Suppose an observer in the train passes the west light just as event W happens. By carefully measuring when the light from event E arrives and calculating the distance between the two lights (by observing how long it takes to travel between the lights at the known speed β), the observer is able to infer when event E actually happened in the train frame. The observer concludes that

A. Event W happened before event E in the train frame.
B. Events W and E were simultaneous in the train frame.
C. Event E happened before event W in the train frame.
D. One cannot determine unambiguously which event occurs first in the train frame.

(*Hint:* Draw a qualitative spacetime diagram.)

R5T.8 A bullet train moving in the $+x$ direction with x-velocity β relative to the ground has lights on the roof of the head and tail cars that blink simultaneously in the train frame. The head car's light happens to be passing by an observer on the ground just as it blinks. The observer *sees* the light from the tail car at a different time, but after correcting for the light travel time from the tail car, the observer concludes that in the ground frame

A. The tail car's light blinked before the head car's light.
B The head car's light blinked before the tail car's light.
C. Both lights blinked simultaneously.
D. One cannot determine unambiguously which event occurs first in the ground frame.

(*Hint:* Draw a qualitative spacetime diagram.)

R5T.9 Two lights are 1000 ns apart along a stretch of railway track. In the ground frame, the west light flashes 600 ns before the east light flashes. Could these flashes be simultaneous in the frame of a train moving along the track at a certain speed (that is less than the speed of light)?

A. Yes, if the train is moving east at the correct speed.
B. Yes, if the train is moving west at the correct speed.
C. Yes, if the train observer happens to be at the right distances from the lights when receiving their flashes.
D. No, the flashes cannot be simultaneous in any frame.

(*Hint:* Draw a qualitative spacetime diagram.)

R5T.10 According to our conventional frame names, the Home Frame is the frame at rest. T or F?

HOMEWORK PROBLEMS

Basic Skills

R5B.1 The spacetime diagram below shows the worldline of an alien spaceship fleeing at a speed of $\beta = \frac{3}{5}$ in the $+x$ direction from space station DS9 after stealing some potentially destructive trilithium crystals. The departure of the ship from DS9 is event A. At event B, DS9 fires a phaser blast (which travels at the speed of light), hoping to disable the vessel. At event C, the fleeing spaceship drops a fuel tank behind, setting the tank at rest relative to DS9 while the spaceship continues on ahead of it (the tank now shields the ship from the point of view of DS9). At event D, the phaser blast hits and destroys the tank, leaving the ship unharmed.

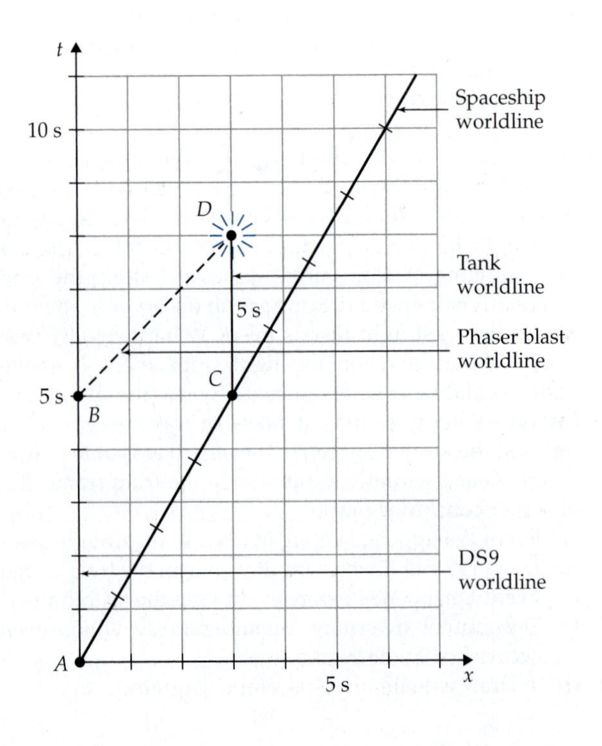

(a) The spaceship frame's t' axis is labeled on the diagram (this is just the worldline of the spaceship itself). This axis has also been calibrated. Check that the calibration is correct (explain how you checked this).

(b) On the diagram (or a copy), draw and calibrate the diagram x' axis for the spaceship frame.

(c) When does the spaceship drop the tank according to its own clock? Explain how you arrived at your answer.

(d) What is the approximate time of event D in the spaceship frame? (Do not use the Lorentz transformation equations. Instead, read the result from the diagram and explain how you arrived at your result.)

(e) Calculate the time of event D in the spaceship's frame, using an appropriate Lorentz transformation equation.

(f) Which event, B or C, occurs first in DS9's frame (or are these events simultaneous)? Which occurs first in the spaceship frame?

(g) Use the appropriate Lorentz transformation equation to compute the x' coordinate of event D. Explain why the sign of your result makes sense.

(h) Event Q is the explosion of a meteor as it collides with DS9's protective shield. According to measurements in the ship's frame, this event occurs at $t' = 3.0$ s, $x' = 2.0$ s. Draw and label this event on the diagram.

(i) Use the appropriate inverse Lorentz transformation equations to compute the Home Frame coordinates of event Q.

R5B.2 A Rigellian spaceship fleeing from battle passes Space Station Delta at an essentially constant velocity of 3/5 in the $+x$ direction: let this be the origin event O. At event B, 80 s after event O as measured in its own frame, Space Station Delta fires a photon torpedo (which travels at the speed of light) toward the fleeing ship. At event C, 100 s after event O in the space station frame, the Rigellians fire a laser flash (which also travels at the speed of light) back toward Space Station Delta. This laser flash destroys the torpedo at event D.

(a) On a sheet of hyperbola graph paper, draw the worldlines of Space Station Delta and the Rigellian ship.
(b) Locate events *O*, *B*, and *C* on the diagram.
(c) Draw and label the torpedo and laser flash worldlines and locate event *D*. What are its coordinates?
(d) Draw the t' and x' axes for the Rigellian frame and use the hyperbolas to calibrate those axes.
(e) Which occurs first in the Rigellian frame, event *B* or *C*?
(f) Use the Lorentz transformation equations to calculate the coordinates of events *B* and *C* in the Rigellian frame and check that your results are consistent with results read from the diagram.

R5B.3 Use a qualitative spacetime diagram to answer the question posed in problem R5T.7 and explain your logic.

R5B.4 Use a qualitative spacetime diagram to answer the question posed in problem R5T.8 and explain your logic.

R5B.5 An event occurs at $t = 6.0$ s and $x = 4.0$ s in the Home Frame. The Other Frame is moving in the $+x$ direction with x-velocity $\beta = 0.5$ relative to the Home Frame.
(a) Use a two-observer spacetime diagram to determine when and where this event occurs in the Other Frame.
(b) Check your work by applying the appropriate Lorentz transformation equation.

R5B.6 An event occurs at $t = 1.5$ s and $x = 5.0$ s in the Home Frame. The Other Frame is moving in the $+x$ direction with x-velocity $\beta = 3/5$ relative to the Home Frame.
(a) Use a two-observer spacetime diagram to determine when and where this event occurs in the Other Frame. (*Hint:* You should find that t' is *negative*.)
(b) Check your work by applying the appropriate Lorentz transformation equation.
(c) Is the time order of this event and the origin event the same in both frames?

R5B.7 An Other Frame moves in the $+x$ direction with x-velocity $\beta = 0.60 = \frac{3}{5}$ with respect to the Home Frame. Other Frame observers observe an event at time $t' = 3.0$ s and position $x' = 1.0$ s.
(a) Use a two-observer spacetime diagram to determine when and where this event occurs in the Home Frame.
(b) Check your work by applying the appropriate Lorentz transformation equation.

R5B.8 An Other Frame moves in the $+x$ direction with x-velocity $\beta = \frac{2}{5}$ relative to the Home Frame. Other Frame observers observe an event at $t' = 30$ s and $x' = 1.0$ s.
(a) Use a two-observer spacetime diagram to determine when and where this event occurs in the Home Frame.
(b) Check your work by applying the appropriate Lorentz transformation equation.

Modeling

R5M.1 Star *A* and star *B* float essentially at rest 500 light-years apart in an inertial reference frame we will take to be the Home Frame. Suppose star *A* (whose position is $x_A = 0$ in the Home Frame) explodes (goes supernova) at a time we define to be $t_A = 0$. Star *B* explodes at time $t_B = 400$ y, both as measured in the Home Frame. A spaceship is moving from star *A* directly toward star *B* at a constant speed $|\beta|$, passing star *A* just as it explodes. Define the $+x$ direction to be the direction the ship is moving.
(a) What is the value of β if the explosions are simultaneous in the spaceship's frame? Answer by constructing a two-observer spacetime diagram, and explain your reasoning.
(b) Where does the explosion of star *B* occur in the spaceship's reference frame? Answer using the two-observer diagram and explain your reasoning. (You may *check* your work using a Lorentz transformation equation.)
(c) When does light from star *B*'s explosion reach the ship? Answer using your two-observer diagram.
(d) Does the answer to part (c) make sense considering your answer for part (b)? Explain.

R5M.2 Consider the question posed in problem R5T.9.
(a) Use a quantitatively accurate spacetime diagram to answer the question and explain your reasoning.
(b) If such a frame is possible, determine its speed relative to the ground frame.

R5M.3 The Federation space cruiser *Execrable* is floating in Federation territory at rest relative to the border of Klingon space, which is 6.0 min away in the $+x$ direction. Suddenly, a Klingon warship flies past the cruiser in the direction of the border at a speed of $\frac{3}{5}$. Call this event *A*, and let it define time zero in both the Klingon and cruiser reference frames. At $t_B = 5.0$ min according to cruiser clocks, the Klingons emit a parting disrupter blast (event *B*) that travels at the speed of light back to the cruiser. The disrupter blast hits the cruiser and disables it (event *C*), and a bit later (according to cruiser radar measurements) the Klingons cross the border into Klingon territory (event *D*).
(a) Draw a two-observer spacetime diagram where the cruiser and the Klingon warship are the Home and Other Frames, respectively. Draw and label the worldlines of the cruiser, the Klingon territory boundary, the Klingon warship, and the disrupter blast. Draw and label events *A*, *B*, *C*, and *D* as points on your diagram.
(b) When does the disrupter blast hit, and when do the Klingons pass into their own territory, according to clocks in the cruiser's frame? Answer by reading the times of these events directly from the diagram.
(c) The Klingon–Federation Treaty states that it is illegal for a Klingon ship in Federation territory to damage Federation property. When the case comes up in interstellar court, the Klingons claim that they are within the letter of the law: according to measurements made in their reference frame, the damage to the *Execrable* occurred *after* they had crossed back into Klingon territory, so they were *not* in Federation territory at the time. Did event *C* (disrupter blast hits the *Execrable*) *really* happen after event *D* (Klingons cross the border) in the Klingons' frame? Answer this question by using your two-observer diagram, and check your work with the Lorentz transformation equations.

R5M.4 Space station DS9 floats in deep space. Let DS9 define the Home Frame's spatial origin. A Ferengi freighter moving at a constant speed of 3/5 in the $+x$ direction passes DS9 (call this event A) at time $t = 0$ in both the freighter frame (the Other Frame) and the DS9 frame. At $t = 4$ h (according to DS9's clocks), Quark (a resident of DS9) sends an encrypted laser message to the freighter (let this be event B). At event C, workers on the Ferengi ship destroy what turns out to be crucial physical evidence in a later investigation. Light from this destruction event reaches DS9 at $t = 8$ h (according to DS9's clocks): let this be event D.

(a) Construct a two-observer spacetime diagram of this situation displaying DS9's worldline, the freighter's worldline, the worldline of Quark's signal, the worldline of the light from the destruction event, and the spacetime locations of events A, B, C, and D.

(b) When and where does event B occur in the Ferengis' frame? Answer by reading the result from your diagram. Check your work by applying the appropriate Lorentz transformation equations.

(c) When and where does event C occur in DS9's frame? Explain your reasoning.

(d) Odo, the DS9 constable, accuses Quark of colluding with the Ferengi and specifically that his encoded message led the Ferengi to destroy the evidence. Is Odo's accusation justified?

R5M.5 Fred sits 65 ns west of the east end (and thus 35 ns east of the west end) of a 100-ns-long train station. Sally operates a reference frame in a train racing east across the countryside at a speed of 0.5. At a certain time (call it $t' = 0$), Sally passes Fred. At that same instant, Fred flashes a strobe lamp (call this event F), which sends bursts of light both east and west. Alan, who is standing at the west end of the station, receives the west-going part of the flash (call this event A), and a bit later (according to clocks in the station) Ellen, who is standing at the east end of the station, receives the east-going flash (call this event E).

(a) When do events A and E occur in the station frame? Who sees the flash first (according to clocks in the station frame), Alan or Ellen?

(b) Draw a two-observer spacetime diagram that displays and labels the worldlines of Sally, Fred, Alan, Ellen, and the two light flashes. Locate and label events F, A, and E as points on the diagram. Carefully draw and calibrate the t' and x' axes for Sally's train frame.

(c) When and where do events A and E occur in Sally's frame? Sally claims that Ellen sees the flash first in her frame. Is this true? Verify your assertions with calculations based on the Lorentz transformation equations.

R5M.6 A Tirillian spaceship fleeing from battle passes space station DS7 at an essentially constant velocity of $\frac{3}{5}$ in the $+x$ direction as measured in DS7's frame. Let the event of the ship passing DS7 be the origin event A in both frames. The Tirillians have a cloaking device that they think makes them invisible to DS7's sensors. However, 40 s after passing DS7 (as measured by the Tirillian clocks) the spaceship passes through a dust cloud that emits a pulse of electromagnetic radiation when disturbed: let this be event B. The instant this pulse (which travels at the speed of light) is received by DS7, the DS7 crew fires a photon torpedo (which also travels at the speed of light) toward the fleeing Tirillians: call this event C. The Tirillians decide 80 s after passing DS7 that they have likely been detected, so they put up their defensive shields (which involves turning off the cloaking device): call this event D.

(a) Use a ruler to draw a complete and carefully constructed two-observer spacetime diagram of the situation, drawing the worldlines of DS7 and the Tirillian spaceship and locating and labeling events A, B, C, and D.

(b) When and where did event B occur in the Home Frame? Use an appropriate Lorentz transformation equation to check what you read from your diagram.

(c) When does event C occur in the Home Frame? Explain how you located this event on the diagram.

(d) When does event C occur in the Tirillian frame? Explain how you can read t'_C from the diagram, and use an appropriate Lorentz transformation equation to verify your result.

(e) Which event, C or D, occurs first in the DS7 frame? Which occurs first in the Tirillian frame? Explain.

(f) Could the Tirillians have made their decision to raise shields as a consequence of observing (somehow) that DS7 had fired a torpedo? Why or why not?

Derivation

R5D.1 Imagine that the physical distance between adjacent marks on the Home Frame t and x axes is some quantity q (for example, q might be 1 cm or 0.25 in.). Argue that the physical distance q' between adjacent marks on the t' and x' axes, as measured *along* those axes, is

$$q' = q\sqrt{\frac{1+\beta^2}{1-\beta^2}} \tag{R5.14}$$

R5D.2 Use the Lorentz transformation equations (equations R5.11) to show that the squared spacetime interval Δs^2 has the same value in all inertial reference frames.

R5D.3 In all of the examples in this chapter, the Other Frame has been moving in the $+x$ direction relative to the Home Frame so that β is positive. What if the Other Frame moves in the $-x$ direction, so that β is negative?

(a) Go through the arguments presented in sections R5.3 through R5.5 with this change in mind, and then construct a complete and calibrated two-observer spacetime diagram for the situation where the Other Frame moves with a speed of $|\beta| = \frac{3}{5}$ in the $-x$ direction with respect to the Home Frame. Describe why you chose to draw and calibrate the diagram as you did.

(b) Event F happens at $t = 6$ s and $x = 2$ s in the Home Frame. Read from your diagram when and where this event occurs in the Other Frame.

(c) Check that the Lorentz transformation equations (with a negative value of β) yield the same result.

R5D.4 Here is one way to derive the Lorentz transformation equations from scratch. For the sake of argument, we will consider events that occur only along the spatial x axis. We will also *assume* that the transformation equations must be linear; that is, they have the form

$$t' = At + Bx \quad \text{and} \quad x' = Ct + Dx \qquad \text{(R5.15)}$$

where A, B, C, and D are unknown constants that do not depend on the coordinates but only on β. (One can show that only linear equations like this transform a constant-velocity worldline in the Home Frame to another constant-velocity worldline in the Other Frame. Since a free particle must follow a constant-velocity worldline in any inertial frame, this linearity is required of any reasonable transformation equations linking the frames.)

(a) The Home Frame coordinates of a light flash moving in the $+x$ direction from the origin event will be such that $x/t = 1$ at all times. Because the speed of light is the same in all frames, the coordinates of this flash in the Other Frame must be such that $x'/t' = 1$ in this case (that is, whenever $x/t = 1$). Argue that the transformation equations R5.15 will deliver the correct result in this special case if and only if $A + B = C + D$.

(b) The Home Frame coordinates of a light flash moving in the $-x$ direction from the origin event will be such that $x/t = -1$. Argue that the transformation equations will deliver the correct result $x'/t' = -1$ in *this* special case if and only if $B - A = C - D$.

(c) The spatial origin of the Other Frame is at $x' = 0$. In the Home Frame, the coordinates of this point are such that $x/t = \beta$, because this point moves with x-velocity β (along with the whole Other Frame). Show that the transformation equations will yield $x' = 0$ whenever $x = \beta t$ if and only if $C = -\beta D$.

(d) These three equations allow us to eliminate B, C, and D in favor of A. Do this and show that the result is

$$t' = A(t - \beta x) \quad \text{and} \quad x' = A(-\beta t + x) \qquad \text{(R5.16)}$$

(e) The squared spacetime interval Δs^2 between any event and the origin event must be the same. Show that this requires that $A = \gamma$, making equations R5.16 the same as the Lorentz transformation equations. Q.E.D.

(f) Alternatively, one can prove that $A = \gamma$ directly from the principle of relativity as follows. The principle requires that the transformation law be the same in both frames (after switching β to $-\beta$, which is the only difference between the frames), we must have

$$t = A^*(t' + \beta x') \quad \text{and} \quad x = A^*(\beta t' + x') \qquad \text{(R5.17)}$$

where A^* is the same function of $-\beta$ that A is of β. Now, we'll see in a moment that the principle also requires that $A^* = A$. Assume this for right now. Transforming a coordinate from the Home Frame to the Other Frame and then back must yield the same thing, so

$$x = A[\beta t' + x'] = A[\beta A(t - \beta x) + A(-\beta t + x)] \qquad \text{(R5.18)}$$

Show that this *requires* that $A = \gamma$.

(g) But how do we know that we *must* have $A = A^*$? Consider performing the following experiment in both frames. Look at the master clock at rest at the spatial origin of the frame that is not your frame. How far has it traveled in your frame when its face registers exactly 1 s after the origin event? If you do this experiment in the Home Frame, then (since the clock in question is at $x' = 0$) the transformation equation says that the distance traveled is $x = A^*(t' + \beta x') = A^*(1\text{ s})$. If you do this experiment in the Other Frame, then the distance traveled is $x' = A(t - \beta x) = A(1\text{ s})$. Explain why the principle of relativity is violated if $A \neq A^*$.

Rich-Context

R5R.1 Each Global Positioning System (GPS) satellite constantly broadcasts a signal that specifies what time the signal is sent (according to an atomic clock on the satellite) as well as information about that satellite's location when the signal is sent. A GPS receiver uses the information sent from multiple satellites to find its location by doing a complicated calculation that accounts for the signal travel time from the satellites, the rotation of the earth, and a variety of effects predicted by both special and general relativity.

To see just some of the effects of special relativity that are involved, consider a starkly simplified GPS system where the satellites fly at a constant speed of β a negligible distance above the x axis on a flat earth. Assume the satellites' atomic clocks are synchronized in the satellites' frame, and that you are standing somewhere along the x axis. At a certain instant, your GPS receiver simultaneously receives a signal from somewhere along the $-x$ direction relative to you from a satellite A and another signal from somewhere along the $+x$ direction from a satellite B. The signals both state the same signal departure time ($t'_A = t'_B = 0$ in the satellite's frame) and specify that the satellites are at locations x'_A and x'_B, respectively, at that time.

(a) Assume that the Galilean transformation equations are true and that both signals have the same speed in the *earth* frame. Show that your position is $x = \frac{1}{2}(x'_B + x'_A)$.

(b) Assume that $\beta = \frac{3}{5}$ and let $x'_A = 1$ ms and $x'_B = 5$ ms (to pick something arbitrary). Carefully construct a spacetime diagram of the situation, and argue from the diagram that $x \neq \frac{1}{2}(x'_B + x'_A)$.

(c) Using this diagram to guide your thinking, develop a general equation that calculates your position along the x axis in the flat earth's frame in terms of β and the reported values of x'_A and x'_B, taking account of special relativity. Check that your proposed equation yields results consistent with your diagram for specific values of β, x'_A, and x'_B given in part (b). (*Hint:* Let event C be the event of your receiver obtaining both signals. Can you find the coordinates of event C in either the spaceship frame or the earth frame? If the former, you can transform to get the coordinates in the earth frame.)

(d) Suppose the satellites' common speed is 3.9 km/s (which is roughly the real GPS satellites' orbital speed) and that their reported positions are $x'_A = -3000$ km and $x'_B = +9000$ km. By how much would the naive calculation of part (a) be in error, according to your relativistic formula? Is this significant?

ANSWERS TO EXERCISES

R5X.1 In the Other Frame, the light flash is measured to have a round-trip time of $2T$ by hypothesis and thus must have taken a time T for each leg of the trip. Since light has speed 1 in all frames, and the light-flash in this case travels up and down the spatial x' axis, the reflection event must have occurred along the spatial x' axis at a distance T from the origin, implying that $x' = T$ for the reflection event.

R5X.2 To isolate t', multiply equation R5.9*b* by β, and then subtract this equation from R5.9*a*, yielding

$$t - \beta x = \gamma(t' + \beta x' - \beta^2 t' - \beta x') = \gamma(1 - \beta^2)t' \quad \text{(R5.19)}$$

Remember now that $\gamma \equiv (1 - \beta^2)^{-1/2}$, so $1 - \beta^2 = 1/\gamma^2$. Substituting in for the above and multiplying both sides by γ, we get

$$\gamma(t - \beta x) = \gamma^2(1 - \beta^2)t' = t' \quad \text{(R5.20)}$$

which is equation R5.10*a*. The proof of equation R5.10*b* is entirely analogous.

R5X.3 **(a)** The graphical solution is shown in figure R5.11 (event A), where it looks like $t' \approx -1.1$ s and $x' \approx 9.6$ s. The Lorentz transformation equations imply that

$$t' = \gamma[t - \beta x] = 1.09[3.0\text{ s} - \tfrac{2}{5}(10.0\text{ s})] = -1.09\text{ s} \quad \text{(R5.21}a\text{)}$$

$$x' = \gamma[-\beta t + x] = 1.09[-\tfrac{2}{5}(3.0\text{ s}) + 10.0\text{ s}] = 9.6\text{ s} \quad \text{(R5.21}b\text{)}$$

(b) The graphical solution is shown in figure R5.11 (event B), where it looks like $t \approx 10.5$ s and $x \approx 7.9$ s. The inverse Lorentz transformation equations tell us that

$$t = \gamma[t' + \beta x'] = 1.09[8.0\text{ s} + \tfrac{2}{5}(4.0\text{ s})] = 10.46\text{ s} \quad \text{(R5.22}a\text{)}$$

$$x = \gamma[\beta t' + x'] = 1.09[\tfrac{2}{5}(8.0\text{ s}) + 4.0\text{ s}] = 7.85\text{ s} \quad \text{(R5.22}b\text{)}$$

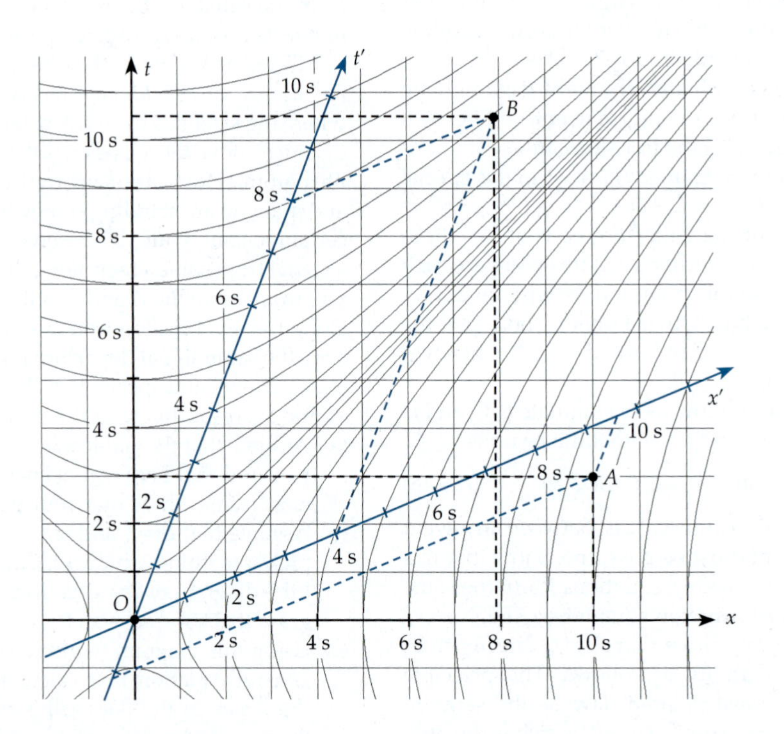

Figure R5.11

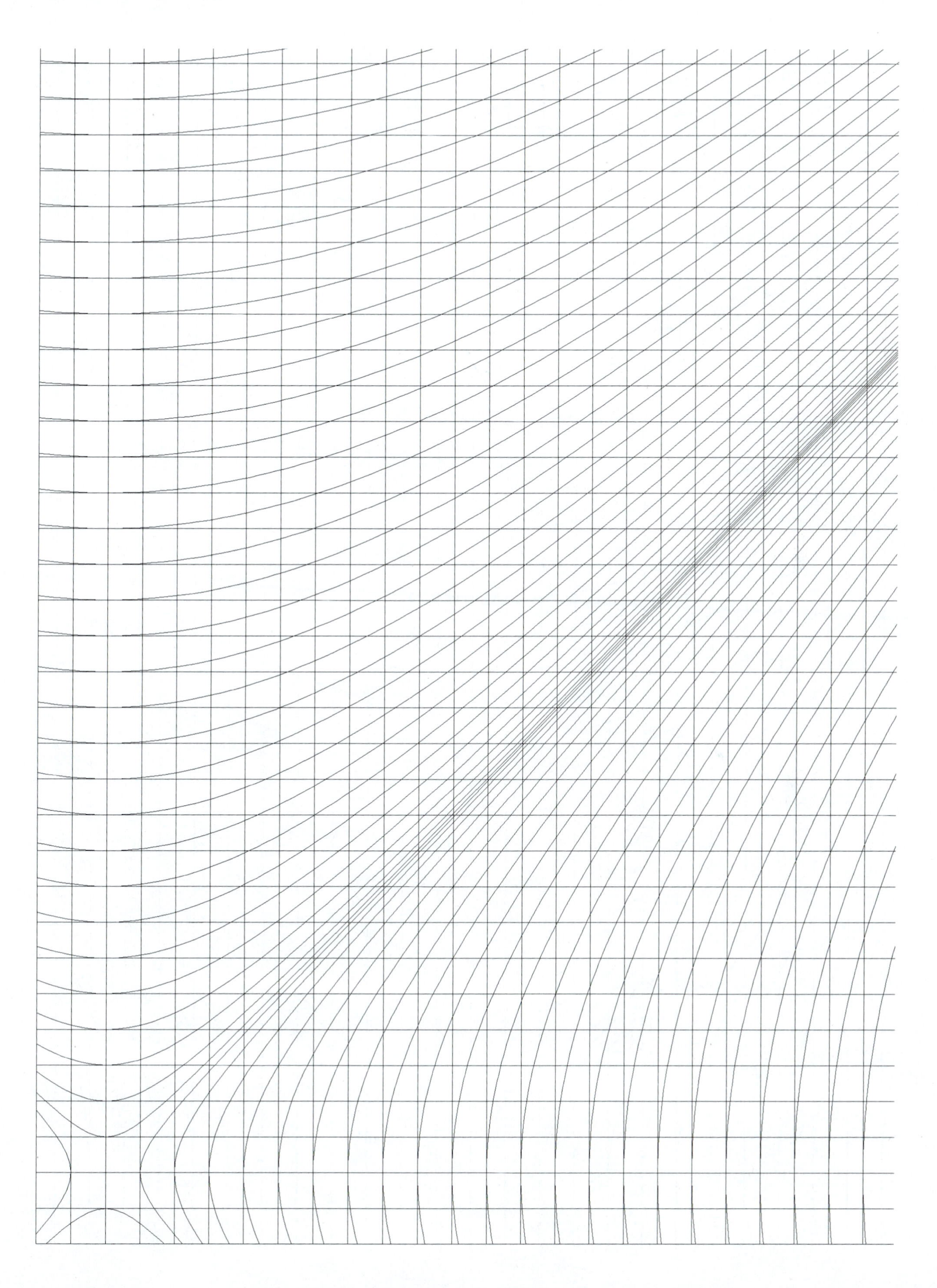

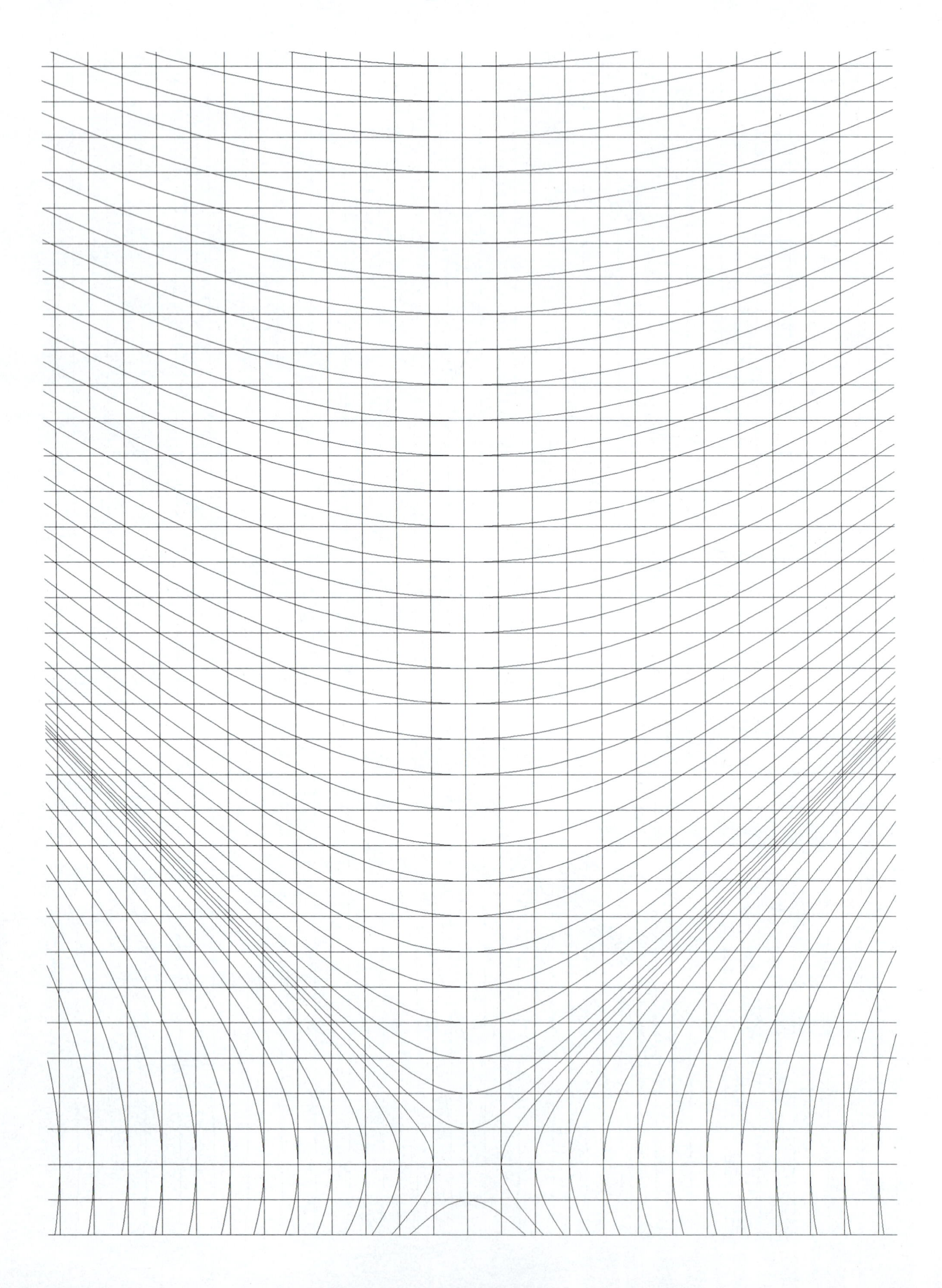

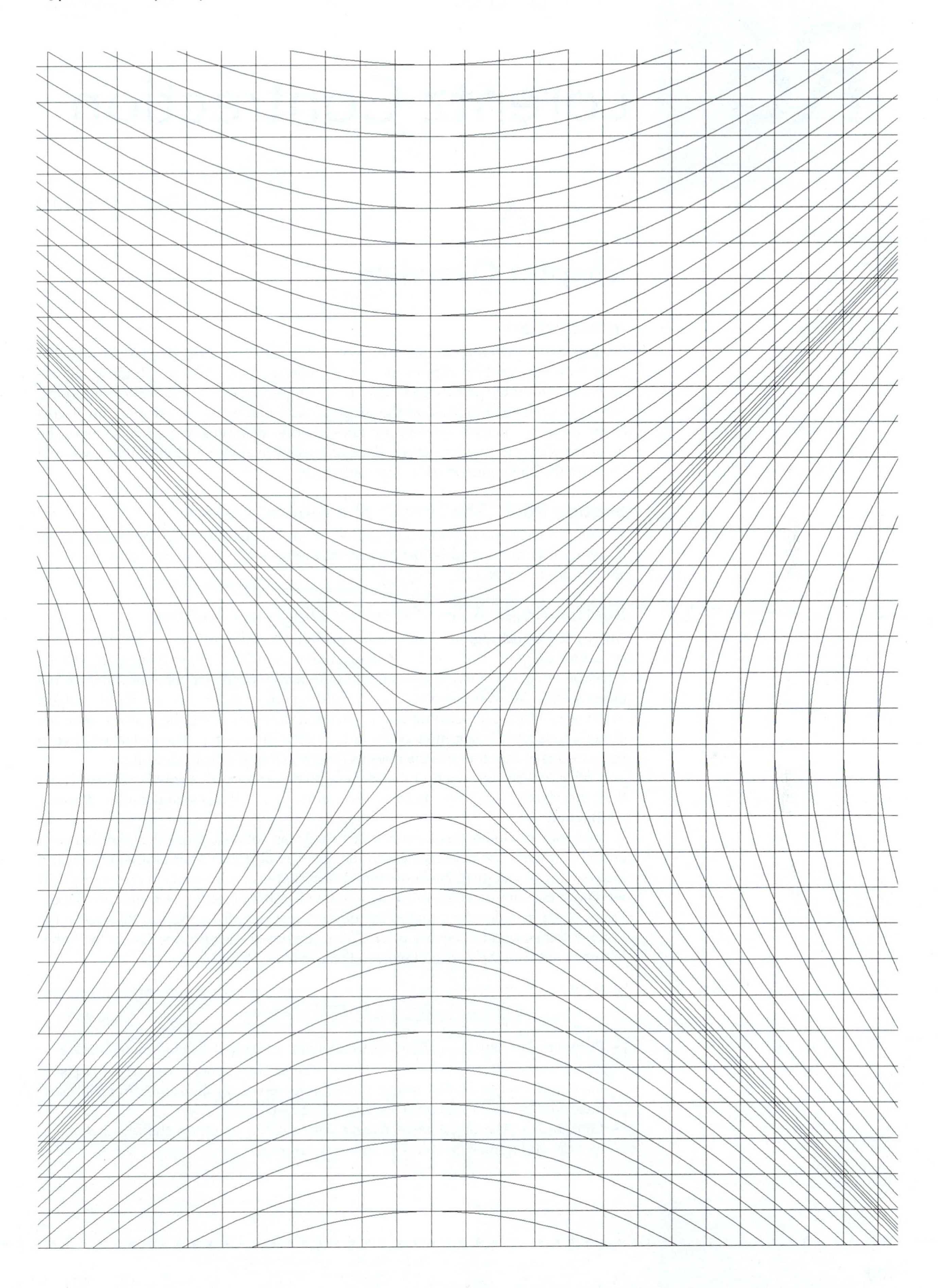

R6 Lorentz Contraction

Chapter Overview

Introduction

Because two-observer spacetime diagrams concisely and vividly display the relationships between events in various inertial reference frames, problem solutions that use such diagrams often prove clearer and more compelling than mathematical solutions, even when we need to back up the diagram with mathematical calculations. In this chapter and chapter R7, we will use two-observer diagrams to explore some of the more peculiar and startling predictions of special relativity. In this chapter, we will focus on the phenomenon of *Lorentz contraction*.

Section R6.1: The Length of a Moving Object

We can precisely and operationally define an object's **length** in a given inertial frame in which it is moving to be *the distance between simultaneous events occurring at the object's ends in that frame*.

Section R6.2: A Two-Observer Diagram of a Stick

An object's **world-region** is the set of worldlines for all particles in the object. On a spacetime diagram, the world-region of a one-dimensional object lying along the $+x$ direction looks like a two-dimensional plane whose edges are the worldlines of the object's ends. We can use a two-observer diagram to determine an object's length in any given frame by locating two events that occur (1) along the worldlines of the object's ends and (2) simultaneously in the frame in question. The distance between those events in that frame is the object's length in that frame, by definition.

When we do this, we find that the length determined this way is always *shorter* than the object's length in its own rest frame: this is the phenomenon of **Lorentz contraction**.

We can calculate this contraction exactly as follows. Consider simultaneous events that mark the object's ends in the frame (call it the Other Frame) in which it moves with x-velocity β. We know that $\Delta t' = 0$ between the events in that frame, and we know that these events must be separated in the object's *rest* frame (the Home Frame) by a distance Δx equal to the object's **rest length** L_R. We can then solve the Lorentz transformation equation R5.12*b* for the object's length $L = \Delta x'$ in the Other Frame in terms of $\Delta t'$ and Δx between these events. The result is

$$L = L_R\sqrt{1-\beta^2} \qquad \text{(R6.2)}$$

- **Purpose:** This equation allows us to calculate a moving object's length L along the direction of its motion in some inertial reference frame, where $|\beta| = |\vec{\beta}|$ is the object's speed in that frame and L_R is its length in the same direction as measured in the inertial frame where it is at rest.
- **Limitations:** The object must have a well-defined length in the direction of motion, and it must be at rest in an inertial frame.

Section R6.3: What Causes the Contraction?

This contraction effect is entirely due to the fact that observers in different frames disagree about clock synchronization. Since different observers will see different pairs of events marking out the object's ends to be simultaneous, they cannot agree on the object's length.

There is an analogy for this in plane geometry. Consider a road with parallel sides. The road's east–west width depends on the orientation of one's coordinate system, even though the road's physical reality is unchanged. Similarly, an object's projection on a frame's spatial axis (which is what its length really is) is frame-dependent, even though its four-dimensional reality is unchanged.

Section R6.4: The Contraction Is Frame-Symmetric

An object is contracted in one frame but not in another. Doesn't this contradict the principle of relativity? No! The principle only requires that the *same experiment* performed in different inertial frames yield the *same result*. Therefore, if an object at rest in frame A is observed to be contracted in frame B, the principle only requires that an object at rest in frame B be similarly contracted when observed in frame A. It is easy to show this is true.

Section R6.5: The Barn and Pole Paradox

Consider a runner who carries a pole (whose rest length is 10 ns) through a barn (whose rest length is 8 ns) at a speed of $|\beta| = \frac{3}{5}$ relative to the barn. In the ground frame, the pole is 8 ns long, so there is an instant when the pole fits entirely inside the barn. In the runner's frame, though, the barn is 6.4 ns long and the pole is 10 ns long. How can a 10-ns pole fit in a 6.4-ns barn?

The solution of this "paradox" is as follows. The phrase "fits inside" really means the event B of the pole's back end entering the barn is simultaneous with the event F of the pole's front end leaving the barn. But events B and F are *not* simultaneous in the runner's frame: event B happens only after enough time has passed after event F for the 10-ns pole to move 3.6 ns beyond the front end of the barn. So the runner *never* sees the barn enclose the pole. The trick with this "paradox" (and many similar ones) is to recognize that events that are simultaneous in one frame are *not* generally simultaneous in another.

Section R6.6: Other Ways to Define Length

We do not *have* to define length as we did in section R6.1. This section, however, argues that several reasonable and well-defined alternative definitions display exactly the same Lorentz contraction effect.

R6.1 The Length of a Moving Object

How can we operationally define the length of something that is moving?

The basic question that will concern us in this chapter can be stated simply as follows: what exactly do we mean by the *length* of a moving object?

As always, we need an *operational definition* of this word if it is to mean anything—that is, we need to describe exactly how we can *measure* an object's length in a given inertial frame. In the particular inertial frame where the object is at rest, it is simple to compare the object to a stationary ruler. But the determination of an object's length in a frame in which it is observed to be moving presents difficulties that need to be handled carefully.

We have defined a reference frame to be an apparatus that measures the spacetime coordinates of *events*. Our first task in the problem of measuring lengths (and indeed most problems in relativity theory) is to rephrase the problem in terms of events. In a given reference frame, how might we characterize the length of an object in terms of events?

Let us consider a concrete example. Suppose we are trying to measure the length of a moving train in the reference frame of the ground. A clock lattice at rest on the ground records the passage of the train through it by describing the motion in terms of events. To be specific, imagine that a certain clock in the lattice records that the *back* end of the train passed at exactly 1:00:00 p.m. (call this event O). Another clock elsewhere in the lattice records that the *front* end of the train passed at exactly 1:00:00 p.m. (call this event A). Therefore, we can say that at exactly 1:00:00 p.m., the train lies between the location of the clock registering event O and that of the clock registering event A. It therefore makes sense to define the train's length to be equal to the distance between those events, as measured in the lattice.

With this image in mind, we operationally *define* an object's length in any inertial reference frame as follows:

One way to define a moving object's length

> An object's **length** in a given inertial frame is defined to be the *distance* between any two *simultaneous events* that occur at its ends.

This expresses the definition of length in the language of events, enabling us to use the tools we have been developing to describe the relationships between events to talk about the *process* of measuring an object's length.

R6.2 A Two-Observer Diagram of a Stick

An object's "world region"

Consider a measuring stick oriented along the spatial x direction and at rest in the Home Frame. How can we represent such an object on a spacetime diagram? To present the full reality of a measuring stick in spacetime, one must plot the worldline of each particle in the stick. Just as a point particle is represented on a spacetime diagram by a curve called a worldline, so a stick is represented by an infinite number of associated worldlines, which one might call a **world-region**. An example of a world-region is shown in figure R6.1.

The definition of length given in section R6.1 yields the expected result when the object in question is at rest. Consider the 4-ns measuring stick of figure R6.2, which is at rest in the Home Frame of the diagram. Events O and A lie at the ends of the measuring stick and are simultaneous in the Home Frame (both occur at $t = 0$ in that frame). According to our definition, then, the length of the measuring stick in the Home Frame is the distance between these two events, which, according to figure R6.2, is simply 4 ns.

Demonstrating Lorentz contraction using a two-observer diagram

Now consider determining the length of this same measuring stick in an Other Frame that is moving with speed β in the $+x$ direction with respect to the Home Frame. An observer in that frame will observe the stick to move

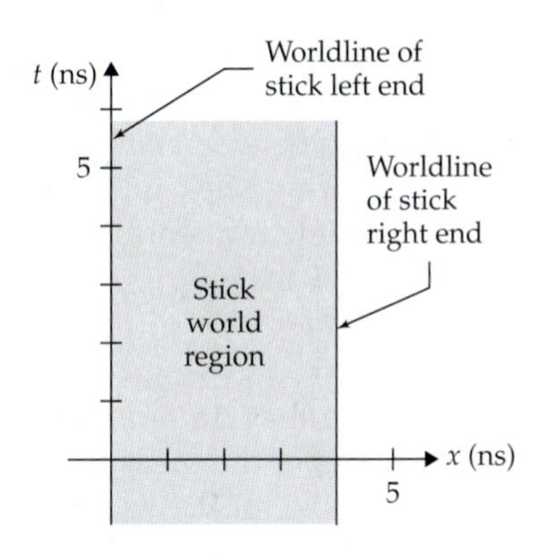

Figure R6.1
Part of the world-region of a 4-ns measuring stick at rest in the Home Frame with one end at $x = 0$ and the other end at $x = 4$ ns. Because it is at rest in the Home Frame, the worldlines of its endpoints are vertical lines, and the worldlines of all the points in between fill in the region of spacetime shown in gray.

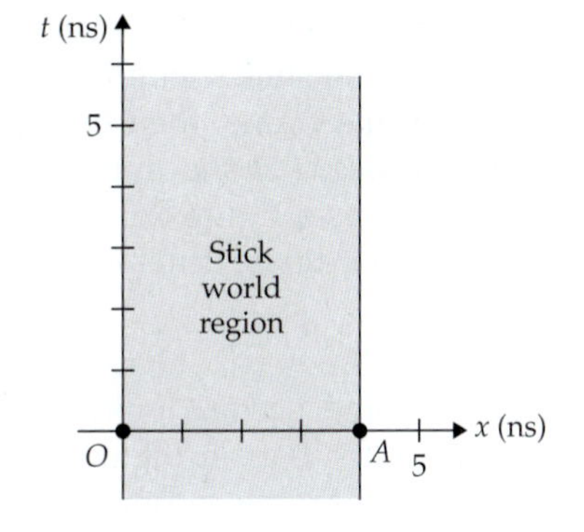

Figure R6.2
The two events O and A lie along the worldlines of the measuring stick's ends and also happen to occur at the same time in the Home Frame. The distance between these events (which is 4 ns in this case) is the measuring stick's length in the Home Frame, by definition.

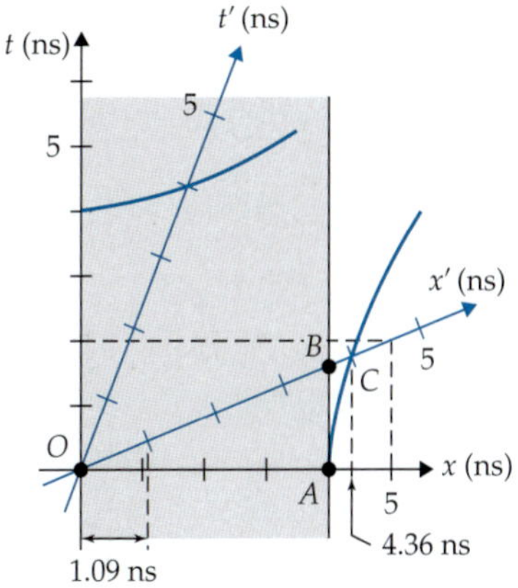

Figure R6.3
Events O and B lie along the worldlines of the measuring stick's ends and occur simultaneously (at $t' = 0$) in an Other Frame moving with $\beta = \frac{2}{5}$ with respect to the Home Frame. According to the diagram, the distance between these events (which is the stick's length in the Other Frame) is ≈ 3.7 ns < 4 ns.

in the $-x$ direction with speed $|\beta|$. Figure R6.3 is a two-observer spacetime diagram showing the Home Frame and Other Frame axes superimposed on the measuring-stick's world-region. (For the sake of concreteness, I have constructed the diagram assuming that $\beta = \frac{2}{5}$.) In the Other Frame, it is not event A but B (as shown on the diagram) that is simultaneous with O and lies at the other end of the measuring stick (both O and B lie on the diagram x' axis, so both occur at $t' = 0$ in the Other Frame). This means that the measuring stick's length in the Other Frame is *defined* to be the distance between events O and B as measured in that frame.

But as the calibrated axes of the Other Frame show, the distance between O and B in that frame is *less* than 4 ns! We can see that this *must* be so as follows. Consider the 4-ns mark on the diagram x' axis (the event labeled C) on the spacetime diagram. This mark must be connected with a hyperbola to the 4-ns mark on the diagram x axis (event A). It is easy to see that the mark on the x' axis is farther along that axis at $t' = 0$ than the stick's right end is at that instant (event B). Therefore, the stick's length in the Other Frame (the distance between events O and B by definition) must be *smaller* than 4 ns.

We can use the other method of axis calibration to see the same thing. The 4-ns mark on the x' axis (event C) must be separated from the origin event by a horizontal displacement $\Delta x_{OC} = \gamma\,\Delta x'_{OC} = (1.09)(4\text{ ns}) \approx 4.36$ ns. This means that event B must be closer to the origin event than the mark event C at $x' = 4.0$ ns, which in turn implies that $\Delta x'_{OB} < 4$ ns!

Either way we look at it, the measuring stick's right end, which is always exactly 4 ns from the spatial origin of the Home Frame, intersects the diagram x' axis at an event B that is *closer* to the origin event than the 4-ns mark on the diagram x' axis, implying that the stick is measured in the Other Frame to have a length of *less* than 4 ns. In fact, you can see that in the particular case shown, the stick has a length of about 3.7 ns in the Other Frame.

We can also easily check this result with the help of the Lorentz transformation equations. In the Other Frame, the measuring stick's length L is *defined* to be the distance $\Delta x'$ between two simultaneous events that occur at the ends of the stick—that is, events for which $\Delta t' = 0$. Assuming we know that the measuring stick's length in the Home Frame is $L_R = 4.0$ ns and that it is at rest in that frame, then the Home-Frame distance between *any* pair of events that occur at the opposite ends of the measuring stick must be $\Delta x = L_R = 4.0$ ns (see figure R6.3). One of the inverse Lorentz transformation equations for coordinate differences (equation R5.12*b*) says that

$$\Delta x = \gamma(\beta \Delta t' + \Delta x') \tag{R6.1}$$

In the case at hand, we are looking for $L = \Delta x'$, knowing that $\Delta x = L_R$ and $\Delta t' = 0$ for events that simultaneously mark out the ends of the measuring stick in the Other Frame. Dropping the $\Delta t'$ term and solving for $\Delta x'$ yields

A general formula for an object's length in a frame in which it is moving

$$L = \Delta x' = \frac{L_R}{\gamma} = L_R\sqrt{1-\beta^2} \tag{R6.2}$$

- **Purpose:** This equation allows us to calculate a moving object's length L along the direction of its motion in some inertial reference frame, where $|\beta| = |\vec{\beta}|$ is the object's speed in that frame and L_R is its length in the same direction as measured in the inertial frame where it is at rest.
- **Limitations:** The object must have a well-defined length along its direction of motion, and it must be at rest in some inertial frame.

Plugging in the relevant numbers in this case, we find that

$$L = (4.0\text{ ns})\sqrt{1-\left(\tfrac{2}{5}\right)^2} = 3.7\text{ ns} \tag{R6.3}$$

in agreement with the result displayed in figure R6.3.

We see that if we accept the quite sensible definition of length given in section R6.1, we are confronted with the fact that an object's length is a *frame-dependent* quantity: its value depends on which inertial frame one chooses to make the measurement. Equation R6.2 implies that an object's length measured in a frame in which the object is moving will always be *smaller* than the value of its length in the frame in which it is at rest. This phenomenon is called **Lorentz contraction**.

The definition of an object's "rest length"

We can use equation R6.2 quite generally to compute the contraction's magnitude, as long as we note that L_R (which we call the object's **rest length**) stands for the object's length measured in the frame in which it is at rest, and L stands for the object's contracted length measured in an inertial frame that moves with speed $|\beta|$ with respect to the object's rest frame.

Exercise R6X.1

Using hyperbola graph paper, estimate the length of a 5-ns measuring stick when it is observed in a frame that is moving with x-velocity $\beta = \frac{4}{5}$ relative to the Home Frame. Use equation R6.2 to check your result.

R6.3 What Causes the Contraction?

Lorentz contraction has its origin in problems of clock synchronization

The previous discussion shows that this contraction effect has nothing to do with some effect of motion that physically compresses a moving object (as, for example, an elastic object such as a balloon would be compressed if it were forced to move rapidly through water). The physical reality of the measuring stick (as represented by its world-region on the spacetime diagram) actually *remains the same*, no matter what reference frame we use to describe it. The fundamental reason why observers in different inertial frames will measure the same object to have different lengths is that the observers disagree about clock synchronization, and therefore disagree about which events mark out the ends of the object "at the same time." For example, observers in the Home Frame of figure R6.3 use events O and A, while observers in the Other Frame use events O and B. We see then that the phenomenon of Lorentz contraction has its origin in the problem of clock synchronization!

A geometric analogy

Nonetheless, the idea that the same object can be measured to have different lengths in different inertial frames may be hard to accept. Yet we are not at all surprised by the analogous behavior of geometric objects on a two-dimensional plane. Let me illustrate. Suppose we wish to determine the east–west width of a road running in a roughly northerly direction on the earth's surface. Two different surveyors set up differently oriented coordinate systems and make this measurement. Is it surprising that they get different results (see figure R6.4)?

The road's east–west width shown in figure R6.4 is greater in the primed coordinate system than in the unprimed system. Has the road magically expanded for the surveyor who laid out the primed coordinate system? Of course not! The road's physical reality does not change just because we change the coordinate system in which we measure it. But because the two surveyors cannot agree on which two points that lie along the road's edges also lie on an east–west line, they will measure the road's east–west width to have different values.

The analogy between a road's "true width" and an object's rest length

We do not find this problematic or even unexpected. Now, we should say that if you are going to measure the road's "true" width, you should measure it by using a coordinate system in which the road runs parallel to the y axis, so that the x axis is *perpendicular* to the road. In that special coordinate system, the width of the road will have its "true" value (which is *shorter* than the value of the same measured in any secondary coordinate system).

Similarly, we might say that to measure an object's "true" length in spacetime, we should measure its length in the inertial frame in which it is at rest. This "true length" (more correctly, the object's **proper length**) will be *longer* than the value measured in any other inertial frame. For clarity's sake, let us always refer to this length as the object's **rest length**.

Here is another way to look at the issue. A real physical object exists in four-dimensional spacetime. Its "length" is its projection on a three-dimensional

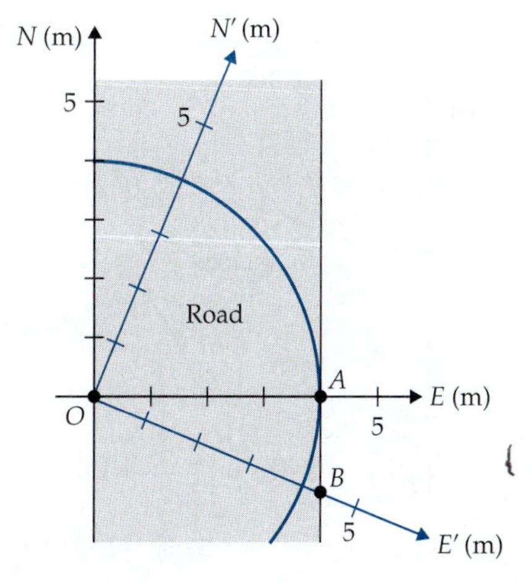

Figure R6.4
Points O and A span the east–west width of the road in the Home Frame coordinate system, while points O and B span the same in the Other Frame coordinate system. The circle shown connects all points that lie 4 meters from the origin. From the picture, it is clear that the road has a greater east–west width in the Other Frame coordinate system than in the Home Frame system. Does this mean the road has magically expanded for the surveyors in the Other Frame coordinate system?

slice of that spacetime that we call "now." An object's "length" is therefore analogous to the width of the shadow that a three-dimensional object casts on a two-dimensional surface—that is, the projection of that object on the surface. If I change the object's orientation with respect to the surface, the shadow may grow or shrink. This is not an illusion: the shadow *really does* change size! But this is not because the three-dimensional object is changing: it is simply because we are viewing it from different angles. Similarly, observers in different inertial reference frames essentially view a four-dimensional object from different "directions," and so conclude that the object's projection on the three-dimensional surface that a particular observer calls "now" changes size. The projection *really does* change size, but the projection is only one perspective on the changeless four-dimensional object.

R6.4 The Contraction Is Frame-Symmetric

Does the Lorentz contraction provide a way of distinguishing a rest frame from a moving frame?

But, you might ask, does this Lorentz contraction effect not violate the principle of relativity? We have seen that an object at rest in the Home Frame is measured to have a shorter length in the Other Frame. Does this not imply that there is a physically measurable distinction between the two frames, a distinction that would violate the requirement that all inertial frames be equivalent when it comes to the laws of physics?

The principle requires that identical experiments yield identical results

The principle of relativity does *not* require that measurements of a specific object or of a set of events have the same values in all reference frames. What the principle *does* require is that if we do exactly the same *physical experiment* in two different inertial reference frames, we get exactly the same result (otherwise, the laws of physics that predict the outcome of the experiment will be seen to be different in the different frames). Now, we have seen that if we take a 4-ns measuring stick at rest in the Home Frame and measure its length in the Other Frame, we will find it to be Lorentz-contracted to 3.7 ns in length. The principle of relativity *does* require that if we perform the *same* experiment in the Home Frame, we get the *same* result—that is, if we take a 4.0-ns measuring stick at rest in the Other Frame and measure its length in a Home Frame moving at a speed of $|\vec{\beta}| = \frac{2}{5}$ with respect to the Other Frame, we should find the stick to be Lorentz-contracted to 3.7 ns.

Figure R6.5 shows that this is indeed so. The worldlines of the ends of a measuring stick at rest in the Other Frame will be parallel to the t' axis, as shown. Events O and D mark out the two ends of the measuring stick at time $t = 0$ in the Home Frame. The distance between these events (i.e., the length of the measuring stick in that frame) is seen to be about 3.7 ns, as expected.

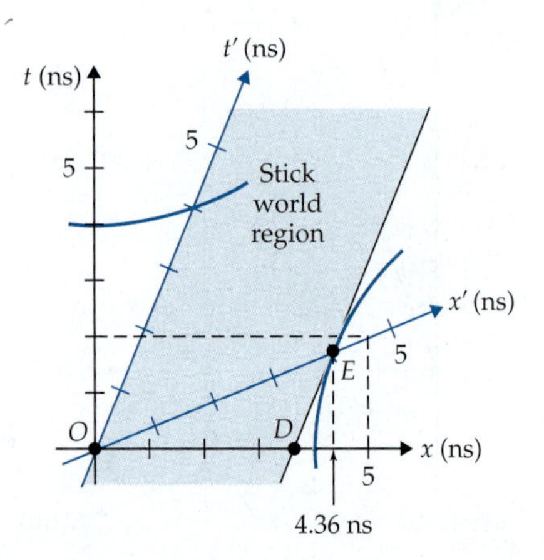

Figure R6.5
This measuring stick has a rest length of 4.0 ns, because events O and E, which occur simultaneously (in the Other Frame) and at opposite ends of the measuring stick, are 4.0 ns apart in this frame. Events O and D occur at opposite ends of the stick and simultaneously in the Home Frame, so the distance between these events is defined to be the length of the stick in that frame. We can see from the diagram that this length is about 3.7 ns.

Again we can also easily check this with the help of the Lorentz transformation equations. In the Home Frame, the length L of the measuring stick is *defined* to be the distance Δx between *simultaneous* events occurring at the ends of the stick—that is, events for which $\Delta t = 0$. Assuming we know that the length of the measuring stick in the Other Frame is $L_R = 4.0$ ns and that it is at rest in that frame, then the distance between *any* pair of events that occur at the opposite ends of the measuring stick must be $\Delta x' = L_R = 4.0$ ns in that frame. One of the Lorentz transformation equations for coordinate differences (equation R5.11*b*) says that

$$\Delta x' = \gamma(-\beta\,\Delta t + \Delta x) \tag{R6.4}$$

In the case at hand, we are looking for $L = \Delta x$, knowing that $\Delta x' = L_R$ and $\Delta t = 0$ for the events in question. Therefore, dropping the Δt term in equation R6.4 and solving for Δx, we get

$$L = \Delta x = \frac{L_R}{\gamma} = L_R\sqrt{1-\beta^2} = (4.0 \text{ ns})\sqrt{1-\left(\tfrac{2}{5}\right)^2} = 3.7 \text{ ns} \tag{R6.5}$$

in agreement with the result displayed by figure R6.5.

Lorentz contraction is frame-symmetric and thus consistent with the principle of relativity

In summary, we see that it doesn't matter whether the stick is at rest in the Home Frame or at rest in the Other Frame. If the stick is observed to have a length L_R in the frame in which it is at rest, it has a length $L = L_R\,(1-\beta^2)^{1/2}$ in any frame that is moving with speed $|\vec{\beta}|$ relative to the stick's rest frame. If the stick is at rest in the Home Frame, it is observed to be contracted in the Other Frame. If it is at rest in the Other Frame, it is observed to be contracted in the Home Frame. Because Lorentz contraction is frame-symmetric in this way, it means it *is* consistent with the principle of relativity.

R6.5 The Barn and Pole Paradox

The predictions of the theory of relativity are counterintuitive enough that it is easy (as a result of fuzzy thinking) to invent situations that at first appear to be paradoxical. We dealt with one of the most famous, the *twin paradox*, in chapter R4. In this section, we will examine another famous apparent paradox, generally known as the *barn and pole paradox*, that is based on a natural but mistaken understanding of the phenomenon of Lorentz contraction.

Description of the barn and pole problem

Consider a pole carried by a pole-vaulter who is running along the ground at a speed $|\vec{\beta}| = \frac{3}{5}$. In the runner's frame, the pole is at rest (of course): let us assume that it has a rest length of 10 ns. An observer on the ground is moving with speed $|\vec{\beta}|$ with respect to the pole's rest frame and so will measure the pole to be Lorentz-contracted to a length of only $L = L_R\,(1-\beta^2)^{1/2} = (10\text{ ns})(1-\frac{9}{25})^{1/2} = (10\text{ ns})(\frac{16}{25})^{1/2} = (10\text{ ns})(\frac{4}{5}) \approx 8$ ns. As the runner presses on, she runs through a barn that also happens to be 8 ns long as measured in the ground frame. Since both the pole and the barn are 8 ns in the ground frame, there is an instant of time in that frame in which *the barn entirely encloses the pole.*

But now look at the situation from the runner's perspective. In her frame, the pole is at rest and has its normal length of 10 ns. She sees the *barn* to be moving relative to her at a speed of $\frac{3}{5}$, and so it is the *barn* that is Lorentz-contracted to $\frac{4}{5}$ of its ground frame length—that is, to $(\frac{4}{5})(8\text{ ns}) = (32/5)\text{ ns} = 6.4$ ns. Thus, the paradox: *how can a barn that is 6.4 ns long ever enclose a 10-ns pole?*

This apparent paradox results from a naive application of the idea that "moving objects are contracted" without really understanding *why* objects are measured to be contracted and exactly how the length of a moving object

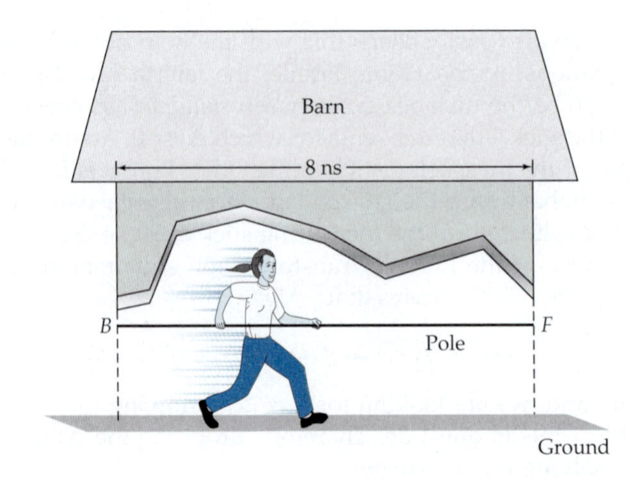

Figure R6.6
A picture of the barn and pole problem as it would be observed in the ground frame. Events F (front of the pole passes through barn's front door) and B (back of pole passes through barn's back door) are simultaneous in the ground frame. We can define the coordinate time of these events to be $t = 0$. Note that at this instant the pole is completely enclosed by the barn.

is measured. We will see that the apparent paradox is resolved if we carefully consider the precise meaning of the words we have used to describe it.

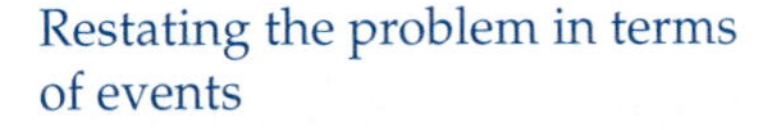
Restating the problem in terms of events

The first step in solving this problem (and virtually every other problem in special relativity) is to rephrase the problem in terms of *events*. Let us call the arrival of the front end of the pole at the front end of the barn event F. Call the arrival of the back end of the pole at the back of the barn event B. To say that there is an instant at which the barn encloses the pole is to say that events F and B are simultaneous in the ground frame (see figure R6.6).

Let us agree to use event B as the origin of both space and time in both frames (that is, B occurs at $x = 0$ and $t = 0$ in the ground frame and $x' = 0$ and $t' = 0$ in the runner's frame). The statement that F and B are simultaneous in the ground frame then means that event F also occurs at $t = 0$ in the ground frame. But when and where does event F occur in the runner's frame?

Figure R6.7 shows a two-observer spacetime diagram for this problem. I have chosen the ground frame to be the Home Frame of the diagram. I have also taken the ground observer's description of the events to be truthful: events B and F do occur simultaneously in the ground frame, and the pole is enclosed by the barn at time $t = 0$ in the ground frame. Notice also that the diagram supports the runner's claim that the barn is 6.4 ns long in her frame: events B and C are simultaneous in the runner's frame and lie at the ends of the barn, so the distance between them is the length of the barn in that frame (by definition), and this distance is indeed about 6.4 ns on the diagram.

The solution to the paradox

So what is the solution to the paradox? The diagram shows that *the runner never observes the pole to be enclosed by the barn*. Event F is *not* simultaneous with event B in the runner's frame: F is simultaneous with event D (note that the line connecting F and D is parallel to the x' axis). This means that event F (front of pole reaches front of barn) occurs about 6 ns *before* event B (back of pole reaches back of barn) in the frame of the runner. At the same time as event F occurs in the runner's frame (that is, at $t' = -6$ ns), you can see from the diagram that the pole's back end is still sticking out behind the barn. When event B finally occurs (at $t' = 0$), the pole's front end is sticking out in front of the barn. (Remember that all events occurring "at the same time" as a given event in the runner's frame lie on a line parallel to the diagram x' axis.)

A check using the Lorentz transformation equations

We can use the Lorentz transformation equations to confirm this picture. In the Home Frame, the coordinate differences between events F and B are $\Delta t_{BF} = 0$ and $\Delta x_{BF} \equiv x_F - x_B = 8\text{ ns} - 0 = 8$ ns. [The factor $\gamma = 1/(1-\beta^2)^{1/2} = 5/4$ here.] Therefore, in the runner's frame,

$$\Delta t'_{BF} = \gamma(\Delta t_{BF} - \beta\Delta x_{BF}) = \frac{5}{4}\left[0 - \frac{3}{5}(8\text{ ns})\right] = -\frac{3}{4}(8\text{ ns}) = -6\text{ ns} \qquad \text{(R6.6)}$$

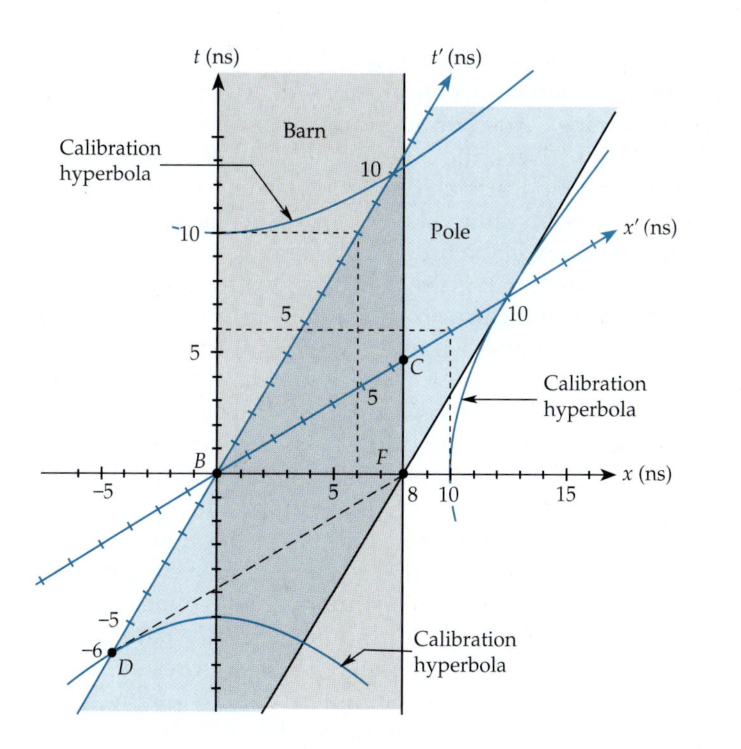

Figure R6.7
Graphical solution to the barn and pole paradox. Events B and C mark the ends of the barn at $t' = 0$ in the runner's frame. These events are roughly 6.4 ns apart according to the diagram, so the barn is indeed about 6.4 ns long as measured in the runner's frame. But note that events B and F are not simultaneous in the runner's frame. Indeed, event F occurs at the same time as event D, or about 6 ns before event B (note that the line connecting events F and D is parallel to the x' axis).

Since $\Delta t'_{BF} = t'_F - t'_B$ and $t'_B = 0$, we have $t'_F = -6$ ns, implying that event F occurs about 6 ns before B in the runner's frame, as we read from the diagram.

To the runner, everything looks exactly as if a 10-ns pole is being carried through a 6.4-ns barn

Now let us think about this for a minute. If *you* were the runner and you were told you were about to run a 10-ns pole through a 6.4-ns barn, what would you *expect* to see? First, you would see the front end of your pole reach the front end of the barn (event F). At this time, your 10-ns pole would stick out 3.6 ns behind the rear of the 6.4-ns barn. After the barn moves backward relative to you another 3.6 ns, the back end of your pole will coincide with the back end of the barn (event B), at which time the front of the pole sticks out 3.6 ns in *front* of the barn (see figure R6.8).

How long before event B should event F occur? The time between these events should be the time required for the barn to move backward a distance of 3.6 ns at a speed of $|\beta| = \frac{3}{5}$ (in the runner's frame), or

$$\Delta t = \frac{\Delta x}{|\beta|} = \frac{3.6 \text{ ns}}{\frac{3}{5}} = (3.6 \text{ ns})\left(\frac{5}{3}\right) = 6 \text{ ns} \tag{R6.7}$$

which is the time between the events indicated on the spacetime diagram in figure R6.7. In fact, you should go over that diagram very carefully and convince yourself that the description given above is indeed exactly what the runner will observe in her reference frame.

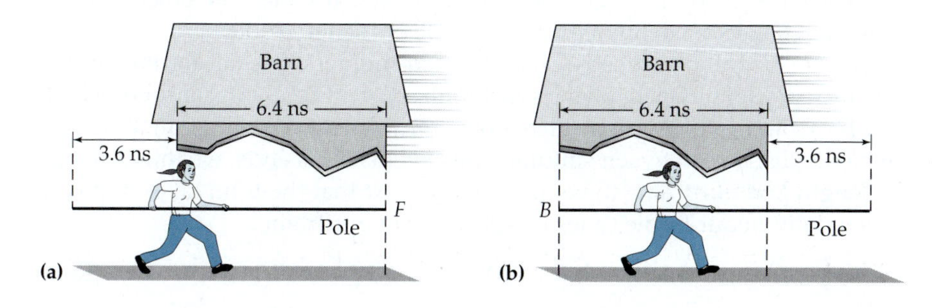

Figure R6.8
(a) The view from the runner's frame. Event F occurs first, at which time the pole sticks out about 3.6 ns behind the barn. (b) Event B occurs next, at which time the pole sticks out 3.6 ns in front of the barn. The time between events F and B is the time that it takes the barn to move a distance of 3.6 ns at its speed of $\beta = \frac{3}{5}$ relative to the runner's frame.

The point is that nothing strange or weird happens in either frame. In the barn frame, observed events are consistent with the interpretation that an 8-ns pole is being carried through an 8-ns barn. In the runner's frame, the time relationship between the same events is, on the other hand, consistent with the interpretation that a 10-ns pole is being carried through a 6.4-ns barn: we do not see anything like "a 10-ns pole enclosed by a 6.4-ns barn." The apparent paradox in the problem as stated is based on the unstated and erroneous assumption that if events F and B were simultaneous (that is, the pole is enclosed in the barn) in the ground frame, the events will *also* be simultaneous in the runner's frame. However, when we recall that the coordinate time measured between two events will *not* in general be the same in different frames, the paradox evaporates. Excepting the phenomenon of Lorentz contraction itself, *nothing* unusual is seen to happen in either frame.

Similar "paradoxes" hinge on the same error about simultaneous events

People have invented a number of apparent paradoxes analogous to the barn and pole paradox (see the homework problems for examples). Such paradoxes almost always involve a hidden assumption that two events that are simultaneous in a given inertial reference frame are simultaneous in all reference frames. We are taken in by the apparent paradox because our intuitive belief in the absolute nature of simultaneity is so natural that we hardly notice when it is assumed. But special relativity teaches us that this assumption is *not* true: because observers in different inertial frames do *not* observe clocks in other frames to be synchronized, they will disagree about whether two given events are simultaneous. Moreover, because we have defined the length of a moving object in a given frame in terms of simultaneous events in that frame, it follows that different observers will disagree about that *length* as well. The frame dependence of simultaneity and the phenomenon of Lorentz contraction are bound together, and indeed they fail to make sense without each other (as these paradoxes show).

Exercise R6X.2

(a) Using figure R6.7, describe how you could verify that at the time of event F in the runner's frame, the pole's back end is about 3.6 ns of distance behind the barn's back door. **(b)** Use the Lorentz transformation equations to verify that at the time of event F in the runner's frame, the pole's back end is about 3.6 ns of distance behind the barn's back door.

R6.6 Other Ways to Define Length

Alternative definitions of length

We have seen that if we define a moving object's length as the distance between simultaneous events occurring at its ends, we get a frame-dependent (Lorentz-contracted) answer. Are there other ways we might define a moving object's length? If so, do these yield different results?

An example of such an alternative definition

The answers to these questions are as follows. Yes, there are other logically reasonable ways of defining a moving object's length, but these definitions, if they *are* logically reasonable, yield the *same* numerical result for the object's length as the definition involving synchronized events presented in section R6.1.

For example, consider again a moving train. Instead of defining its length to be the distance between simultaneous events at its ends, we might define its length L in our frame to be the total time Δt that the train takes to pass a given point in our frame times its speed $|\beta|$ in our frame:

$$L \equiv |\beta|\,\Delta t \tag{R6.8}$$

The value of $|\beta|\,\Delta t$ should yield the distance that the train moves as it passes the point, which is a perfectly reasonable definition of its length.

This definition yields the same result as the original

This definition, however, yields exactly the same (Lorentz-contracted) result as our original definition. To see this, let us apply this formula to determine the length in the runner's frame of the barn described in section R6.5. Event B is the event of the barn's back end passing the point $x' = 0$ (that is, the diagram t' axis) in the runner's frame, while event E is the event of the barn's front end passing this point. You can see from figure R6.7 (the relevant portion of which is repeated in figure R6.9) that it takes the barn roughly $\Delta t' \approx 10.7$ ns to pass the point $x' = 0$ in the runner's frame. Since the barn is traveling backward at a speed of $|\beta| = \frac{3}{5}$ in the runner's frame, the barn's length according to our new definition must be

$$L' \approx \tfrac{3}{5}(10.7 \text{ ns}) \approx 6.4 \text{ ns} \tag{R6.9}$$

which is roughly the same (contracted) result we found before.

We can (as usual) use a Lorentz transformation equation to find the barn's exact length according to this definition. Events B and E occur at $x' = 0$ in the runner's frame by definition, so $\Delta x' = 0$ for these events. Since these events occur at opposite ends of the barn, they are separated by $\Delta x = 8$ ns in the barn's frame. One of the inverse Lorentz transformation equations says that

$$\Delta x = \gamma(\beta\Delta t' + \Delta x') \tag{R6.10a}$$

so (since $\Delta x' = 0$)

$$\Delta t' = \frac{\Delta x}{\gamma\beta} = \frac{8 \text{ ns}}{(5/4)(\frac{3}{5})} = \frac{4}{3}(8 \text{ ns}) = \frac{32}{3} \text{ ns} = 10.67 \text{ ns} \tag{R6.10b}$$

According to equation R6.8, the barn's length in the runner's frame is thus

$$L' = |\beta|\,\Delta t' = \frac{3}{5}\left(\frac{32}{3} \text{ ns}\right) = \frac{32}{5} \text{ ns} = 6.4 \text{ ns} \tag{R6.11}$$

which is exactly the same as the result we found before.

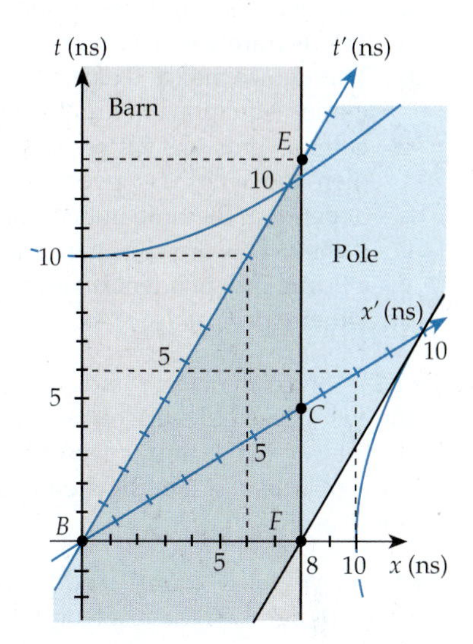

Figure R6.9
Events B and E are the events where the barn's back and front ends respectively pass the spatial origin of the runner's frame. The time this takes in the runner's frame is about 10.7 ns.

This is just one example of an alternative definition of length (see problem R7A.1 for discussion of another). *All* known reasonable definitions of the length of a moving object yield the same result as equation R6.2: no matter how you compute it, the length of an object determined in a frame where it is moving is Lorentz-contracted from its rest length by the factor $(1-\beta^2)^{1/2}$.

. . . as do other definitions

Exercise R6X.3

Using equation R6.8, find the length of the runner's pole in the ground frame **(a)** by reading Δt from figure R6.9 and **(b)** by computing Δt with the help of an appropriate Lorentz transformation equation.

TWO-MINUTE PROBLEMS

R6T.1 A moving object's length in a given frame is defined to be the distance between two events that occur at opposite ends of the object and that are simultaneous in that frame. *Why* is it crucial that the events we use to define a moving object's length be *simultaneous*?
A. This is purely conventional: there is no other reason.
B. This choice makes it easier to use the Lorentz transformation equations to find the length.
C. If the events are *not* constrained to be simultaneous, then the length is poorly defined: its value would depend on the time interval between the events.
D. If the events are simultaneous, then the length will be a frame-independent quantity.
E. Other (specify)

R6T.2 Since an object's ends do not move in its rest frame, the events used to mark out an object's length in that frame do not *have* to be simultaneous: the distance between them is the object's rest length whether they are simultaneous or not. T or F?

R6T.3 An object of rest length L_R moving at one-half the speed of light will have a length equal to:
A. $\frac{1}{2}L_R$
B. $\frac{3}{4}L_R$
C. $\left(\frac{1}{2}\right)^{1/2}L_R$
D. $0.87L_R$
E. $\frac{1}{4}L_R$
F. Other (specify)

R6T.4 An object is at rest in the Home Frame. Imagine an Other Frame moving at a speed of $|\beta| = \frac{4}{5}$ with respect to the Home Frame. The object's length in the Other Frame is measured to be 15 ns. What is its length as observed in the Home Frame?
A. 15 ns
B. 12 ns
C. 9 ns
D. 19 ns
E. 25 ns
F. Other (specify)

R6T.5 An object's length would be negative in a frame where it travels faster than the speed of light. T or F?

R6T.6 Suppose an object is in a frame where it is moving at speed $|\vec{\beta}_0|$ and its length is L_0 at that speed. If we double the speed ($|\vec{\beta}_1| = 2|\vec{\beta}_0|$), then the object's length is compressed by a factor of two ($L_1 = \frac{1}{2}L_2$). T or F?

R6T.7 The most important reason an object is observed to be shorter in a frame where it is moving than in a frame where it is at rest is that
A. The force of motion strongly compresses an object that is moving at relativistic speeds.
B. "Simultaneity" is not a frame-independent concept.
C. The measuring sticks used by the moving observer are Lorentz-contracted.
D. The clocks used by the moving observer run slower.

R6T.8 In the pole and barn problem, the barn never actually encloses the pole in the ground frame. T or F?

R6T.9 We can define a moving object's length to be its speed times the time it takes to pass a given point. T or F?

HOMEWORK PROBLEMS

Basic Skills

R6B.1 How fast must an object be moving in a given frame if its measured length in that frame is one-half its rest length?

R6B.2 How fast must an object be moving in a given frame if its measured length in that frame is to be significantly different from its rest length? (Assume you can measure the object's length to 1 part in 10,000—that is, to four significant figures.)

R6B.3 An Other Frame moves with a speed of 0.80 relative to the Home Frame. An object at rest in the Home Frame has a length of 30 ns. What is the object's length in the Other Frame?

R6B.4 An Other Frame moves with a speed of 0.80 relative to the Home Frame. An object at rest in the Other Frame has a length of 30 ns as measured in the Home Frame. What is the object's length in the Other Frame?

R6B.5 Suppose an object with a rest length of 10 ns is at rest in a frame that is moving with a speed of 0.50 relative to the Home Frame. Draw a two-observer spacetime diagram of this situation, and use it to determine the length of the object in the Home Frame. Check your result by using equation R6.2.

R6B.6 Suppose an object with a rest length of 5 ns is at rest in the Home Frame. The Other Frame is moving with a speed of 0.50 relative to the Home Frame. Draw a two-observer spacetime diagram of this situation, and use it to determine the length of the object in the Other Frame. Check your result using equation R6.2.

R6B.7 An observer at rest with respect to the sun measures the earth's diameter (in the direction of its motion) as it swings by in its orbit. How many centimeters shorter is this diameter in this frame than its rest diameter of 12,760 km? (*Hint:* Use the binomial approximation.)

R6B.8 About how many femtometers shorter than its rest length is a car's length measured in the ground frame if that car is traveling at 30 m/s (66 mi/h) in that frame? Assume the car's rest length is 5.0 m. (1 fm = 10^{-15} m is about the size of an atomic nucleus.)

Modeling

R6M.1 Imagine an alien spaceship traveling so rapidly that it crosses our galaxy (whose rest diameter is 100,000 ly) in only 100 y of spaceship time. Observers at rest in the galaxy say that this is possible because the ship's speed $|\beta|$ is so close to 1 that the proper time it measures between its entry into and departure from the galaxy is much shorter than the galaxy-frame coordinate time between those events (about 100,000 y). But how does this look to the aliens? To them, the galaxy moves backward relative to them at the speed $|\beta| \approx 1$, and so is Lorentz-contracted to a bit less than 100 ly across. *This* is what makes it possible for the whole galaxy to fly by them in only 100 y.

(a) Find the exact value of the speed $|\beta|$ that the aliens must achieve to cross the galaxy in 100 y.

(b) Find the galaxy's diameter in the aliens' frame, and verify that a galaxy with such a diameter moving at speed $|\beta|$ will completely pass the aliens' ship in 100 y.

R6M.2 In the experiment described in problem R3M.11, particles travel at a speed of $|\vec{v}| = 0.866$ between detectors 2.08 km apart. This takes 8.0 μs as measured by laboratory clocks. Since the half-life of the particles involved is 2.00 μs, we might naively expect only about one-sixteenth of the particles to survive the trip. But that problem's solution shows that the particles' clocks actually only measure 4.00 μs for the trip between the detectors, and thus about one-fourth survive.

But now consider how this all looks to an observer traveling with one of the particles. In the particle's frame, the laboratory and the detectors appear to be moving past at a speed of 0.866. In 4.00 μs (as measured by the particle's clock), the laboratory will only move by a distance of 1.04 km at that speed, so the particles will only see *one-half* of the distance between the detectors go by. But laboratory observers claim that by the time the particles' clocks read 4.00 μs, they have covered the *full* distance between the detectors. Is this not a paradox?

Resolve the apparent paradox by considering Lorentz contraction. How far apart are the detectors in the particle frame? How does this resolve the paradox? (*Hint:* Problem R6M.1 discusses a similar situation.)

R6M.3 As discussed in section R3.4, muons created in the upper atmosphere can sometimes reach the earth's surface. Imagine that one such muon travels the 60 km from the upper atmosphere to the ground (in the earth's frame) in one muon half-life of 1.52 μs (in the muon's frame). How thick is the portion of the earth's atmosphere from the muon's creation point to the ground in the *muon's* frame?

R6M.4 How fast would you have to be moving relative to our Milky Way galaxy such that the galaxy (whose rest diameter is 100,000 ly) is only 3.65 light-days across in your reference frame (enabling you to cross it in about 3.65 days according to your clock)? Express your answer in the form $|\beta| = 1 - \delta$. (*Hint:* See problem R6D.1.)

R6M.5 Imagine a cube 30 cm on a side. About how fast would this cube have to be moving relative to you if in your frame it was as thin as a sheet of paper in the direction of its motion? (*Hints:* You may find the approximation discussed in problem R6D.1 useful. Also, can you think of an easy way to estimate the thickness of a sheet of paper?)

Derivation

R6D.1 *A useful approximation.* Consider a speed $|\beta|$ that is very close to the speed of light: $|\beta| = 1 - \delta$, where $\delta \ll 1$. Use the binomial approximation to show that under these circumstances $1/\gamma = (1-\beta^2)^{1/2} \approx (2\delta)^{1/2}$. How accurate is this approximation when $|\beta| = 0.9$? When $|\beta| = 0.99$?

R6D.2 Prove mathematically that the alternative definition of length given in section R6.6 *always* yields the same result as equation R6.2, as follows. Consider an object of rest length L_R at rest in a frame that we can choose to call the Other Frame. Let β be the x-velocity at which the Other Frame (and thus the object) is moving relative to the Home Frame. Let Δt be the Home-Frame time between the events of the object's front end passing a certain point in the Home Frame and its rear end passing that same point. Since these events occur at the same place in the Home Frame, $\Delta x = 0$ between these events in that frame. But since these events occur at opposite ends of the object, the distance between them in the Other Frame is $|\Delta x'| = L_R$. Use an appropriate Lorentz transformation equation to determine Δt in terms of Δx, $\Delta x'$, and β, and then use the result to prove that $L \equiv |\beta|\,\Delta t = L_R(1-\beta^2)^{1/2}$.

R6D.3 We can use the metric equation to derive equation R6.2 as follows. An object with rest length L_R moving in the $+x$ direction with x-velocity β passes a clock at rest in your inertial frame. Let event F be the object's front end passing that clock, and let event B be the object's back end passing that clock.

(a) Argue that in the object's frame, the coordinate time between these events is equal to $L_R/|\beta|$.

(b) What is the *distance* between these events in the object's frame?

(c) Define the object's length in *your* reference frame to be the distance the object travels in the time it takes to pass by your clock—that is, $L = |\beta|\Delta t$, where Δt is the time measured between events F and B by your clock. Use the metric equation and the information in parts (a) and (b) above to arrive at equation R6.2. (*Hint:* Your clock is present at both events.)

R6D.4 *Transformation of angles.* Consider a meterstick at rest in a given inertial frame (make this the Other Frame) oriented in such a way that it makes an angle of θ' with respect to the x' direction in that frame. In the Home Frame, the Other Frame is observed to move in the $+x$ direction with an x-velocity of β.

(a) Keeping in mind that the distances measured *parallel* to the line of relative motion are observed to be Lorentz-contracted in the Home Frame while distances measured perpendicular to the line of motion are not, show that the angle θ we will observe this meterstick to make with the x direction in the Home Frame is given by

$$\theta = \tan^{-1}\left(\frac{\tan\theta'}{\sqrt{1-\beta^2}}\right) \qquad \text{(R6.12)}$$

(b) What would the meter stick's length be as measured in the Home Frame?

(c) Assume the meterstick makes an angle of 30° with the x' direction in the Other Frame. How fast would that frame have to be moving with respect to the Home Frame for the meterstick to be observed in the Home Frame to make an angle of 45° with the x direction?

Rich-Context

R6R.1 *The space wars paradox.** Two spacecraft of equal rest length L_R = 100 ns pass very close to each other as they travel in opposite directions at a relative speed of $|\beta| = \frac{3}{5}$. The captain of ship O has a laser cannon at the tail of her ship. She intends to fire the cannon at the instant her bow is lined up with the tail of ship O'. Since ship O' is Lorentz-contracted to 80 ns in the frame of ship O, she expects the laser burst to miss the other by 20 ns, as shown in figure R6.10a (she intends the shot to be "across the bow"). However, to the observer in ship O', it is ship O that is contracted to 80 ns. Therefore, the observer on O' concludes that if the captain of O carries out her intention, the laser burst will strike ship O' about 20 ns *behind* the bow, with disastrous consequences (figure R6.10b).

Assume the captain of O carries out her intentions exactly as described, according to measurements in her own frame, and analyze what *really* happens as follows:

(a) Construct a carefully calibrated two-observer spacetime diagram of the situation described. Define event A to be the coincidence of the bow of ship O and the tail of ship O' and event B to be the firing of the laser cannon. Choose A to define the origin event in both frames, and locate B according to the description of the intention of O' above. When and where does this event occur as measured in the O' frame, according to the diagram? (You may assume that the ships pass each other so closely that the travel time of the laser burst between the ships is negligible.)

(b) Verify the coordinates of B, using the Lorentz transformation equations.

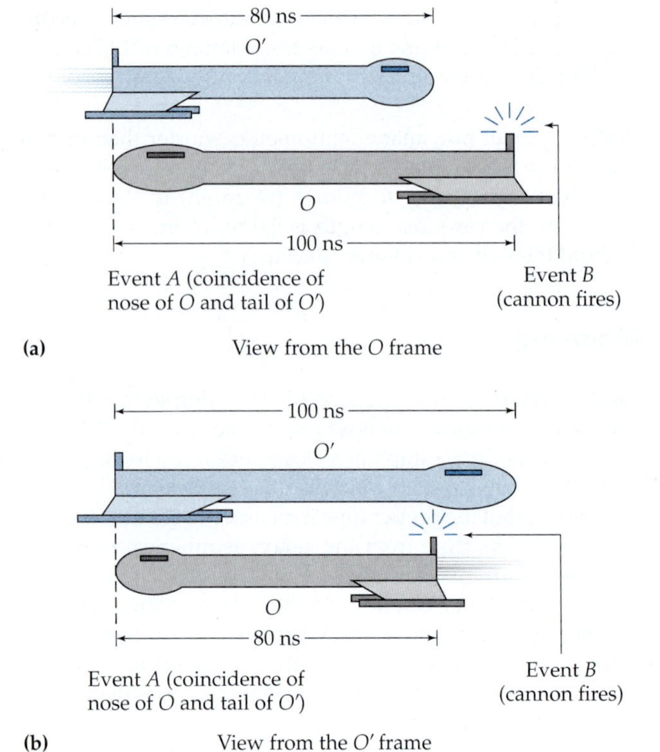

Figure R6.10
The situation described in problem R6R.1.

(c) Write a short paragraph describing whether the cannon burst really hits or not, according to the results you found above. Discuss the hidden assumption in the statement of the apparent paradox, and point out how one of the drawings in figure R6.10 is misleading.

R6R.2 *The bullet hole paradox.*† Two guns are mounted a distance of 40 ns apart on the embankment beside some railroad tracks. The barrels of the guns project outward toward the track so that they almost brush a speeding express train as it passes by. The train moves with a speed of $|\beta| = \frac{3}{5}$ with respect to the ground. Suppose the two guns fire simultaneously (as measured in the ground frame), leaving two bullet holes in the train.

(a) Let event R be the firing of the rear gun and event F the firing of the front gun. These events occur 40 ns apart and at the same time in the ground frame. Draw a carefully constructed two-observer diagram of the situation, taking the ground frame to be the Home Frame and taking R to be the origin event. Be sure to show and label the axes of the ground and train frames, the worldlines of the guns, the worldlines of the bullet holes that they produce, and events R and F.

*Adapted from E. F. Taylor and J. A. Wheeler, *Spacetime Physics*, San Francisco: Freeman, 1966, pp. 70–71.

†Adapted from B. M. Casper and R. J. Noer, *Revolutions in Physics*, New York: Norton, 1972, pp. 363–364.

(b) Argue, using your diagram, that the bullet hole worldlines are about 50 ns apart as measured in the train frame. Verify this by using the Lorentz transformation equations to show that events R and F occur 50 ns apart in the train frame.

(c) In the ground frame, the guns are 40 ns apart. In the train frame, the guns are moving by at a speed of $|\beta| = \frac{3}{5}$, and the distance between them is Lorentz-contracted to less than 40 ns. Use the Lorentz contraction formula to show that the guns are in fact 32 ns apart in the train frame. Describe how this same result can be read from your spacetime diagram.

(d) Doesn't this lead to a contradiction? How can two guns that are 32 ns apart in the train frame fire simultaneously and yet leave bullet holes 50 ns apart in the train frame? Write a paragraph in which you carefully describe the logical flaw in the description of the "contradiction" given in the last sentence. (*Hint:* Focus on the word *simultaneously*.) Describe what *really* is observed to happen in the train frame, and thus how it is perfectly natural for guns that are 32 ns apart to make holes that are 50 ns apart.

R6R.3 *The space cadets paradox.** A very long measuring stick is placed in empty space at rest in an inertial frame we'll call the "stick frame." A spaceship of rest length L_R travels along the measuring stick in the $+x$ direction with an x-velocity $\beta = \frac{4}{5}$ relative to it. Two space cadets P and Q with knives and synchronized watches are stationed at rest on the ship frame in the ends of the spaceship. At a prearranged time, each cadet simultaneously reaches through a porthole and slices through the measuring stick.

(a) How long is the spaceship according to the cadets?

(b) How long is the spaceship according to observers along the measuring stick (that is, observers at rest in the "stick frame")?

(c) Use the Lorentz transformation equations to show that observers along the measuring stick would conclude that the measuring stick's cut portion has length $\frac{5}{3}L_R$.

(d) Since the cutting events occur simultaneously in the spaceship frame, they do *not* occur simultaneously in the stick frame. Use the Lorentz transformation equations to find the time separation of the two cutting events as viewed in the stick frame.

(e) Explain in a short paragraph how it is that two cadets who are only $\frac{3}{5}L_R$ apart (as measured in the stick frame) can cut a hunk of measuring stick $\frac{5}{3}L_R$ long if they really cut simultaneously according to their synchronized watches.

R6R.4 Albert and Becky are passengers on a train that is moving through a long straight tunnel at a velocity of $\vec{\beta} = 3/5$ in a direction we will call the $+x$ direction. Albert and Becky are 80 ns of distance apart as measured in the tunnel (Home) frame. At a pre-arranged instant of time in the train frame that we will call $t' = 0$, Albert reaches through his window and paints a mark on the tunnel wall (event A) and Becky also reaches through her window and paints a mark on the wall (event B). Let Albert's position define the spatial origin $x' = 0$ in the train frame.

(a) Construct a calibrated two-observer spacetime diagram of the situation described. Draw Albert and Becky's worldlines, locate events A and B, and draw the worldlines of the marks. (Remember that the marks do not exist before events A and B.)

(b) You should see that events A and B are 100 ns apart in the train frame. Check this by considering the fact that the train is Lorentz contracted in the tunnel frame.

(c) Read from your diagram the displacement Δx between the marks in the tunnel frame, and show with lines and labels on the diagram how you arrived at that result.

(d) Check your work for part (b) using an appropriate Lorentz transformation equation. (*Hint:* $\Delta x > 100$ ns.)

(e) How is it possible for Albert and Becky, who are 80 ns apart in the tunnel frame, to paint their marks simultaneously and yet leave marks on the tunnel that are more than 100 ns apart? Explain in quantitative detail how everything works in the tunnel frame, assuming that Albert and Becky really do make their marks at the same time in the train frame.

R6R.5 Rachel and Fred are standing in the ground frame (the Home Frame) along a train track that runs in the $+x$ direction. Fred is 200 ns further up the track than Rachel. A train traveling in the $+x$ direction at an x-velocity of $\beta = 3/5$ is passing by. At a pre-arranged time according to their previously synchronized watches (call this $t = 0$), Rachel and Fred reach out and mark the side of the train with a magic marker. Let the event of Rachel's marking the train be R, and the event of Fred marking the train be F.

(a) Construct a carefully calibrated two-observer spacetime diagram of the situation described. In addition to the reference frame axes for both frames, draw Fred and Rachel's worldlines, locate events F and R, and draw the worldlines of the marks. (Remember that the marks do not exist before events F and R.)

(b) What is the distance between the marks in the train frame? Indicate the distance on your diagram.

(c) Check that your answer for part (b) is correct by using an appropriate Lorentz transformation equation. (*Hint:* The result should be greater than 200 ns.)

(d) What is the separation between Rachel and Fred in the train frame? Indicate this separation on the diagram.

(e) Check that your answer for part (b) is correct by using an appropriate inverse Lorentz transformation equation. (*Hint:* The result should be less than 200 ns.)

(f) Describe clearly and quantitatively how the train's conductor would explain how Rachel and Fred can be less than 200 ns apart in the train frame and yet draw marks that are more than 200 ns apart. (Don't just state the events' time coordinates in the train frame: explain exactly what the conductor observes, how far the conductor sees Rachel and Fred move between the events if that is relevant, and how everything looks completely normal in the train frame. A few well-chosen sentences will suffice.)

*Thanks to W. F. Titus of Carleton College.

Advanced

R6A.1 *The radar method*. Imagine using the radar method to measure the length of an object moving at an x-velocity of $-\beta$ with respect to your own frame (which is the Other Frame). At a certain time (event A) you send forth a light flash from the master clock at the spatial origin of your frame. This flash bounces off a mirror at the far end of the object (event R) and then returns to your clock (event B). If you time this all just right so that the near end of the object passes your clock (event O) at exactly the time halfway between the emission event A and the reception event B (as measured by that clock), then you know that event O and the reflection event R are simultaneous. This means that at that instant, the object lies exactly between the clock and the light flash as it bounces off the mirror. The length L of the object in your frame is thus equal (in SR units) to the time it takes the light to come back from event R, since light travels at a speed of 1 in all frames.

Now imagine viewing this measurement process from a Home Frame in which your frame is moving in the x-direction with an x-velocity β. Because the object is at rest in that frame, the distance between its ends in that frame is L_R. Draw a two-observer diagram of the situation as viewed by observers in the Home Frame (let O be the origin event in both frames). Argue that the coordinate distance between events O and A in the Home Frame is $\Delta x = |\beta|\Delta t$, where Δt is the coordinate time measured between those events in the Home Frame. Also argue (using similar triangles on the diagram) that $\Delta t = L_R$. Then, use the metric equation to relate the time measured between events O and A measured in *your* frame (which is equal to L) to the coordinate time Δt measured between the events in the Home Frame, and show that you end up with the same result as that given by the Lorentz contraction equation R6.2.

R6A.2 *Light clocks*. Consider a light clock as shown in figure R3.3, except imagine the light clock to be laid on its side so that the light flash travels along the clock's direction of motion. Show that the only way this sideways light clock will measure the correct spacetime interval between events A and B (as any decent clock should) is if the distance between its mirrors is Lorentz-contracted by the amount stated by equation R6.2.

ANSWERS TO EXERCISES

R6X.1 A redrawn version of figure R6.3 is shown below.

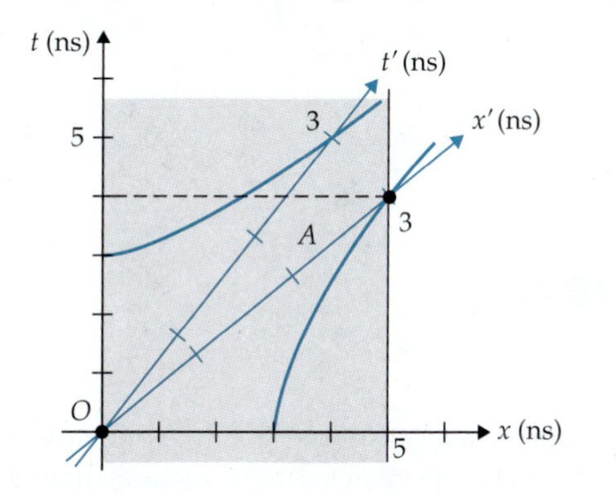

Events O and A occur simultaneously in the Other Frame and thus mark out the length of the stick in that frame. From the diagram, it looks as if the stick is about 3 ns long in the Other Frame. Equation R6.2 says that $L = L_R(1-\beta^2)^{1/2} = (5\text{ ns})\left(1-\frac{16}{25}\right)^{1/2} = (5\text{ ns})\left(\frac{3}{5}\right) = 3$ ns.

R6X.2 **(a)** Event P shown in the figure below (a slightly modified version of figure R6.7) occurs at the barn's back end at the same time as event F ($t'_F = -6$ ns).

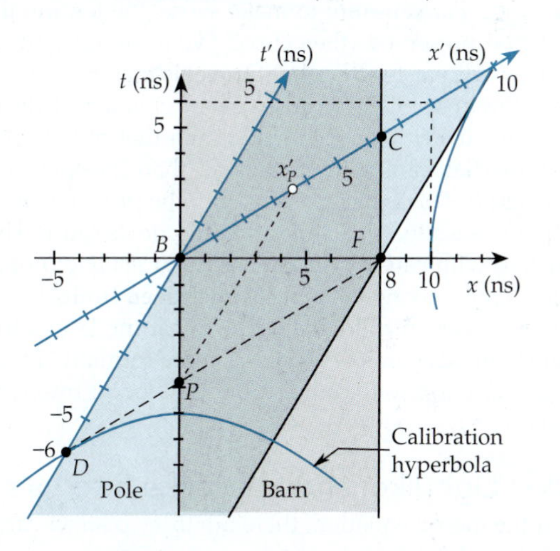

To find out *where* this event occurs in the runner's frame, we draw a line parallel to the t' axis up to the x' axis, as shown. We see from this line that the barn's back end is at $x' = 3.6$ ns when event F occurs, while the pole's back end is (always) at $x' = 0$. Therefore, the pole's back end is indeed about 3.6 ns behind the barn's back end in the runner's frame at this time.

(b) The barn's back end is always at $x_P = 0$ in the Home Frame. So, according to one of the inverse Lorentz transformation equations, we have

$$x_P = \gamma(\beta t'_P + x'_P) \tag{R6.13}$$

$$\Rightarrow \quad x'_P = \frac{x_P}{\gamma} - \beta t'_P = 0 - \frac{3}{5}(-6\text{ ns}) = 3.6\text{ ns} \tag{R6.14}$$

Since the pole's back end is always at $x' = 0$, it is sticking out 3.6 ns behind the barn at this time.

R6X.3 **(a)** The figure below (a slightly modified copy of figure R6.9) implies that it takes about 13.4 ns (the time between events F and E) in the Home Frame for the pole to pass the fixed point $x = 8$ ns in the Home Frame.

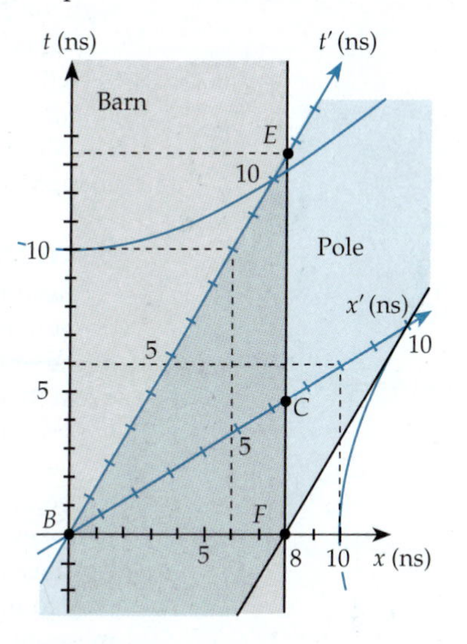

Since the pole's speed is $|\beta| = \frac{3}{5}$ in this frame, this implies that the pole has a length $L = |\beta|\,\Delta t = \left(\frac{3}{5}\right)(13.4\text{ ns}) = 8.0$ ns.
(b) Events F and E occur at the same position ($x = 8$ ns) in the Home Frame, so they are separated by $\Delta x = 0$ in that frame. Since these events occur at opposite ends of the pole in the runner's frame, they are separated by $\Delta x' = 10$ ns in that frame (*negative* because $\Delta x' \equiv x'_E - x'_F = 0\text{ ns} - 10\text{ ns} = -10$ ns). So using the Lorentz transformation equation $\Delta x' = \gamma(-\beta\,\Delta t + \Delta x)$, we find that

$$\Delta t = \frac{\Delta x}{\beta} - \frac{\Delta x'}{\gamma\beta} = 0 - \frac{-10\text{ ns}}{\left(\frac{5}{4}\right)\left(\frac{3}{5}\right)}$$
$$= \frac{40}{3}\text{ ns} = 13.3\text{ ns} \tag{R6.15}$$

This means that $L = |\beta|\,\Delta t = \frac{3}{5}(40/3\text{ ns}) = 8$ ns exactly.

R7 The Cosmic Speed Limit

Chapter Overview

Introduction

One of the most fundamental and surprising consequences of the principle of relativity is that nothing can travel faster than the speed of light in a vacuum. In this chapter, with the help of some two-observer diagrams, we will explore why this must be.

This chapter also provides an appropriate context for discussing the Einstein velocity transformation equations that replace the Galilean velocity transformation equations (equations R1.3), since the latter do allow for faster-than-light speeds. The Einstein velocity transformation equations provide important background for chapters R8 and R9.

Section R7.1: Causality and Relativity

A **causal influence** is any effect (particle, object, wave, or message) produced by one event that can cause another event. For causality to make sense, the **temporal order** of the events in question must be preserved (the caused event cannot precede the event that causes it). The problem with faster-than-light travel is that it *violates causality*. We can see this as follows. Suppose event P causes event Q and that the causal influence moving from P to Q moves with a speed faster than that of light. In such a case, we can always find an inertial frame moving slower than the speed of light where event Q occurs before event P. This is inconsistent with the principle of relativity, since it violates the concept of causality for an effect to precede its cause. The only way to make relativity consistent with causality is to insist that *nothing* (not even a message) can travel faster than light, making it a real **cosmic speed limit**.

The concept of causality expresses in colloquial terms what the law of increase of entropy (the **second law of thermodynamics**) expresses more formally. To say that we must preserve causality is really to say that the second law of thermodynamics must obey the principle of relativity.

Section R7.2: Timelike, Lightlike, and Spacelike Intervals

Because of the minus signs in the metric equation, there are three distinct categories of spacetime interval:

1. **Timelike:** when $\Delta s^2 > 0$
2. **Lightlike:** when $\Delta s^2 = 0$
3. **Spacelike:** when $\Delta s^2 < 0$

These categories are frame-independent, since the value of Δs^2 is frame-independent.

How can we measure the spacetime interval between the two events? If Δs^2 is timelike, we can find an inertial frame where the events occur at the same *place*, and we measure Δs with the frame's clock at that location. If Δs^2 is *spacelike*, we can find a frame where the events occur at the same *time*: the distance that we measure with a ruler between the events in that frame is the **spacetime separation** $\Delta\sigma \equiv (-\Delta s^2)^{1/2}$ between the events.

Section R7.3: The Causal Structure of Spacetime

Events separated by a spacelike interval *cannot* be causally connected, since the causal influence would have to travel faster than the speed of light. Because the categories of spacetime interval are frame-independent, *all* observers will agree about (1) which events have a timelike (or lightlike) interval with a given event P and occur *after* it in all frames, (2) which events have a timelike (or lightlike) interval with P and occur *before* it in all frames, and (3) which events have a spacelike interval with P. Events in the first category may be causally influenced by P, so we say such events lie in the **future** of P. Events in the second category may causally influence P, so we say such events lie in the **past** of P. Events in the third category cannot be causally related to P.

In a spacetime diagram with two spatial axes, P's past and future look like cones whose points touch at P and whose surfaces are rings of light converging on or expanding from P: this structure is P's **light cone**.

Section R7.4: The Einstein Velocity Transformation

We can use a two-observer spacetime diagram to determine a particle's velocity in one inertial frame if we know its velocity in another and the two frames' relative velocity. We do this by (1) setting up a calibrated two-observer spacetime diagram showing axes for both frames; (2) drawing the particle's worldline with the correct slope relative to the axes of the frame in which we know its velocity; and (3) measuring that worldline's slope according to the other frame's axes.

Alternatively, one can use the Lorentz transformation equations to compute the velocity of the object in either frame, using its velocity components measured in the other. The result is

$$v'_x = \frac{v_x - \beta}{1 - \beta v_x} \qquad v'_y = \frac{v_y\sqrt{1-\beta^2}}{1 - \beta v_x} \qquad v'_z = \frac{v_z\sqrt{1-\beta^2}}{1 - \beta v_x} \tag{R7.14}$$

$$v_x = \frac{\beta + v'_x}{1 + \beta v'_x} \qquad v_y = \frac{v'_y\sqrt{1-\beta^2}}{1 + \beta v'_x} \qquad v_z = \frac{v'_z\sqrt{1-\beta^2}}{1 + \beta v'_x} \tag{R7.8}$$

- **Purpose:** The first set of equations describes how to calculate an object's velocity components v'_x, v'_y, and v'_z measured in the Other Frame from its velocity components v_x, v_y, and v_z measured in the Home Frame; the second set tells you how to do the reverse, where β is the x-velocity of the Other Frame relative to the Home Frame.
- **Limitations:** These equations assume that the two frames are inertial and that they are in standard orientation with respect to each other.
- **Notes:** Equations R7.14 are the **(direct) Einstein velocity tranformation equations**, and equations R7.8 are the **inverse Einstein velocity transformation equations**. These equations replace the Galilean transformation equations (equations R1.3).

R7.1 Causality and Relativity

"Nothing can go faster than the speed of light." This statement is a well-known consequence of special relativity. But why *must* this statement be true? Do loopholes exist that might make faster-than-light travel possible?

What is causality?

In sections R3.2 and R4.2, we saw how the metric equation and the proper time equation fail if we apply them to a clock moving faster than light: both equations imply that the time registered between two events by such a clock would be an imaginary number, which is absurd. In both cases, this absurdity results from the violation of the $\Delta t^2 > |\Delta\vec{d}|^2$ restriction necessary for the derivation of the metric equation. Thus, neither equation really says anything useful about what a clock traveling faster than light would measure.

In this section, we will see that there is a deeper problem with traveling faster than light: *it violates causality*. What do I mean by *causality?* In physics (and more broadly, in daily life), we know that certain events *cause* other events to happen (see figure R7.1). For example, even couch potatoes know that if you press the appropriate button on the remote control, the TV channel will change. **Causally connected** events *must* happen in a certain order in time: the event being caused must follow the event that causes it. For example, we would be deeply disturbed if the TV channel changed just *before* we pressed the remote control button!

Consider two distinct events (call them P and Q) such that event P *causes* event Q, or more precisely, Q happens as a direct consequence of the reception of some kind of information that P has occurred. This information can be transmitted from P to Q in any number of ways: via some mechanical effect (such as the movement of an object or the propagation of a sound wave), via a light flash, via an electric signal, via a radio message, etc. Basically, the information can be carried by *any* object or effect that can move from place to place and is detectable.

Let's consider the TV remote control again as a specific example. Suppose you press a button on your TV remote control handset (event P). The information that the button has been pressed is sent to the TV set in some manner, and in response, the TV set changes channels (event Q). Keep this basic example in mind as we go through the argument that follows.

Why faster-than-light causal influences are absurd

Now let us pretend that the **causal influence** that connects event P to event Q *can* flow between them at a constant speed $|\vec{v}_{ci}|$ *faster* than the speed of light as measured in your inertial frame, which we will call the Home

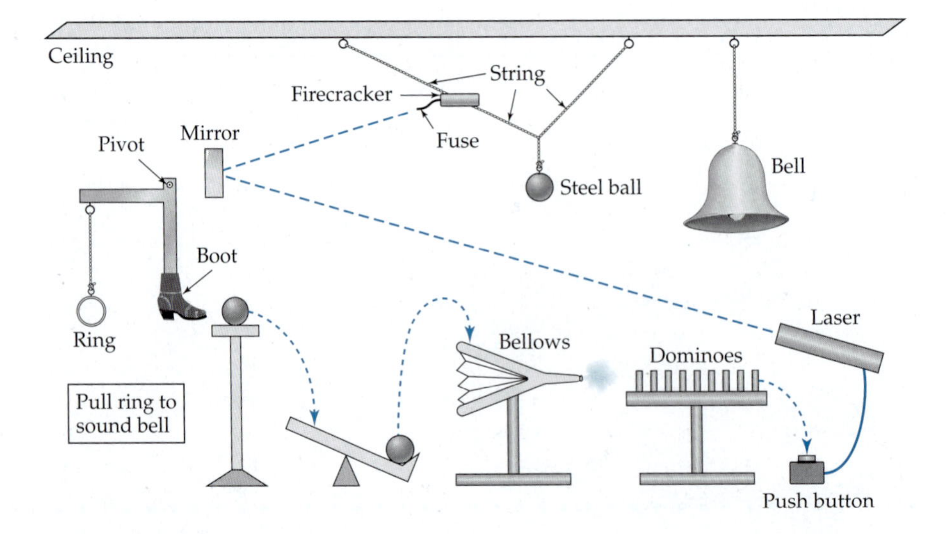

Figure R7.1
Some causal connections.

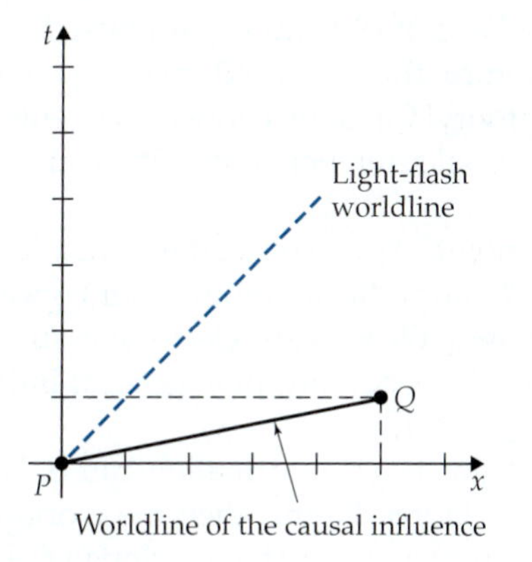

Figure R7.2
Suppose that events P and Q are connected by a hypothetical causal influence traveling with a speed $|\vec{v}_{ci}|$ faster than the speed of light (specifically, 5 times the speed of light for the sake of concreteness).

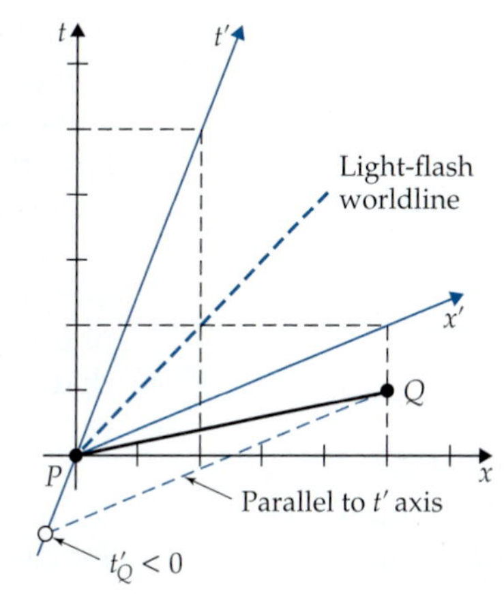

Figure R7.3
This two-observer spacetime diagram shows the same situation as in figure R7.2, but with added axes for an Other Frame moving with $\beta = \frac{2}{5}$ with respect to the Home Frame. Note that in the Other Frame, event Q occurs *before* event P.

Frame. (Perhaps the TV manufacturer has found some way to convey a signal from the remote to the TV using "Z waves" that travel faster than light.) We will show that this leads to a logical absurdity. Choose event P to be the origin event in that frame, and choose the spatial x axis of the frame so that both events P and Q lie along it. (We can always do this: it is just a matter of choosing the origin and orientation of our reference frame. Choosing the frame to be oriented in this way is purely a matter of convenience.)

Figure R7.2 shows a spacetime diagram (drawn by an observer in the Home Frame) of a pair of events P and Q fitting the description above. Note that if the causal influence flows from P to Q faster than the speed of light, its worldline on the diagram will have a slope $1/|\vec{v}_{ci}| < 1$, which is *less* than the slope of the worldline of a light flash leaving event P at the same time (which is also shown on the diagram for reference).

Now consider figure R7.3. In this two-observer spacetime diagram, I have drawn the t' and x' axes for an Other Frame that travels with a speed $|\beta| = \frac{2}{5}$ in the $+x$ direction relative to the Home Frame. Note that according to section R5.5, the slope of the diagram x' axis in such a diagram is β. Note also that since the slope of the causal influence worldline is $1/|\vec{v}_{ci}| < 1$, it is *always* possible to find a value of β such that $1/|\vec{v}_{ci}| < \beta < 1$, meaning that it is always possible to find a reference frame moving slower than the speed of light relative to the Home Frame whose x' axis lies *between* the light-flash worldline and the causal influence worldline, as shown. In such a frame, event Q will be measured to occur *before* event P, as one can see by reading the time coordinates of these events from the diagram.

Faster-than-light influences imply reversal of cause and effect in some frames

Thus, in such an Other Frame, event P is observed to occur *after* event Q does. But this is absurd: event P is supposed to *cause* event Q. How can an event be measured to occur before its cause? This is not merely a semantic issue, nor is it mere appearance. According to any and every physical measurement that one might make in the Other Frame, event Q will really be observed to occur *before* its "cause" P.

To vividly illustrate the absurdity, consider our TV remote example. If the signal could go from your remote control to the TV faster than light, in certain inertial reference frames, you would observe the TV set changing channels *before* the button was pushed. If this were to happen in your reference frame, you would consider this a violation of the laws of physics (presuming your TV set was not broken). But the laws of physics are supposed to hold in *every* inertial reference frame. Therefore, this observed inversion of cause and effect violates the principle of relativity! Causally connected events must have the same **temporal order** in *all* inertial reference frames if we are to preserve the concept's meaning.

We have only three options at this point. We can reject the principle of relativity and start over at square one. We can radically modify our conception of causality in a way that is yet unknown. Or we can reject the assumption that got us into this trouble in the first place, namely, that a causal influence can flow from P to Q faster than light ($|\vec{v}_{ci}| > 1$).

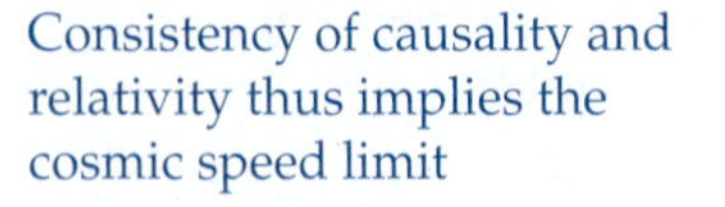

Consistency of causality and relativity thus implies the cosmic speed limit

The last option is clearly the least drastic. If information can only flow from P to Q with a speed $|\vec{v}_{ci}| \leq 1$ in the Home Frame, then the worldline of the causal influence connecting event P to event Q will have a slope $1/|\vec{v}_{ci}| > 1$. Any Other Frame must travel with $|\beta| < 1$, by this hypothesis (since the parts of the reference frame, like any material object, could in principle be the agent of a causal influence). The slope β of the Other Frame's diagram x' axis on a spacetime diagram will always be less than the slope $1/|\vec{v}_{ci}|$ of the causal connection worldline connecting P and Q, and thus Q will occur after P in *every* other inertial reference (figure R7.4).

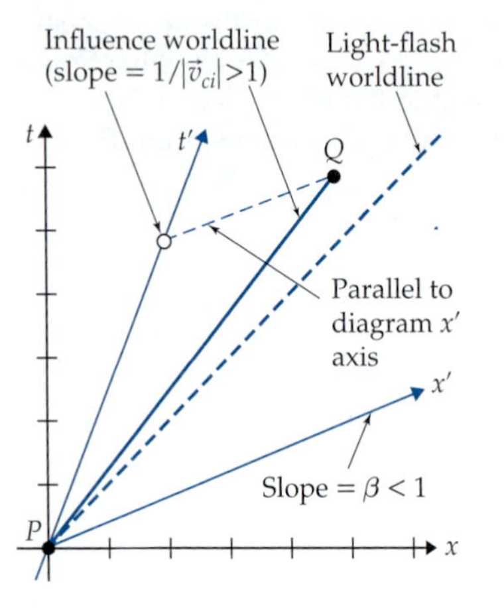

Figure R7.4
The slope β of the x' axis can never have a magnitude large enough to be above the causal influence worldline if the speed of that influence is less than 1.

So, if the speed of reference frames and causal influences is limited to some $|\vec{v}_{ci}| < 1$, then effects will occur *after* their causes in every inertial reference frame, which is necessary if the idea of causality is to be consistent with the principle of relativity.

> **THEOREM: The Cosmic Speed Limit:** In order for causality (that is, the idea that one event can *cause* another event to happen) to be consistent with the principle of relativity, information (that is, *any* effect representing a causal connection between two events) *cannot* travel between two events with a speed $|\vec{v}_{ci}| > 1$.

Since anything movable and detectable can carry information (that is, cause things to happen), this consequence of the principle of relativity applies not only to all physical objects (waves, particles, and macroscopic objects) but indeed to *any* trick or means of conveying a message that exists or might be imagined (for example, instantaneous changes in a gravitational or electric field, telepathy, or magic).

Causality and the second law of thermodynamics

Now, at the most basic physical level, the physical law that defines the temporal order of cause and effect is the **second law of thermodynamics**, which requires that the entropy of the universe always increase (or at least remain the same) during any physical process. This law thus implies that events in certain physical processes can occur in one temporal order but *cannot* occur in the reverse order. Therefore, if the second law of thermodynamics is to be true in all inertial frames (as required by the principle of relativity), then the temporal order of all events that might be linked by that law must be preserved in all inertial frames. Thus, the cosmic speed limit really follows from the assumption that the second law of thermodynamics is consistent with the principle of relativity. "Cause and effect" is really just an intuitive and colloquial way to talk about the invariant temporal order imposed on events by the second law.

Summary

So, with a straightforward argument using two-observer spacetime diagrams, we have proved the existence of a cosmic speed limit, an idea having profound physical and philosophical implications. As usual, this prediction is amply supported by experiment. No particle, object, or signal of any kind has ever been definitely observed to travel at faster than the speed of light in a vacuum. Science fiction fans and space travel buffs who hope for the discovery of faster-than-light travel may hope in vain: both the argument (based as it is on the firmly accepted and fundamental ideas of the principle of relativity and the physical reality of causality) and the experimental evidence present a pretty ironclad case for this cosmic speed limit.*

Exercise R7X.1

Suppose a causal influence moves between events P and Q at 3 times the speed of light relative to the Home Frame. How fast would an Other Frame have to move relative to the Home Frame for event Q to occur *before* event P in the Other Frame?

R7.2 Timelike, Lightlike, and Spacelike Intervals

We are now in a position to understand more fully the true physical nature of the spacetime interval between *any* two events in spacetime. In section R3.2, we saw that for two events whose coordinate differences in a given inertial reference frame are Δt, Δx, Δy, and Δz, the quantity $\Delta s^2 = \Delta t^2 - \Delta x^2 - \Delta y^2 - \Delta z^2$ has a frame-independent value that is equal to the time registered by an inertial clock present at both events. But to make the proof of the metric equation work, we had to assume that $|\Delta\vec{d}| \equiv (\Delta x^2 + \Delta y^2 + \Delta z^2)^{1/2}$ was *smaller* than Δt so that there was more than sufficient time for a light flash to travel from one event to the other along the length of the light clock. The purpose of this section is to investigate the meaning of the spacetime interval when $|\Delta\vec{d}| > \Delta t$ (that is, when this condition is violated).

Δs^2 between two arbitrary events can be positive, negative, or zero

We have exploited the analogy between spacetime geometry and Euclidean plane geometry extensively in the last few chapters. We have noted, though, that the minus signs in the metric equation (which do not appear in the corresponding Pythagorean relation) lead to some subtle differences between spacetime geometry and Euclidean geometry. One of these differences is the following. In Euclidean geometry, the squared distance $|\Delta\vec{d}|^2$ between two points on a plane is necessarily positive:

$$|\Delta\vec{d}|^2 = \Delta x^2 + \Delta y^2 \geq 0 \tag{R7.1}$$

But the metric equation allows the squared spacetime interval between two events to be positive, zero, or negative, depending on the relative sizes of the coordinate separations $|\Delta\vec{d}|$ and Δt between those events:

$$\Delta s^2 = \Delta t^2 - |\Delta\vec{d}|^2$$

Therefore,

$$\text{If} \quad |\Delta\vec{d}| > \Delta t \quad \text{then} \quad \Delta s^2 < 0! \tag{R7.2}$$

We see that while there is only *one* kind of distance between two points on a plane, the possible spacetime intervals between two events in spacetime fall

*This does not mean that interstellar travel is out of the question. Remember that the ship time measured in a spaceship traveling close to the speed of light (relative to the galaxy) is much shorter than the coordinate time we would measure (at rest in the galaxy). Therefore, a trip to distant stars can be made as short as desired for the *passengers* by simply constraining the ship's speed to be sufficiently close to the speed of light. But (at least if special relativity is true) there appears to be no way to make a trip of 1000 ly in our galaxy in less than 1000 y *in the frame of the galaxy,* no matter how short this might seem to the passengers. This does put some severe limits on the possibilities of interstellar commerce! We will examine other difficulties associated with interstellar travel in chapter R9.

into *three* distinct categories depending on the sign of Δs^2. These categories are as follows:

The three categories for the spacetime interval

1. If $\Delta s^2 > 0$, we say that the interval between the events is **timelike**.
2. If $\Delta s^2 = 0$, we say that the interval between the events is **lightlike**.
3. If $\Delta s^2 < 0$, we say that the interval between the events is **spacelike**.

The reasons for these names will become clear shortly.

The peculiar category here is the *spacelike* category—there is nothing corresponding to it in ordinary plane geometry (where the squared distance between two events is always positive). What does it mean for two events to have a *spacelike* spacetime interval between them?

Spacelike spacetime intervals exist, but cannot be measured with a clock

First, note that events separated by spacelike spacetime intervals certainly do exist. For example, consider the case of two events that occur at the same *time* but at different *locations* in a given inertial reference frame. Since the time separation between these events is zero in that frame, we have $\Delta s^2 = 0 - |\Delta\vec{d}\,|^2 = -|\Delta\vec{d}\,|^2 < 0$, so the interval between these events is necessarily spacelike. Therefore, we do need the spacelike interval classification.

As we have already discussed, the squared spacetime interval between two events Δs^2 that appears in the metric equation $\Delta s^2 = \Delta t^2 - |\Delta\vec{d}\,|^2$ has been linked with the frame-independent time measured by an inertial clock present at both events *only* in the case where $\Delta t^2 > |\Delta\vec{d}\,|^2$. For two events for which $\Delta t^2 < |\Delta\vec{d}\,|^2$, it is not clear how one can directly measure the squared spacetime interval between the events at all. For example, for an inertial clock to be present at both events where $|\Delta\vec{d}\,| > \Delta t$, it would have to travel at a speed $|\vec{v}\,| > 1$ in that frame. We have just seen that this is impossible; thus, a spacelike spacetime interval *cannot* be measured by a clock or anything else that travels between the events. Since the proof of the metric equation given in section R3.2 does not handle the case of spacelike intervals, it is not even obvious the squared spacetime interval $\Delta s^2 = \Delta t^2 - |\Delta\vec{d}\,|^2$ is frame-independent when it is less than zero.

Δs^2 is frame-independent even when it is negative

In fact, the squared spacetime interval Δs^2 *does* have a frame-independent value, no matter what its sign is. This can easily be demonstrated by using the Lorentz transformation equations for coordinate differences, given by equations R5.11. The argument goes like this. Let Δt, Δx, Δy, Δz be the coordinate separations of two events measured in the Home Frame, and let $\Delta t'$, $\Delta x'$, $\Delta y'$, $\Delta z'$ be the coordinate separations of the same two events measured in an Other Frame moving in the $+x$ direction with x-velocity β with respect to the Home Frame. Then, equations R5.11 imply that

$$\begin{aligned}
&(\Delta t')^2 - (\Delta x')^2 - (\Delta y')^2 - (\Delta z')^2 \\
&\quad = [\gamma(\Delta t - \beta\,\Delta x)]^2 - [\gamma(-\beta\,\Delta t + \Delta x)]^2 - \Delta y^2 - \Delta z^2 \\
&\quad = \gamma^2(\Delta t^2 - 2\beta\,\Delta t\,\Delta x + \beta^2\,\Delta x^2) - \gamma^2(\beta^2\,\Delta t^2 - 2\beta\,\Delta t\,\Delta x + \Delta x^2) - \Delta y^2 - \Delta z^2 \\
&\quad = \gamma^2(\Delta t^2 + \beta^2\,\Delta x^2 - \beta^2\,\Delta t^2 - \Delta x^2) - \Delta y^2 - \Delta z^2 \\
&\quad = \gamma^2(1 - \beta^2)(\Delta t^2 - \Delta x^2) - \Delta y^2 - \Delta z^2 \\
&\quad = \Delta t^2 - \Delta x^2 - \Delta y^2 - \Delta z^2
\end{aligned} \tag{R7.3}$$

where in the last step, I used $\gamma \equiv 1/\sqrt{1-\beta^2}$. The sign of $\Delta s^2 = \Delta t^2 - |\Delta\vec{d}\,|^2$ is irrelevant to this derivation, so we will find that the squared interval Δs^2 has the same frame-independent value in *every* inertial reference frame, whether Δs^2 is spacelike, timelike, or lightlike.

How can we measure the value of the spacetime interval between two events separated by a spacelike interval? We cannot use a clock, as we have noted already. In fact, we measure a spacelike spacetime interval with a *ruler*, as we will shortly see.

The spacetime separation

Let us define the **spacetime separation** $\Delta\sigma$ of two events in this way:

$$\Delta\sigma^2 \equiv |\Delta\vec{d}|^2 - \Delta t^2 = -\Delta s^2 \tag{R7.4}$$

The spacetime separation, so defined, is conveniently real whenever the interval between the events is spacelike. Now note that *if* we can find an inertial reference frame where the events are simultaneous ($\Delta t = 0$), we have

$$\Delta\sigma^2 = |\Delta\vec{d}|^2 \quad \Rightarrow \quad \Delta\sigma = |\Delta\vec{d}| \quad \text{(in a frame where } \Delta t = 0) \tag{R7.5}$$

Now, I claim that we can *always* find a frame in which $\Delta t = 0$ if the events are separated by a spacelike interval. Suppose two events occur with coordinate differences Δt and $|\Delta\vec{d}|$ ($>\Delta t$) as measured in the Home Frame. Reorient and reposition the axes of the Home Frame so the events in question both occur along the spatial x axis, with the later event located in the $+x$ direction relative to the earlier event. (This can be done without loss of generality: we are always free to choose the orientation of our coordinate system to be whatever we find convenient.) Once this is done, $|\Delta\vec{d}| = \Delta x$ in the Home Frame.

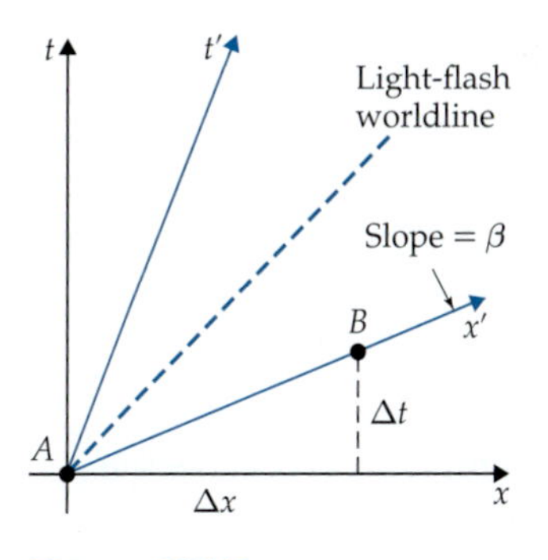

Figure R7.5
Given any pair of events A and B separated by a spacelike interval ($\Delta x > \Delta t$) in the Home Frame, we can find an Other Frame where the two events are simultaneous. The speed of this Other Frame simply must have the right value so that the diagram x' axis can connect both points. Since this axis has slope β, this means β must be equal to $\Delta t/\Delta x$.

Now, consider an Other Frame in standard orientation with respect to the Home Frame and traveling in the $+x$ direction with x-velocity β with respect to the Home Frame. According to equation R5.11*a*, the time-coordinate difference between these events in the Other Frame is then

$$\Delta t' = \gamma(\Delta t - \beta\Delta x) \tag{R7.6}$$

These events will be simultaneous in the Other Frame (that is, $\Delta t'$ will be equal to zero) if and only if the relative speed of the frames is chosen to be $|\beta| = |\Delta t/\Delta x| = |\Delta t|/|\Delta\vec{d}|$. This relative speed $|\beta|$ will be less than 1 since $|\Delta\vec{d}| > \Delta t$ for our events by hypothesis. In short, given *any* pair of events that are separated by a spacelike interval in some inertial frame (which we are calling the Home Frame), it is possible to find an inertial Other Frame moving with positive x-velocity $\beta < 1$ with respect to the Home Frame in which observers will find the two events to be simultaneous (see figure R7.5).

In short, if the spacetime interval between two events is spacelike, then

How to measure a spacelike spacetime interval

1. We can find an inertial frame where these events occur at the *same time*.
2. The spacetime separation $\Delta\sigma$ is the *distance* between the events in that special frame. We can measure this with a ruler stretched between the events in that frame.
3. If observers in any other inertial frame use equation R7.4 to calculate $\Delta\sigma$, they will get the same value as measured directly in the special frame.

These statements are directly analogous to statements that can be made about events separated by a timelike spacetime interval. If the spacetime interval between two events is *timelike*, then

1. We can find an inertial frame in which these events occur at the *same place* (this is the frame of the inertial clock that is present at both events).
2. The *time* between the events in this special frame is Δs. We can measure this with a clock present at both events and at rest in this frame.
3. If observers in any other inertial frame use the ordinary metric equation to calculate Δs, they will get the same value as measured directly in the special frame.

Thus, there is a fundamental symmetry between spacelike and timelike spacetime intervals, a symmetry that arises because both reflect the same underlying physical truth: we can describe the separation of *any* two events in space and time with a frame-independent quantity Δs^2 (which we will call the **squared spacetime interval**) analogous to the *squared distance* between two points in plane geometry. It is simply a peculiarity of the geometry of spacetime that the quantity in spacetime that corresponds to ordinary (unsquared) distance on the plane comes in three distinct flavors (the spacetime *interval* Δs if $\Delta s^2 > 0$ for the events, the spacetime *separation* $\Delta\sigma$ if $\Delta s^2 < 0$, and the lightlike *interval* $\Delta s = \Delta\sigma = 0$ when $\Delta s^2 = 0$), which are measured in different ways using different tools. But it is important to realize that these three quantities are only different aspects of the same basic frame-independent concept Δs^2.

Why the categories of spacetime interval have the names they do

We see that we directly measure timelike intervals with a *time*-measuring device (an inertial *clock* present at both events), while we directly measure spacelike intervals with a *space*-measuring device (a *ruler* stretched between the events in the particular inertial frame where the events are simultaneous). This is why these interval classifications have the names *timelike* and *spacelike*: the names tell us whether we should measure the interval with a clock (because the interval is timelike) or with a ruler (because it is spacelike). The lightlike interval classification stands between the other two. When the interval between two events is lightlike, we have $|\Delta\vec{d}| = \Delta t$, which implies that these events could be connected by a flash of light.

Exercise R7X.2

Consider the events shown in the drawing below. Classify the spacetime interval between each *pair* of events as being timelike, spacelike, or lightlike.

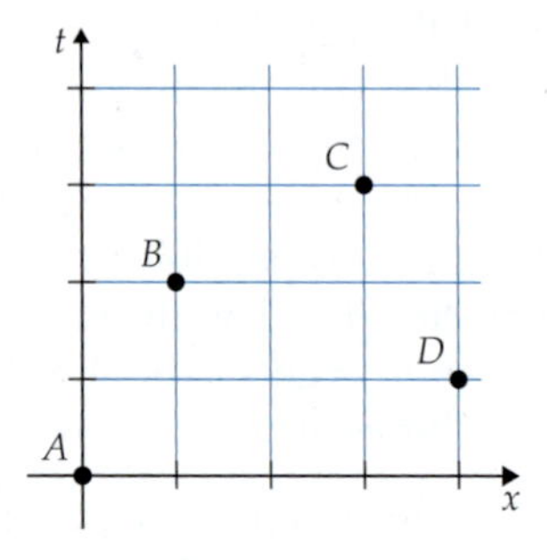

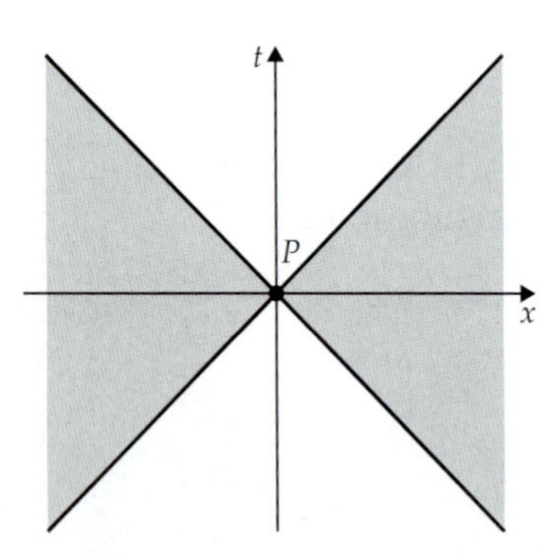

Figure R7.6
The frame-independent regions of spacetime associated with event *P*. The spacetime interval between *P* and any event in the white region is *timelike*, any event in the gray region is *spacelike*, and any event along the black diagonal lines is *lightlike*.

R7.3 The Causal Structure of Spacetime

Now, *because* it is true that the value of the squared spacetime interval Δs^2 is frame-independent no matter what its sign, all inertial observers will agree as to whether the interval between a given pair of events is timelike, lightlike, or spacelike (since if they all agree on the *value* of Δs^2, they will surely all agree on its sign). This means that the spacetime around any event *P* can be divided up into the distinct regions shown in figure R7.6, and every observer will agree about which events in spacetime belong to which region.

Because we can define these regions in a frame-independent manner, they plausibly reflect something absolute and physical about the geometry of spacetime. In fact, *these regions distinguish those events that can be causally connected to P from those that cannot*. Remember that in section R7.1, we found

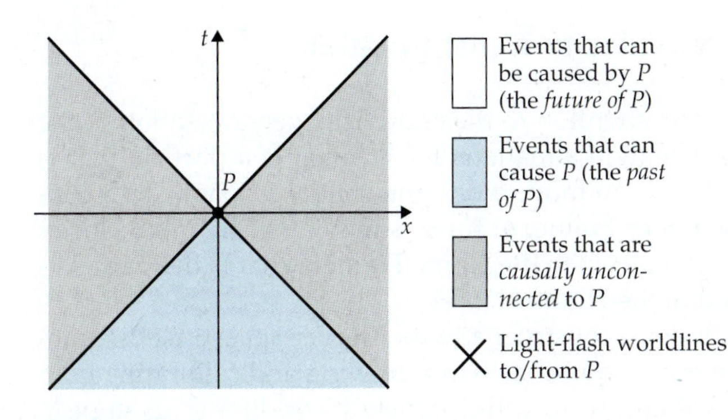

Figure R7.7
The causal structure of spacetime relative to event P.

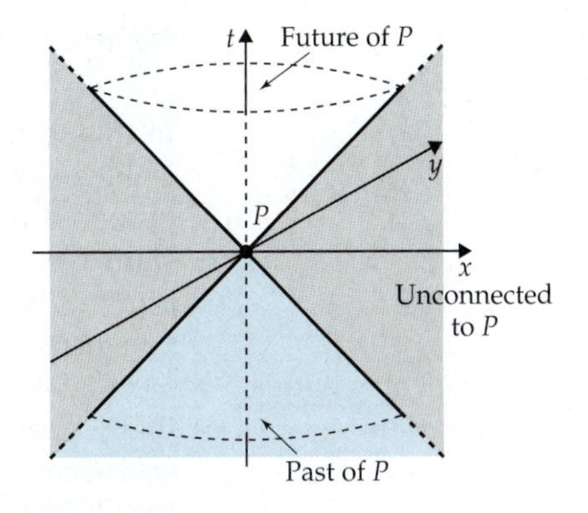

Figure R7.8
The *light cone* of P (shown on a spacetime diagram having two spatial dimensions).

that two events can be causally linked only if $|\Delta\vec{d}| \leq \Delta t$ between them; otherwise, the causal influence would have to travel between the events faster than the speed of light. This means that every event that can be causally linked with P must have a timelike (or perhaps lightlike) interval with respect to P: such events will lie in the white regions shown in figure R7.6.

Understanding the causal structure of spacetime

We can be more specific yet. Since the temporal order of events is preserved in all inertial frames if $|\Delta\vec{d}| \leq \Delta t$ (see figure R7.3 and the surrounding text), all events in the *upper* white region in figure R7.6 will occur *after* P in every frame (and thus could be caused by P), and all events in the lower white region of figure R7.6 will occur *before* P in every frame (and thus could cause P). We refer to these regions as the **future** and **past** of P, respectively.

Events whose spacetime interval with respect to P is spacelike ($|\Delta\vec{d}| > \Delta t$) cannot influence P or be influenced by it. We say that these events (which inhabit the shaded region of figure R7.6) are causally *unconnected* to event P.

With this in mind, we can relabel the regions in figure R7.6 as shown in figure R7.7. Because every observer agrees on the value of the spacetime interval between event P and any other event, every observer agrees as to which event belongs in which classification. The structure illustrated is thus an intrinsic, frame-independent characteristic of the geometry of spacetime.

The light cone associated with an event

Now, the boundaries of the regions illustrated in figure R7.7 are light-flash worldlines. If we consider two spatial dimensions instead of one, an omnidirectional light flash is seen as an ever-expanding ring, like the ring of waves formed by the splash of a stone into a still pool of water. If we plot the growth of such a ring on a spacetime diagram, we get a cone. The boundaries between the three regions described are then two tip-to-tip cones, as shown in figure R7.8. Physicists call this boundary surface the **light cone** for the given event P.

To summarize, the point of this section is that the spacetime interval classifications, which are basic, frame-independent features of the geometry of spacetime, have in fact a deeply physical significance: the sign of the squared spacetime interval between two events unambiguously describes whether these events can be causally connected or not. The light cone shown in figure R7.8 effectively illustrates this geometric feature of spacetime.

R7.4 The Einstein Velocity Transformation

In this section, we turn our attention to the relativistic generalization of the Galilean velocity transformation equations R1.3. Imagine a particle that is observed in the Other Frame to move along the spatial x' axis with a constant x-velocity v_x'. The Other Frame, in turn, is moving with an x-velocity β in the $+x$ direction with respect to the Home Frame. What is the particle's x-velocity v_x as observed in the Home Frame?

Transforming velocities with a two-observer diagram

Figure R7.9 shows how to construct a two-observer spacetime diagram that we can use to answer this question. After drawing and calibrating both sets of coordinate axes, we simply draw the particle's worldline so its slope in the Other Frame is $1/v_x'$. We can then find its x-velocity v_x in the Home Frame by taking the inverse slope of that line in the Home Frame, which we can do by picking an arbitrary "rise" (5 ns in figure R7.9), determining the worldline's "run" for that rise, and then calculating the inverse of the rise/run = run/rise. In the case shown in figure R7.9, where $\beta = \frac{3}{5}$ and $v_x' = \frac{3}{5}$, we find that the value of v_x is about 0.86, and *not* $\frac{3}{5} + \frac{3}{5} = \frac{6}{5}$ that the Galilean velocity transformation equations would predict.

How to derive the inverse Einstein velocity transformation equations

Now let us see if we can derive an exact equation that (like the diagram) allows us to find v_x in terms of v_x' and β. Consider two infinitesimally separated events along the particle's worldline (which we will assume is moving along the spatial x axis). Let the coordinate differences between these events as measured in the Home Frame be dt and dx. Let the coordinate differences between the same two events as measured in the Other Frame be dt' and dx'. The particle's x-velocity as it travels between these events is

$$v_x \equiv \frac{dx}{dt} = \frac{\gamma(\beta\, dt' + dx')}{\gamma(dt' + \beta\, dx')} \tag{R7.7a}$$

where I have used the infinitesimal limit of the difference version of the inverse Lorentz transformation equations (equations R5.12). Dividing the right side top and bottom by dt' and using $dx'/dt' \equiv v_x'$, we get the following relativistically exact equation:

$$v_x = \frac{\beta + v_x'}{1 + \beta v_x'} \tag{R7.7b}$$

In a similar fashion, you can derive the y and z component equations as well. The complete set of equations for v_x, v_y, and v_z are

$$v_x = \frac{\beta + v_x'}{1 + \beta v_x'} \qquad v_y = \frac{v_y'\sqrt{1-\beta^2}}{1 + \beta v_x'} \qquad v_z = \frac{v_z'\sqrt{1-\beta^2}}{1 + \beta v_x'} \tag{R7.8}$$

- **Purpose:** These equations describe how to compute an object's velocity components v_x, v_y, and v_z measured in the Home Frame from its velocity components v_x', v_y', and v_z' measured in the Other Frame, where β is the x-velocity of the Other Frame relative to the Home Frame.
- **Limitations:** These equations assume that the two frames are inertial and that they are in standard orientation with respect to each other.

Exercise R7X.3

Use the method described above to derive the formula for v_z.

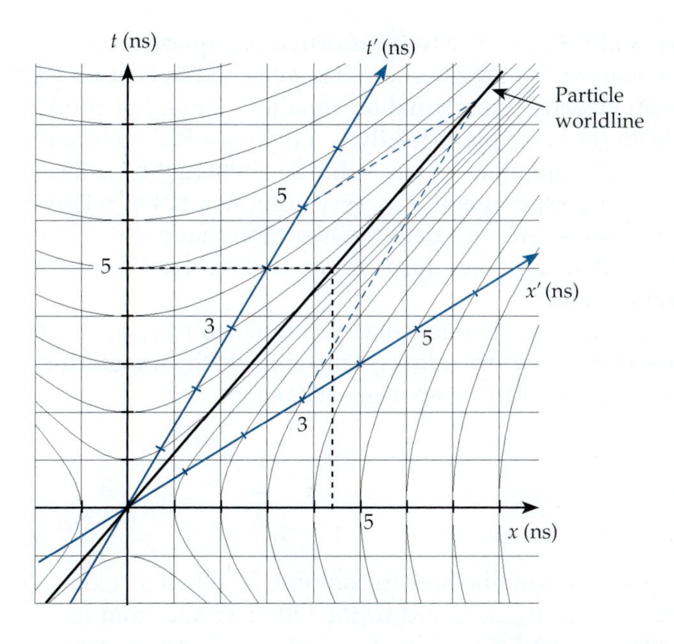

Figure R7.9
We can use a two-observer diagram to find a particle's x-velocity in one frame, given its x-velocity in another and the two frames' relative velocity. For example, if we know that the particle's Other Frame x-velocity is $v'_x = \frac{3}{5}$ and that the Other Frame moves at $\beta = \frac{3}{5}$ relative to the Home Frame, then we can construct an appropriate two-observer diagram for this β, draw the particle's worldline with a slope of $\frac{5}{3}$ relative to the Other Frame axes, and read the slope of this line relative to the Home Frame axes (about $4.3/5 \approx 0.86$ in this case). If we knew the particle's Home Frame x-velocity, we could just reverse the process.

An example application

As an example of the use of equation R7.8*a*, consider the particular problem illustrated by figure R7.9, where we have $v'_x = \beta = \frac{3}{5}$. The final speed of the particle in the Home Frame (according to equation R7.8*a*) is

$$v_x = \frac{3/5 + 3/5}{1 + (3/5)(3/5)} = \frac{6/5}{34/25} = \frac{30}{34} \approx 0.88 \tag{R7.9}$$

This is close to the result we read from figure R7.9.

These equations reduce to the Galilean equations at low speed . . .

We call equations R7.8 the **inverse Einstein velocity transformation equations**: they express algebraically what a two-observer diagram like figure R7.9 expresses graphically. Note that the result is different from what the *Galilean* velocity transformation predicts: solving equation R1.3*a* for v_x yields

$$v_x = \beta + v'_x \quad \text{from the Galilean velocity transformation} \tag{R7.10}$$

Note that when the velocities β and v'_x are very small, the factor $\beta v'_x$ that appears in the denominator of equation R7.8 becomes *very* small compared to 1. In this limit, then, equation R7.8*a* reduces to the Galilean equation R7.10:

$$v_x = \frac{\beta + v'_x}{1 + \beta v'_x} \approx \frac{\beta + v'_x}{1} = \beta + v'_x \quad \text{in low-velocity limit} \tag{R7.11}$$

The same kind of argument applies to the other two component equations as well. The Galilean transformation equations are therefore reasonably accurate for everyday velocities, but only represent an *approximation* to the true velocity transformation law expressed by equations R7.8.

. . . and are consistent with the cosmic speed limit and the invariant speed of light

Equations R7.8 never yield a Home Frame x-velocity that exceeds the speed of light, even if both β and v'_x are 1 (their maximum possible value):

$$v_x = \frac{1+1}{1+1\cdot 1} = \frac{2}{2} = 1 \tag{R7.12}$$

Moreover, equation R7.8*a* (unlike the Galilean equation R1.3*a*) is consistent with the idea that the speed of light is equal to 1 in all frames: if the x-velocity of a light flash in the Other Frame is $v'_x = 1$, its speed in the Home Frame will be

$$v_x = \frac{\beta + 1}{1 + \beta \cdot 1} = \frac{\beta + 1}{1 + \beta} = 1 \tag{R7.13}$$

independent of the value of β.

In short, the inverse Einstein velocity transformation equations (equations R7.8) provide the answer to the question that we raised in chapter R1 about how the Galilean velocity transformation equations should be modified to be consistent with the principle of relativity. Equations R7.8 reduce to the Galilean transformation equations at low velocities (where the Galilean transformation is known by experiment to be very accurate), but at relativistic velocities they are consistent with both the assertion that nothing can be measured to go faster than light and the assertion that light itself has the same speed in all inertial frames.

Equations R7.8 convert Other Frame velocity components v'_x, v'_y, v'_z to Home Frame components v_x, v_y, v_z. The (direct) **Einstein velocity transformation equations** transform the velocity components the other way.

$$v'_x = \frac{v_x - \beta}{1 - \beta v_x} \qquad v'_y = \frac{v_y\sqrt{1-\beta^2}}{1 - \beta v_x} \qquad v'_z = \frac{v_z\sqrt{1-\beta^2}}{1 - \beta v_x} \tag{R7.14}$$

- **Purpose:** These equations describe how to compute an object's velocity components v'_x, v'_y, and v'_z measured in the Other Frame from its velocity components v_x, v_y, and v_z measured in the Home Frame, where β is the x-velocity of the Other Frame relative to the Home Frame.
- **Limitations:** These equations assume that the two frames are inertial and they are in standard orientation with respect to each other.

These equations can be derived by solving equations R7.8 for the Other Frame components or by using the direct Lorentz transformation equations R5.11 as we used equations R5.12 in the derivation of equations R7.8. But one needn't do all this work: simply note that the only difference between the Home Frame and the Other Frame is the sign of β. If you compare equations R7.14 with equations R7.8, you will see they are the same, except that I have changed β to $-\beta$. Again, these equations themselves are laws of physics that satisfy the principle of relativity.

TWO-MINUTE PROBLEMS

R7T.1 Consider the events shown in the figure below:

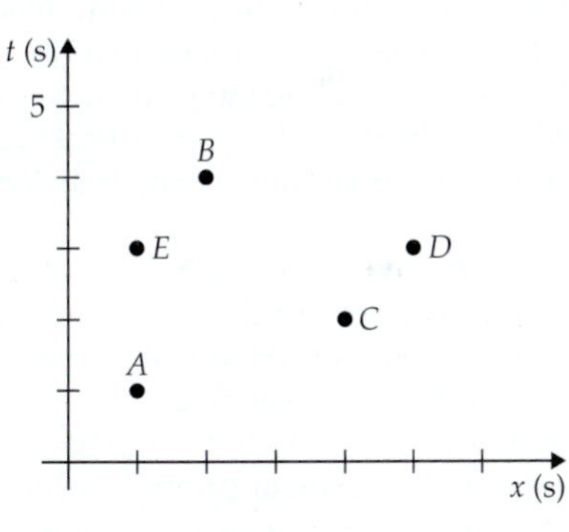

For each of the ten event pairs in this spacetime diagram, classify the spacetime interval between them.
A. The interval is timelike.
B. The interval is lightlike.
C. The interval is spacelike.

R7T.2 Which pairs of events shown in the spacetime diagram for problem R7T.1 could in principle be causally connected? Which could not? (For each of the ten pairs, answer T if they could be causally connected, and F if not.)

R7T.3 Suppose that event A is the origin event in the Home Frame and that event B occurs at $t = 1$ ns and $x = 10$ ns in that frame. What would be the minimum x-velocity β that an Other Frame would have to have relative to the Home Frame if B occurred first in that frame?
A. 10
B. 1
C. 0.60
D. 0.40
E. 0.10
F. B can't occur first.

R7T.4 Two blinking warning lights are 3000 m apart along a railroad track. Suppose that in the ground frame the west light blinks (event W) 5 μs before the east light blinks (event E). Now imagine that both lights are observed by a passenger on the train who passes the west light just as it blinks. It is possible for the train to be moving fast enough that the two blinks are *observed* (not necessarily *seen*) by the passenger to be simultaneous. T or F?

R7T.5 Two blinking warning lights are 1000 ns apart along a railroad track. A train moves from west to east at a speed of 0.5. In the train frame, an observer registers flashes from each light, and after correcting for light travel time, concludes that the east light flashed 200 ns before the west light flashed. Which light flashed first in the ground frame? (*Hint:* Draw a spacetime diagram.)
A. The east light flashes before the west light.
B. The west light flashes before the east light.
C. The lights flash simultaneously.
D. One does not have enough information to say.

R7T.6 Suppose an explosion occurs at $x = 0$ in the Home Frame (event A). Light from the explosion is detected by a detector at position $x = -100$ ns (event B) and by a detector at position $x = +50$ ns (event C). Events B and C are causally connected. T or F?

R7T.7 A laser beam is emitted on earth (event A), bounces off a mirror placed on the moon by Apollo astronauts (event B), and then returns to a detector on earth (event C). The detector is 12 ns of distance from the laser.
(a) The spacetime interval between events A and B is:
A. Timelike
B. Lightlike
C. Spacelike
(b) The spacetime interval between events A and C is:
A. Timelike
B. Lightlike
C. Spacelike
(c) Events A and C are definitely causally connected. T or F?

R7T.8 If the spacetime interval between two events is timelike, then the temporal order of the two events is the same in *every* inertial reference frame. T or F?

R7T.9 If the spacetime interval between two events A and B is spacelike and event A occurs before event B in some Home Frame, then it is *always* possible to find an Other Frame where the events occur in the other order. T or F?

R7T.10 An object moves with speed $v_x = +0.9$ in the Home Frame. In an Other Frame moving at $\beta = 0.60$ relative to the Home Frame, the object's x-velocity v'_x is
A. $v'_x = 1.5$
B. $1 < v'_x < 1.5$
C. $0.9 < v'_x < 1$
D. $0.3 < v'_x < 0.9$
E. $v'_x < 0.3$
F. $v'_x < 0$

R7T.11 Harry Potter points his wand at the sun and cries "Spotificus!" Within 8 minutes, observers all over the earth see a large string of sunspots form on the sun. Did Harry create the sunspots?
A. Yes. Harry is a very powerful wizard.
B. Yes, if magical effects travel at the speed of light.
C. No. This is not physically possible, even with magic.

R7T.12 An unidentified spaceship cruises past Outpost 11, a space station floating in deep space, at a speed of $\frac{3}{5}$, without acknowledging requests for identification. Ten seconds after the ship passes (according to Outpost 11 clocks), Outpost 11 sends a message to the departing ship, warning that it will commence firing if the ship does not respond. Fifteen seconds after that, Outpost 11 sensors indicate that the ship has raised its shields. This is clearly a response to the warning. T or F? (*Hint:* Draw a spacetime diagram.)

HOMEWORK PROBLEMS

Basic Skills

R7B.1 At 11:00:00 a.m. a boiler explodes in the basement of the Museum of Modern Art in New York City (call this event A). At 11:00:00.0003 a.m., a similar boiler explodes (call this event B) in the basement of a soup factory in Camden, New Jersey, a distance of 150 km from event A.

(a) Why is it impossible for the first event to have caused the second event?

(b) An alien spaceship cruising in the direction of Camden from New York measures the Camden event to occur at the same time as the New York event. What is the approximate speed of the spaceship relative to earth?

R7B.2 Two balls are simultaneously ejected (event A) from the point $x = 0$ in some inertial frame. One rolls in the $+x$ direction with speed 0.80 and eventually hits a wall at $x = 8.0$ ns (event B). The other rolls in the $-x$ direction with speed 0.40, eventually hitting a wall at $x = -8.0$ ns (event C). Is the spacetime interval between B and C spacelike or timelike? Could these events be causally connected?

R7B.3 A meteor strikes the moon (event A), causing a large and vivid explosion. Exactly 0.47 s later (as measured in an inertial reference frame attached to the earth), a radio telescope receiving signals from the moon goes on the fritz. Could these events be causally related? Explain.

R7B.4 Suppose event B happens 3.0 ns after and 5.0 ns east of event A in the Home Frame. In the Other Frame, the events happen at the same time.

(a) How fast is the Other Frame moving eastward with respect to the Home Frame if this is true?

(b) What is the distance between the events in that frame?

R7B.5 Do problem R7T.4. Hand in your spacetime diagram and explain your reasoning.

R7B.6 Do problem R7T.12. Hand in your spacetime diagram and explain your reasoning.

R7B.7 An object moves with velocity $v'_x = \frac{2}{5}$ in an inertial frame attached to a train, which in turn moves with x-velocity $\beta = \frac{4}{5}$ in the $+x$ direction with respect to the ground. What is the object's velocity v_x with respect to the ground? Evaluate this by reading the velocity from a carefully constructed two-observer spacetime diagram. Check your answer by using the appropriate Einstein velocity transformation equation.

R7B.8 An object moves at a speed of 0.80 in the $+x$ direction, as measured in the Home Frame. What is its x-velocity in an Other Frame that is moving at a speed of 0.60 in the $+x$ direction, also as measured in the Home Frame?

R7B.9 Rocket A travels to the right and rocket B to the left at speeds of $\frac{3}{5}$ and $\frac{4}{5}$, respectively, relative to the earth. What is the velocity of A measured by observers in rocket B? Answer by reading the velocity from a carefully constructed two-observer diagram. Check your answer, using an appropriate Einstein velocity transformation.

R7B.10 Two trains approach each other from opposite directions along a linear stretch of track. Each has a speed of $\frac{3}{4}$ relative to the ground. What is the speed of one train relative to the other? Answer this question by using an appropriate Einstein velocity transformation equation. As part of your solution, explain carefully which object you are taking to be the Home Frame, which object you are taking to be the Other Frame, and which is the object whose speed you are measuring in both frames.

R7B.11 Two cars travel in the same direction on the freeway. Car A travels at a speed of 0.90, while car B can only muster a speed of 0.60. What is the relative speed of the cars? As part of your solution, explain carefully which object you are taking to be the Home Frame, which object you are taking to be the Other Frame, and which is the object whose speed you are measuring in both frames.

Modeling

R7M.1 Suppose we set up a laser that is spinning at a speed of 100 rotations per second around an axis perpendicular to the beam that the laser creates.

(a) How far away must you place a screen so the spot of light the laser beam produces on the screen sweeps along it at a speed faster than that of light?

(b) Would a spot speed greater than that of light violate the cosmic speed limit? (*Hint:* If you stand at one edge of the screen and a friend stands at the other, can you send a message to your friend using the sweeping spot? If so, how? If not, why not?)*

R7M.2 You are on a jury for a terrorism trial. The facts of the case are these. On June 12, 2047, at 2:25:06 p.m. Greenwich Mean Time (GMT), the earth–Mars shuttle *Ares* exploded as it was being refueled in low earth orbit. (Fortunately, no passengers were aboard, and the refueling was handled by robots.) At 2:27:18 p.m., police record a holoscene (a 3D holographic movie) of a raid on a hotel room on Mars conducted on the basis of an unrelated anonymous tip. In the holo, the defendant is shown with a radio control transmitter in hand. Forensic experts have testified that the *Ares* was blown up by remote control, and that the reconstructed receiver was consistent with the transmitter in the defendant's possession. Just before the explosion, a

*Adapted from E. F. Taylor and J. A. Wheeler, *Spacetime Physics*, San Francisco: Freeman, 1966, p. 62.

caller predicted the blast and took responsibility on behalf of the Arean Liberation Army; the defendant has links to that organization. Cell-net records show that the defendant spoke with someone near the earth less than 10 min before. Hall monitors in the hotel showed that the defendant entered the hotel room at 2:23:12 p.m. and was not carrying the transmitter at that time. The defendant has taken the Fifth Amendment and has offered no defense other than a plea of not guilty. A fragment of the trial transcript follows.

Prosecutor: Is the time shown floating at the top of the holo-scene of the raid from a clock internal to the camera?

Police witness: No, I am told that the indicated time is computed from signals originating from the master clock on earth that is part of the Solar System Positioning System (SSPS) that defines a solar-system-wide inertial frame fixed on the sun. According to the manual (*pulls out the manual*), "The signal from the earth master clock is suitably corrected in the camera for the motions of the earth and Mars and the light travel time from the earth, so that the time displayed is exactly as if it were from the clock at the camera's location, at rest in the solar system frame, and synchronized with the earth-based master clock." We do this deliberately so as to be able to correlate events on a solar-system-wide basis.

Prosecutor: There is no chance that *this* time (*freezes holo-scene*), which shows the police yanking the device from the defendant here at exactly 2:27:20 p.m. GMT, is in error?

Police witness: No, the camera was checked two days previously as part of a normal maintenance program.

Prosecutor: We have been told that the *Ares* blew up at exactly 2:25:06 p.m. GMT. Was that time determined using the SSPS also?

Police witness: Yes.

Prosecutor: You have testified that the hall monitors show the defendant entering the room at exactly 2:23:12 p.m. This time was also determined using the SSPS?

Police witness: Yes.

Prosecutor: So the defendant was alone in the room at the time that the *Ares* exploded?

Police witness: Yes, in the SSPS frame.

Prosecutor: So this holoscene shows you capturing the defendant red-handed just after the destruction of the *Ares*, with the incriminating transmitter still in hand . . .

Defense: Objection, your Honor!

Judge, sighing: Sustained.

Guilty or not guilty? Write a paragraph justifying your reasoning very carefully.

R7M.3 Starbase Alpha coasts through deep space at a velocity of 0.60 in the $+x$ direction with respect to earth. Let the event of the starbase traveling by the earth define the origin event in both frames. Suppose that at $t = 8.0$ h a giant accelerator on earth launches you toward the starbase at 10 times the speed of light, relative to earth. After you get to the starbase, you use a similar accelerator to launch you back toward earth at 10 times the speed of light, relative to the starbase. Using a carefully constructed (full-page) two-observer diagram, show that in such a case you will return to the earth before you left. (This is yet another way to illustrate the absurdity of faster-than-light travel.)

R7M.4 A flash of laser light is emitted by the earth (event A) and hits a mirror on the moon (event B). The reflected flash returns to earth, where it is absorbed (event C).

(a) Is the spacetime interval between events A and B spacelike, lightlike, or timelike?

(b) What about the spacetime interval between B and C?

(c) What about the interval between events A and C?

Support your answers by describing your reasoning.

R7M.5 A solar flare (see the picture of an actual flare below) bursts through the sun's surface at 12:05 p.m. GMT, as measured by an observer in an inertial frame attached to the sun. At 12:11 p.m., as measured by the same observer, the Macdonald family's home computer fries a circuit board. Could these events be causally connected? Explain.

R7M.6 The first stage of a multistage rocket boosts the rocket to a speed of 0.1 relative to the ground before being jettisoned. The next stage boosts the rocket to a speed of 0.1 relative to the final speed of the first stage, and so on. How many stages does it take to boost the payload to a speed in excess of 0.95? (Credit: NASA)

R7M.7 Suppose that in the Home Frame, two particles of equal mass m are observed to move along the x axis with equal and opposite speeds $|\vec{v}| = \frac{3}{5}$. The particles collide and stick together, becoming one big particle which remains at rest in the Home Frame. Now imagine observing the same situation from the vantage point of an Other Frame that moves in the $+x$ direction with an x-velocity of $\beta = \frac{3}{5}$ with respect to the Home Frame.

(a) Find the velocities of all the particles as observed in the Other Frame, using the appropriate Einstein velocity transformation equations.

(b) We have defined the momentum of a particle with mass m and velocity $\vec{v}$ to be $\vec{p} = m\vec{v}$. Is the system's total momentum conserved in the Home Frame?

(c) Is the system's total momentum conserved in the Other Frame? Is this a problem? (We'll talk more about momentum in chapter R8.)

R7M.8 Suppose that in a particle physics experiment in the Home Frame, a particle of mass m_0 moving at a velocity $\vec{v}_0 = \frac{3}{5}$ in the $+x$ direction suddenly decays into a particle of mass m_1 moving at a speed of $\vec{v}_1 = \frac{4}{5}$ in the $+x$ direction and a particle of m_2 at rest.

(a) We have defined the momentum of a particle with mass m and velocity $\vec{v}$ to be $\vec{p} = m\vec{v}$. Assuming that the system's total momentum (defined this way) is conserved in this decay process in the Home Frame, what must the masses m_1 and m_2 be?

(b) Now let's look at this decay process in an Other Frame moving along with the initial particle. According to the Einstein velocity transformation, what are the final x-velocities of the decay products in this frame?

(c) Is the system's total momentum conserved in this frame? Is this a problem? (We'll talk more about momentum in chapter R8.)

R7M.9 A train travels in the $+x$ direction with an x-velocity of $\frac{4}{5}$ relative to the ground. At a certain time, two balls are ejected, one traveling in the $+x$ direction with an x-velocity of $\frac{3}{5}$ relative to the train, and the other traveling in the $-x$ direction with an x-velocity of $-\frac{2}{5}$ relative to the train.

(a) What are the balls' x-velocities relative to the ground?

(b) What is the x-velocity of the first ball relative to the second?

R7M.10 A particle moves with a speed of $\frac{4}{5}$ in a direction 60° away from the spatial $+x'$ axis toward the spatial y' axis, as measured in a frame (the Other Frame) that is moving in the $+x$ direction with an x-velocity of $\beta = \frac{1}{2}$ relative to the Home Frame. What are the magnitude and direction of the particle's velocity in the Home Frame?

R7M.11 A train travels in the $+x$ direction with a speed of $\frac{4}{5}$ relative to the ground. At a certain time, two balls are ejected so that they travel with a speed of $\frac{3}{5}$ (as measured in the train frame) in opposite directions *perpendicular* to the train's direction of motion.

(a) What are the balls' speeds relative to the ground?

(b) What is the angle that the path of each ball makes with the x axis in the ground frame?

Derivation

R7D.1 Derive the first of equations R7.14 by solving equation R7.7*b* for v'_x.

R7D.2 Show that when $|\beta| << 1$, the Lorentz transformation equations R5.11*a* and R5.11*b* reduce to the Galilean transformation equations R1.2*a* and R1.2*b*. (*Discussion:* The problem is trivial except for equation R5.11*a*. In that equation, how can we justify dropping the βx term when we need to keep the $-\beta t$ term in equation R5.11*b*? Think about typical magnitudes of quantities in an everyday experiment.)

R7D.3 Here is a quick argument that no material object can go faster than the speed of light. Consider an object traveling in the $+x$ direction with respect to some inertial frame (call this the Home Frame) at a speed $|\vec{v}| > 1$ in that frame. Show, using a two-observer spacetime diagram, that it is possible to find an Other Frame moving in the $+x$ direction at an x-velocity $\beta < 1$ with respect to the Home Frame in which the object's worldline lies along the diagram x' axis; and find the value of β (in terms of $|\vec{v}|$) that makes this happen. Why is it absurd for the worldline of any object to coincide with the diagram x' axis?

R7D.4 Show, using the Einstein velocity transformation equations R7.8, that a particle traveling in any arbitrary direction at the speed of light will be measured to have the speed of light in all other inertial frames.

Rich-Context

R7R.1 You are the captain of a spaceship that is moving through an asteroid belt on impulse power at a speed of $\frac{4}{5}$ relative to the asteroids. Suddenly you see an asteroid dead ahead a distance of only 24 s away, according to sensor measurements in your ship's reference frame. You immediately shoot off a missile, which travels forward at a speed of $\frac{4}{5}$ relative to your ship. The missile hits the asteroid and detonates, pulverizing the asteroid into gravel. However, you learned in Starfleet Academy that it is not safe to pass through such a debris field (even with shields on full) sooner than 8 s (measured in the asteroid frame) after the detonation. Are you safe?

R7R.2 A spaceship travels at a speed of 0.90 along a straight-line path that passes 300 km from a small asteroid. Exactly 1.0 ms (in the asteroid frame) in time before reaching the point of closest approach, the ship fires a photon torpedo which travels at the speed of light. This torpedo is fired perpendicular to the ship's direction of travel (as measured in the ship's frame) on the side closest to the asteroid. Will this torpedo hit the asteroid? If not, does it pass the asteroid on the near side or far side (relative to the ship)? Carefully explain your reasoning.

R7R.3 Your school gets a letter from one Kent C. M. Tugadett, a wealthy alumnus who will give a large gift to the

Physics Department if someone can successfully explain to him how to resolve the following paradox, which has bothered him since his college physics class. "A train is moving along a straight railroad track at a speed $|\vec{v}|$ close to the speed of light. Since moving clocks run slow, the train's clocks all run slow compared to ground-frame clocks. Now, a runner runs inside the train at the same speed $|\vec{v}|$ relative to the train but in the opposite direction. Since the runner is moving relative to the train, the runner's watch must run more slowly than the train clocks and so doubly slowly compared to clocks on the ground. But even the Einstein velocity transformation (check this) says that the runner will be at rest relative to the ground. So the runner's watch must run at the *same rate* as the ground clocks. How can this be, since the runner's watch is slower than the train clocks which are slower than the ground clocks?"

It falls to you to answer this question. Carefully and *diplomatically* explain to Mr. Tugadett how to resolve the apparent paradox, and so earn the gift for your school.* (*Hint:* This is yet another situation where the "moving clocks run slow" idea is misleading.)

Advanced

R7A.1 Imagine that in its own reference frame, an object emits light uniformly in all directions. Suppose this object moves in the $+x$ direction with respect to the Home Frame at an x-velocity of β.

(a) Show that the portion of the light that is emitted in the forward hemisphere in the object's own frame is observed in the Home Frame to be concentrated in a cone that makes an angle of

$$\phi = \sin^{-1}\frac{1}{\gamma} \tag{R7.15}$$

with respect to the x axis.

(b) Show that if $\beta = 0.99$, the angle within which this portion (which amounts to one-half of the object's light) is concentrated is only 8.1°.

(This forward concentration of the radiation emitted by a moving object is called the *headlight effect*.)

R7A.2 Consider a very long pair of scissors. If you close the scissor blades fast enough, you might imagine that you could cause the *intersection* of the scissor blades (that is, the point where they cut the paper) to travel from the near end of the scissors to the far end at a speed faster than that of light, without causing any *material part* of the scissors to exceed the speed of light. Argue that (1) this intersection can indeed travel faster than the speed of light in principle, but that (2) if the scissors blades are originally open and at rest and you decide to send a *message* to a person at the other end of your scissors by suddenly closing them, you will find that the intersection (and thus the message) *cannot* travel faster than the speed of light. (*Hint:* The information that the handles have begun to close must travel through the metal from the handles to the blades and then down the blades to cause the intersection to move forward. What carries this information?)†

*Adapted from a problem in Taylor and Wheeler, *Spacetime Physics*, 2nd edition, Freeman, 1992.

†See M. A. Rothman, "Things that Go Faster than Light," *Sci. Am.*, vol. 203, p. 142, July 1960.

ANSWERS TO EXERCISES

R7X.1 Let us take event P to be the origin event in the Home Frame. Now, consider figure R7.3. For event Q to occur below the x' axis in the Other Frame, the x' axis must have a slope greater than that of the worldline of the causal influence connecting events P and Q. If the causal influence moves at a speed of 3, then its slope on a two-observer diagram like figure R7.3 will be $\frac{1}{3}$. The slope of the Other Frame x' axis must be therefore greater than $\frac{1}{3}$. Since the slope of this axis is equal to β, the speed at which the Other Frame moves with respect to the Home Frame, the speed of the Other Frame must be greater than $\frac{1}{3}$.

R7X.2 The spacetime interval between A and B is timelike, between A and C is lightlike, and between A and D is spacelike. The interval between B and C is spacelike. The interval between D and B is spacelike, but the interval between D and C is timelike.

R7X.3 According to the inverse Lorentz transformation equations, we have

$$v_z = \frac{dz}{dt} = \frac{dz'}{\gamma(dt' + \beta dx')} = \frac{dz'/dt'}{\gamma[1 + \beta(dx'/dt')]}$$

$$= \frac{v_z'}{\gamma(1 + \beta v_x')} = \frac{v_z'\sqrt{1-\beta^2}}{1 + \beta v_x'} \tag{R7.16}$$

R8 Four-Momentum

Chapter Overview

Introduction

Our goal in this final subdivision of the unit is to make the laws of conservation of momentum and energy compatible with the principle of relativity. This chapter introduces a redefinition of the concept of momentum (which turns out to include energy as an integral part), and chapter R9 explores the implications of this idea.

Section R8.1: A Plan of Action

In this chapter, we will see that

1. Conservation of a system's total **Newtonian momentum** $\not{\vec{p}} \equiv m\vec{v}$ is not consistent with the principle of relativity. (The slash notation indicates an outmoded definition of momentum.)
2. *Four-momentum* is a natural relativistic redefinition of momentum.
3. Conservation of four-momentum is consistent with the principle of relativity.
4. The conserved fourth component of four-momentum is energy.

Section R8.2: Newtonian Momentum Isn't Conserved

Consider a collision that conserves Newtonian momentum in the Home Frame. If we use the Einstein velocity transformation equations to calculate the velocities of the colliding objects in another inertial frame, we find that Newtonian momentum is *not* conserved in the second frame. Conservation of Newtonian momentum is therefore incompatible with the principle of relativity.

Section R8.3: The Four-Momentum Vector

An object's **four-momentum $\boldsymbol{p}$** is a four-dimensional vector quantity (a **four-vector**) whose four components in a given inertial reference frame are

$$[p_t, p_x, p_y, p_z] = \left[m\frac{dt}{d\tau},\ m\frac{dx}{d\tau},\ m\frac{dy}{d\tau},\ m\frac{dz}{d\tau}\right] \quad \text{(R8.8)}$$

- **Purpose:** This equation defines (in a given inertial reference frame) the components $[p_t, p_x, p_y, p_z]$ of a particle's four-momentum $\boldsymbol{p}$ in terms of its mass m and the rate of change of its spacetime coordinates $[t, x, y, z]$ (in that frame) with respect to the particle's own proper time τ.
- **Limitations:** This equation is a definition and so has no limitations.
- **Notes:** The four-momentum treats time and space coordinates evenhandedly and points tangent to the particle's worldline in spacetime.

Because $d\tau = (1-|\vec{v}|^2)^{1/2}dt$, the spatial components of $\boldsymbol{p}$ are indistinguishable from the components of Newtonian momentum $\not{\vec{p}}$ when the particle's speed $|\vec{v}| \ll 1$. (Equation R8.10 in figure R8.4 shows how to calculate $[p_t, p_x, p_y, p_z]$ from m and $\vec{v}$.)

Section R8.4: Properties of Four-Momentum

The components of the four-momentum transform as follows:

$$\begin{bmatrix} p'_t \\ p'_x \\ p'_y \\ p'_z \end{bmatrix} = \begin{bmatrix} \gamma(p_t - \beta p_x) \\ \gamma(-\beta p_t + p_x) \\ p_y \\ p_z \end{bmatrix} \tag{R8.13}$$

- **Purpose:** This equation describes how we can calculate a particle's four-momentum components $[p'_t, p'_x, p'_y, p'_z]$ in the Other Frame, given its components $[p_t, p_x, p_y, p_z]$ in the Home Frame, where β is the Other Frame's x-velocity relative to the Home Frame and $\gamma \equiv (1 - \beta^2)^{-1/2}$.
- **Limitations:** This equation assumes that both frames are inertial and that they are in standard orientation.
- **Note:** These equations are *very* similar to the Lorentz transformation equations for coordinate differences.

The particle's mass m is the frame-independent **four-magnitude** of its four-momentum: $m = |\boldsymbol{p}| = (p_t^2 - p_x^2 - p_y^2 - p_z^2)^{1/2}$ (see equation R8.31 in figure R8.4).

Section R8.5: Four-Momentum and Relativity

Equation R8.13 implies quite generally that if a system's total four-momentum is conserved in any inertial frame, it will be conserved in *all* inertial frames. Therefore, a law of conservation of four-momentum *is* compatible with the principle of relativity.

Section R8.6: Relativistic Energy

This conservation law only works, however, if p_t is conserved along with p_x, p_y, and p_z. In the limit that $|\vec{v}| \ll 1$, $p_t \approx m + \frac{1}{2}m|\vec{v}|^2$, so conservation of p_t in that limit implies that a system's total mass plus its total kinetic energy is conserved. We call p_t the particle's **relativistic energy** E. Note that even a particle at rest has a **rest energy** $E_{\text{rest}} = m$ ($= mc^2$ in SI units), and that conservation of relativistic energy allows conversion of this form of energy into other forms. We define a particle's **relativistic kinetic energy** K to be $K \equiv E - m$: this is *approximately* equal to $\frac{1}{2}m|\vec{v}|^2$ when $|\vec{v}| \ll 1$.

We define a particle's **relativistic momentum** $|\vec{p}|$ to be the magnitude of the spatial components of its four-momentum: $|\vec{p}| \equiv (p_x^2 + p_y^2 + p_z^2)^{1/2}$. Note that a particle's speed in a given frame is $|\vec{v}| = |\vec{p}|/E$. Figure R9.5 summarizes virtually everything you need to know about doing calculations with four-momentum components.

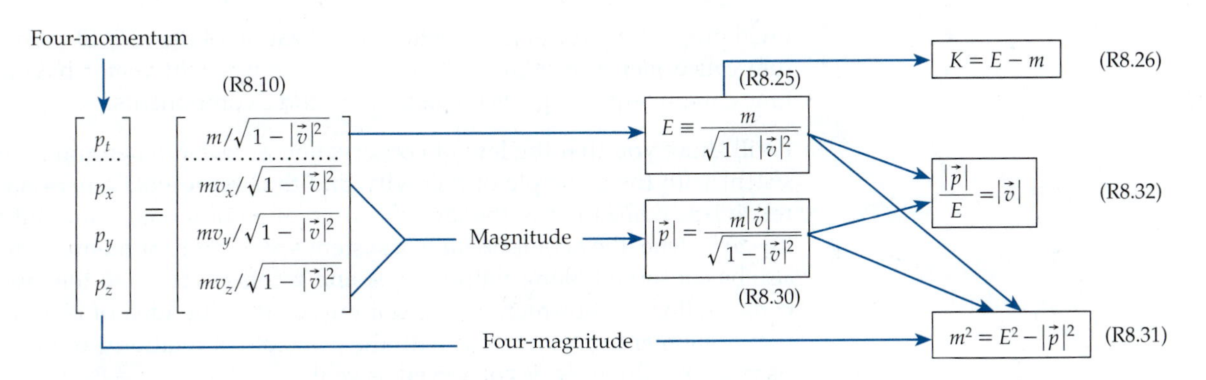

Figure R8.4

Virtually everything you need to know about four-momentum.

R8.1 A Plan of Action

Up to this point, we have been studying **relativistic kinematics** — how we describe and measure the motion of objects in special relativity. But physicists are also interested in *dynamics*, which comprises the study of how interactions between objects *determine* the objects' motions. In this volume's final subsection, we will explore the basic principles of **relativistic dynamics**.

The laws of Newtonian physics are not consistent with relativity

The basic principles of Newtonian dynamics are Newton's three laws of motion, which in turn are based on the laws of conservation of momentum and energy. In section R1.5, we saw that various laws of Newtonian dynamics were consistent with the principle of relativity *if* the Galilean velocity transformation equations (equations R1.2) are true. But in chapter R7 we saw that these equations are *not* true: they only represent the low-velocity limit of the relativistically correct Einstein velocity transformation equations (equations R7.14). *This means that the laws of Newtonian dynamics are NOT generally consistent with the principle of relativity.* Those laws likewise represent only low-velocity approximations to the laws of *relativistic* dynamics, laws that are the same in *all* inertial frames (as the principle of relativity requires).

Well, what are these laws of relativistic dynamics, and how can we find them? We *could* address this question by searching for a relativistic generalization of Newton's second law, then Newton's third law, then the law of universal gravitation, and so on. But it turns out that trying to do this is trickier than it looks, and leads to ugly equations that are not really illuminating.

To find the laws of relativistic dynamics, we start with conservation of momentum

Things work much better if we instead start with the more basic law of *conservation of momentum* and make this law consistent with the principle of relativity first. Not only is the correct adaptation of this law fairly easy to find, but it also proves to be very illuminating and rich in implications and applications. Indeed, just as we found for Newtonian dynamics in unit C, we can learn virtually everything that is useful to know about relativistic dynamics by closely examining the law of conservation of momentum.

An overview of the argument to be found in this chapter

The basic argument in this chapter can be outlined as follows.

1. I will show you that conservation of **Newtonian momentum** $\not{\vec{p}} \equiv m\vec{v}$ is *not* consistent with the principle of relativity: if an isolated system's total Newtonian momentum is conserved in one inertial frame, it is *not* conserved in other frames. This means that we need to redefine momentum so that its conservation law is consistent with the principle of relativity.

2. I will propose a natural relativistic generalization of the idea of momentum called *four-momentum*, which is a *four*-component vector having a time component as well as the usual x, y, and z components.

3. I will show you that the law of conservation of *four*-momentum *is* consistent with the principle of relativity, and thus represents a reasonable relativistic realization of the law of conservation of momentum. But for this to be true, the t component of a system's total four-momentum must *also* be conserved along with its x, y, and z components. So the law of conservation of four-momentum not only makes the idea of conservation of momentum consistent with the principle of relativity, but it tells us that something *else* is conserved as well.

4. This fourth conserved quantity turns out to be a relativistic version of the concept of *energy*. Thus, the law of conservation of four-momentum actually *unifies* two of the three great conservation laws discussed in unit C.

I mean for the slash notation in the equation $\not{\vec{p}} \equiv m\vec{v}$ to indicate that from this point on, this Newtonian definition of momentum is now obsolete. From

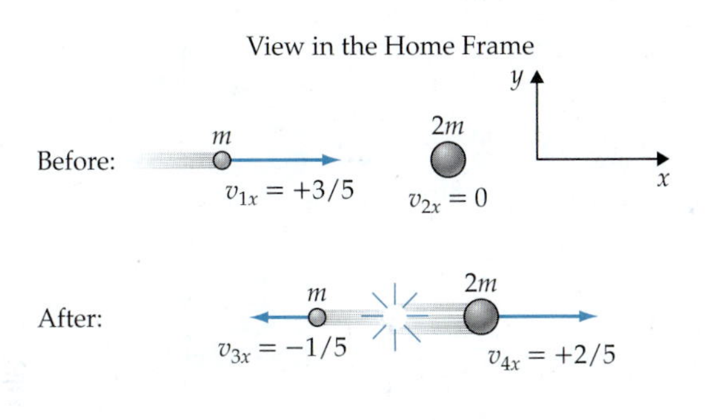

Figure R8.1
A hypothetical collision of two particles as observed in the Home Frame. The total Newtonian momentum of this isolated system is conserved in this frame.

now on (and essentially in units E and Q as well), $\vec{p}$ will refer to the relativistic momentum we are about to define.

R8.2 Newtonian Momentum Isn't Conserved

In this section, I will argue that the law of conservation of Newtonian momentum is *not* consistent with the principle of relativity and thus cannot be a valid law of physics as stated. To illustrate the problem, it is sufficient to demonstrate a single instance of the inconsistency. For the sake of simplicity, I will illustrate the problem using a simple one-dimensional collision.

A hypothetical collision

Figure R8.1 shows such a collision as observed in the Home Frame. In this frame, an object with mass m is moving in the $+x$ direction with an x-velocity $v_{1x} = +\frac{3}{5}$. It then strikes an object of mass $2m$ at rest ($v_{2x} = 0$). Let's assume the objects are isolated and the collision is elastic. If so, the Newtonian equations for one-dimensional elastic collisions (see section C14.2) imply that the lighter mass will rebound from the collision with an x-velocity of $v_{3x} = -\frac{1}{5}$, while the heavier object will rebound with an x-velocity of $v_{4x} = +\frac{2}{5}$.

Newtonian momentum is conserved in the Home Frame for this collision

We can easily verify that the system's total Newtonian x-momentum is conserved in the Home Frame for the collision as described:

$$\text{Total } \not{p}_x \text{ before:} \quad mv_{1x} + 2mv_{2x} = m\left(+\tfrac{3}{5}\right) + 2m(0) = +\tfrac{3}{5}m \tag{R8.1a}$$

$$\text{Total } \not{p}_x \text{ after:} \quad mv_{3x} + 2mv_{4x} = m\left(-\tfrac{1}{5}\right) + 2m\left(+\tfrac{2}{5}\right) = +\tfrac{3}{5}m \tag{R8.1b}$$

How the collision looks in the Other Frame

Now consider how this collision looks when observed in an Other Frame that is moving in the $+x$ direction with an x-velocity $\beta = \frac{3}{5}$. Since this frame essentially moves along with the lightweight object, that object appears to be at rest in the Other Frame: $v'_{1x} = 0$. Because the larger object is at rest in the Home Frame, and the Home Frame is observed to be moving backward with respect to the Other Frame at a speed of $\frac{3}{5}$, the x-velocity of the more massive object must also be $v'_{2x} = -\frac{3}{5}$. The objects' final x-velocities are not so easy to intuit, so we need to use the Einstein velocity transformation equation R7.14*a* to compute these velocities:

$$v'_{3x} = \frac{v_{3x} - \beta}{1 - \beta v_{3x}} = \frac{-\frac{1}{5} - \frac{3}{5}}{1 - \left(\frac{3}{5}\right)\left(-\frac{1}{5}\right)} = \frac{-\frac{4}{5}}{\frac{28}{25}} = -\frac{20}{28} = -\frac{5}{7} \tag{R8.2a}$$

Similarly, you can show that

$$v'_{4x} = -\frac{5}{19} \tag{R8.2b}$$

Exercise R8X.1

Verify equation R8.2*b*.

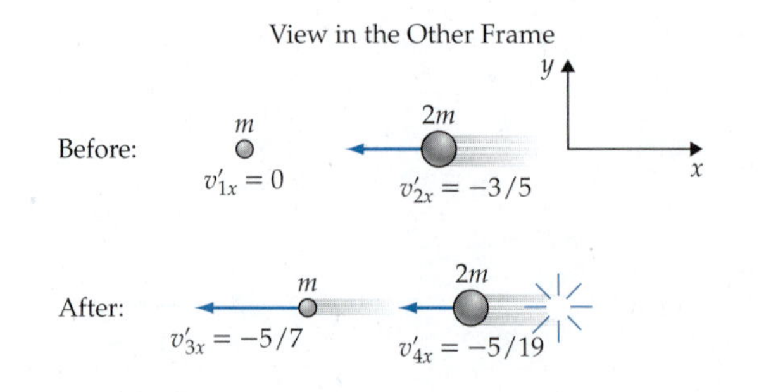

Figure R8.2
The same collision as observed in the Other Frame. Newtonian momentum is *not* conserved in this frame.

Note that these *must* be the objects' final velocities if the Einstein velocity transformation equations are true and figure R8.1 is accurate. In the Other Frame, then, the collision process must go as shown in figure R8.2.

Newtonian momentum is *not* conserved in the Other Frame

In this frame, though, the system's total Newtonian x-momentum is *not* conserved:

$$\text{Total } p_x \text{ before:} \quad mv'_{1x} + 2mv'_{2x} = m(0) + 2m(-\tfrac{3}{5}) = -\tfrac{6}{5}\,m \tag{R8.3a}$$

$$\text{Total } p_x \text{ after:} \quad mv'_{3x} + 2mv'_{4x} = m(-\tfrac{5}{7}) + 2m(-\tfrac{5}{19}) = -\tfrac{165}{133}\,m \tag{R8.3b}$$

(Note that $165/133 = 1.24 > 6/5 = 1.20$.) The law of conservation of Newtonian momentum therefore does *not* hold in the Other Frame, even though it did hold in the Home Frame. (Note that the law of conservation of momentum requires that *each* component of the system's total momentum be conserved separately, so if even *one* component, the x component in this case, is not conserved, then momentum as a whole is not conserved.)

The principle of relativity requires that the laws of physics be the same in all inertial reference frames. The conclusion is inescapable: if the Einstein velocity transformation equations are true, then the law of conservation of Newtonian momentum is *not* consistent with the principle of relativity.

I hope you can see that the root of the problem is the Einstein velocity transformation equations. If the Galilean velocity transformation equations were true, the final velocities would be $v'_{3x} = -\frac{4}{5}$ and $v'_{4x} = -\frac{1}{5}$, and momentum *would* be conserved in the Other frame as well as the Home Frame, because $mv'_{3x} + 2mv'_{4x} = m(-\frac{4}{5}) + 2m(-\frac{1}{5}) = -\frac{6}{5}\,m$.

R8.3 The Four-Momentum Vector

Review of the definition of Newtonian momentum

In unit C, we defined an object's Newtonian momentum $\vec{p}$ as follows:

$$\vec{p} \equiv m\vec{v} = m\frac{d\vec{r}}{dt} \tag{R8.4}$$

This is the *Newtonian* definition of momentum, which we need to modify to make the law of conservation of momentum consistent with the principle of relativity. But *how* might we modify this definition? It helps to look at the Newtonian definition very closely to try to understand its meaning.

The vector $d\vec{r}$ in equation R8.4 represents an infinitesimal displacement in space, which we divide by an infinitesimal time interval dt to get the object's velocity vector $\vec{v}$. The components of the infinitesimal displacement vector $d\vec{r}$ are $[dx, dy, dz]$, so the components of the Newtonian momentum are

$$p_x \equiv mv_x = m\frac{dx}{dt} \qquad p_y \equiv m\frac{dy}{dt} \qquad p_z = m\frac{dz}{dt} \tag{R8.5}$$

Notice that the $\vec{p}$ vector is *parallel* to the infinitesimal displacement $d\vec{r}$, and so will be tangent to the object's path through space (see figure R8.3a).

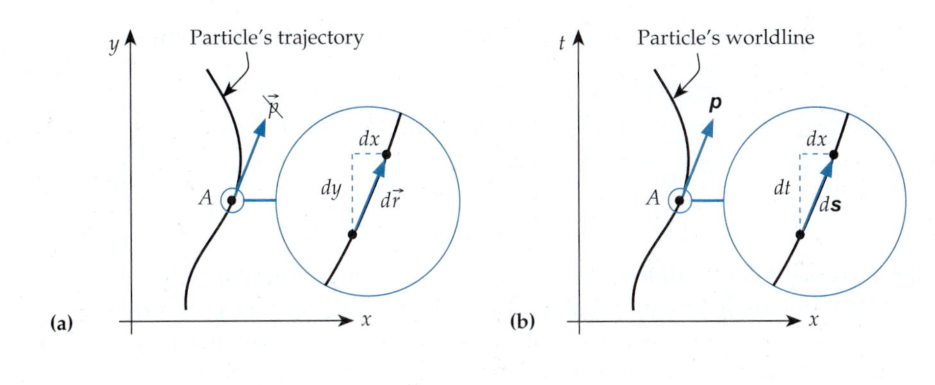

Figure R8.3
(a) A graph showing a particle's trajectory through space. The particle's Newtonian momentum $\not{\vec{p}}$ at a point A is a vector parallel to the displacement $d\vec{r}$ that connects two infinitesimally separated points surrounding A. (b) A spacetime diagram showing a particle's worldline in spacetime. The object's four-momentum $\boldsymbol{p}$ at event A is an arrow parallel to the four-displacement $d\boldsymbol{s}$ that connects two infinitesimally separated events surrounding A.

How can we revise this definition to make it more "relativistic"?

How can we arrive at a relativistic generalization of this process? In special relativity, space and time are considered to be equal parts of the unitary whole we call spacetime, so we describe the motion of an object not merely by describing its path through space but rather by describing its *worldline through spacetime*. The appropriate relativistic generalization of an "infinitesimal displacement in space" $d\vec{r}$ between two infinitesimally separated points on an object's path in space is a displacement $d\boldsymbol{s}$ in *spacetime* between two infinitesimally separated *events* on the object's *worldline* (see figure R8.3b). Note that in any given inertial reference frame, the displacement $d\boldsymbol{s}$ in spacetime between two events is specified by *four* numbers:

$$d\boldsymbol{s} = \begin{bmatrix} dt \\ dx \\ dy \\ dz \end{bmatrix} \tag{R8.6}$$

Including the time displacement dt on an equal footing with the spatial displacements dx, dy, and dz makes the displacement $d\boldsymbol{s}$ a four-component vector that physicists call a **four-vector**.

Given the components $[dt, dx, dy, dz]$ of the displacement four-vector $d\boldsymbol{s}$ between two infinitesimally separated events on the worldline of our object, how do we define its relativistic momentum? By analogy with the Newtonian momentum, we want to divide $d\boldsymbol{s}$ by a quantity that expresses the *time* between the events, and then multiply by the object's mass. In Newtonian mechanics, time is universal and absolute, so the time between the events does not depend on how it is measured. The most important flaw in the definition of Newtonian momentum from the *relativistic* viewpoint, however, is that time is *not* universal and absolute. The time dt measured between the events in the frame where we measure dx, dy, and dz is not the same as the time dt' measured in some other inertial frame, which in turn is not generally the same as the proper time $d\tau$ between the events measured by a clock traveling with the particle. Which of these times should we choose?

Dividing by dt makes the spatial components of $m\,d\boldsymbol{s}/dt$ the same as the spatial components of ordinary momentum, which we know doesn't work. Using the object's *own* proper time (which is uniquely and unambiguously linked to its motion) makes much more sense. Moreover, $d\tau$ has the advantage of being frame-independent. So our proposed relativistic redefinition of an object's Newtonian momentum $\not{\vec{p}}$ is the **four-momentum $\boldsymbol{p}$**:

Definition of a particle's four-momentum vector

$$\boldsymbol{p} \equiv m\frac{d\boldsymbol{s}}{d\tau} \tag{R8.7}$$

Note that when drawn as an arrow in a spacetime diagram, $\boldsymbol{p}$ is tangent to the particle's worldline, just as $\not{\vec{p}}$ was tangent to its spatial trajectory.

The four-momentum is a four-dimensional vector having components

$$\boldsymbol{p} = \begin{bmatrix} p_t \\ p_x \\ p_y \\ p_z \end{bmatrix} = \begin{bmatrix} m(dt/d\tau) \\ m(dx/d\tau) \\ m(dy/d\tau) \\ m(dz/d\tau) \end{bmatrix} \tag{R8.8}$$

- **Purpose:** This equation defines (in a given inertial reference frame) the components $[p_t, p_x, p_y, p_z]$ of a particle's four-momentum $\boldsymbol{p}$ in terms of its mass m and the rate of change of its spacetime coordinates $[t, x, y, z]$ (in that frame) with respect to the particle's own proper time τ.
- **Limitations:** This is a definition, so it has no limitations.
- **Notes:** The four-momentum is defined to have *four* components so that the time and space coordinates are treated evenhandedly and so that the resulting four-vector always points tangent to the particle's worldline in spacetime.

I should note that physicists doing relativity research commonly use ***boldface sans-serif letters*** to represent four-vectors such as $\boldsymbol{p}$ and $d\boldsymbol{s}$. I will follow that convention here. However, I recognize that this is difficult to distinguish $\boldsymbol{p}$ from p in handwriting. I suggest the following notations for handwriting symbols for four-vectors: either put a squiggle under the four-vector's letter (for example, p̰) or put a box around it (for example, $\boxed{\text{p}}$). The squiggle is a traditional proofreader's mark for "make this letter boldface." You can think of the box as saying "this quantity has four components," just as the box has four sides. Use whichever works for you and your instructor.

Expressing the particle's four-momentum in terms of its mass and velocity

We can express the components of the four-momentum in a given inertial frame in terms of the object's ordinary velocity measured in that frame. As we saw in chapter R4, the proper time between two infinitesimally separated events measured by a clock traveling between them at a speed $|\vec{v}|$ in a given inertial frame is related to the coordinate time dt measured in that frame between those events by

$$d\tau = \sqrt{1 - |\vec{v}|^2}\, dt \tag{R8.9}$$

This means that

$$p_t \equiv m\frac{dt}{d\tau} = \frac{m\,dt}{\sqrt{1-|\vec{v}|^2}\,dt} = \frac{m}{\sqrt{1-|\vec{v}|^2}} \tag{R8.10a}$$

$$p_x \equiv m\frac{dx}{d\tau} = \frac{m}{\sqrt{1-|\vec{v}|^2}}\left(\frac{dx}{dt}\right) = \frac{mv_x}{\sqrt{1-|\vec{v}|^2}} \tag{R8.10b}$$

$$p_y \equiv m\frac{dy}{d\tau} = \frac{m}{\sqrt{1-|\vec{v}|^2}}\left(\frac{dy}{dt}\right) = \frac{mv_y}{\sqrt{1-|\vec{v}|^2}} \tag{R8.10c}$$

$$p_z \equiv m\frac{dz}{d\tau} = \frac{m}{\sqrt{1-|\vec{v}|^2}}\left(\frac{dz}{dt}\right) = \frac{mv_z}{\sqrt{1-|\vec{v}|^2}} \tag{R8.10d}$$

These equations tell us how to calculate the components of the four-momentum of an object in a given frame, given the object's velocity vector $\vec{v}$ in that frame.

The low-velocity limit of the four-momentum components

When an object's speed $|\vec{v}|$ becomes very small compared to the speed of light ($|\vec{v}| \ll 1$), the square roots in the denominators in equations R8.10 become almost equal to 1, and we have

$$p_t \approx m \tag{R8.11a}$$

$$p_x \approx mv_x \tag{R8.11b}$$

$$p_y \approx mv_y \tag{R8.11c}$$

$$p_z \approx mv_z \tag{R8.11d}$$

Thus, in the limit where the velocity of an object is very small, the spatial components of the four-momentum reduce to being the same as the corresponding components of the object's Newtonian momentum. Therefore, in everyday circumstances, we are unable to tell whether it is really Newtonian momentum or four-momentum that is the conserved quantity.

Note also that since velocity in SR units is unitless, all four components of an object's four-momentum have units of *mass* in SR units. An object exactly at rest has four-momentum components $p_t = m$, $p_x = p_y = p_z = 0$.

Example R8.1

Problem: Suppose an object with mass $m = 1.0$ kg moves with a velocity such that $v_x = 0$, $v_y = \frac{4}{5}$, and $v_z = 0$ in some inertial reference frame. What are the components of its four-momentum in that frame?

Solution Note that this object's speed is $|\vec{v}| = \frac{4}{5}$, so $\sqrt{1 - |\vec{v}|^2} = \sqrt{1 - \frac{16}{25}} = \sqrt{\frac{9}{25}} = \frac{3}{5}$. The object's four-momentum is therefore

$$p = \frac{m}{\sqrt{1 - |\vec{v}|^2}} \begin{bmatrix} 1 \\ v_x \\ v_y \\ v_z \end{bmatrix} = \frac{1.0 \text{ kg}}{3/5} \begin{bmatrix} 1 \\ 0 \\ 4/5 \\ 0 \end{bmatrix} = \begin{bmatrix} \frac{5}{3} \text{ kg} \\ 0 \text{ kg} \\ \frac{5}{3}\frac{4}{3} \text{ kg} \\ 0 \text{ kg} \end{bmatrix} = \begin{bmatrix} 1.67 \\ 0 \\ 1.33 \\ 0 \end{bmatrix} \text{kg} \tag{R8.12}$$

R8.4 Properties of Four-Momentum

A particle's four-momentum vector is tangent to its worldline in spacetime

Why define an object's relativistic four-momentum in this way? The definition has several attractive features. One feature has already been mentioned: *on a spacetime diagram, the four-momentum is represented by an arrow tangent to the object's worldline through spacetime,* just as the ordinary momentum vector is an arrow tangent to the object's trajectory through space.

It puts space and time on an equal footing

It is also nice that the definition of the four-momentum treats the time coordinate in the same manner as the spatial coordinates: all four coordinate displacements dt, dx, dy, dz appear on an equal footing in the definition of the four-momentum given by equations R8.8. We have already seen how it is important in relativity theory to treat time and space as being equal participants in the larger geometric whole that we call spacetime. The definition of four-momentum given earlier maintains this symmetry.

It has a very straightforward transformation law

All this symmetry has a certain beauty about it which an intuitive physicist like Einstein might take as corroborating evidence that we are on the right track with this definition. But we will see that the most important feature of how the four-momentum vector is defined is that given its components in one inertial frame, we can calculate its components in any other inertial frame, using a very straightforward (and already familiar) set of equations.

The components of an object's four-momentum are *frame-dependent* quantities, because the values of the coordinate differences dt, dx, dy, dz

that appear in the numerators of equations R8.8 are frame-dependent. The differential proper time $d\tau$ appearing in the denominator, on the other hand, is *frame-independent*. In this text, we will also consider the mass m of the object to be a *frame-independent* measure of the amount of "stuff" in the object.*

Suppose we know the components $[p_t, p_x, p_y, p_z]$ of a given object's four-momentum $\boldsymbol{p}$ in the Home Frame, and we want to find the corresponding components in an Other Frame moving in the $+x$ direction with x-velocity β relative to the Home Frame. We can calculate the time component p_t' of the object's four-momentum in the Other Frame as follows:

$$p_t' = m\frac{dt'}{d\tau} = m\frac{\gamma(dt - \beta dx)}{d\tau} = \gamma m\frac{dt}{d\tau} - \gamma\beta m\frac{dx}{d\tau} = \gamma(p_t - \beta p_x) \qquad \text{(R8.13}a\text{)}$$

where I have used the Lorentz transformation (specifically equation R5.12*a*) to express dt' as measured in the Other Frame in terms of dt and dx as measured in the Home Frame. Similarly, you can show that the transformation equation for the four-momentum x component is

$$p_x' = \gamma(-\beta p_t + p_x) \qquad \text{(R8.13}b\text{)}$$

We also have

$$p_y' = m\frac{dy'}{d\tau} = m\frac{dy}{d\tau} = p_y \qquad \text{(R8.13}c\text{)}$$

$$p_z' = m\frac{dz'}{d\tau} = m\frac{dz}{d\tau} = p_z \qquad \text{(R8.13}d\text{)}$$

Exercise R8X.2

Verify that equation R8.13*b* is correct.

The transformation has the same form as the Lorentz transformation equations

Compare these equations to Lorentz transformation equations:

$$\Delta t' = \gamma(\Delta t - \beta\Delta x),\ \Delta x' = \gamma(-\beta\Delta t + \Delta x),\ \Delta y' = \Delta y,\ \Delta z' = \Delta z \qquad \text{(R8.14)}$$

Equations R8.13 are the *same* as these equations except that the four-momentum components p_t, p_x, p_y, and p_z have been substituted for the coordinate displacement components Δt, Δx, Δy, and Δz, respectively. *Thus, the components of the four-momentum transform from frame to frame according to the Lorentz transformation equations,* just as coordinate differences do!

The transformation equations for the four-momentum come out so nicely because (1) the time coordinate appears on an equal footing with the spatial components in the definition of the four-momentum, (2) we have divided the displacement by the frame-independent differential proper time $d\tau$ instead of the frame-dependent differential coordinate time dt, and (3) we define the object's mass m to be a frame-independent quantity.

Formal definition of a four-vector quantity

In fact, the technical definition of a *four-vector* requires this kind of transformation law. Physicists define a **four-vector** to be a physical quantity represented by a vector whose four components transform according to the Lorentz transformation equations (that is, just as the coordinate differences Δt, Δx, Δy, Δz do) when we go from one inertial reference frame to another.

*You may have heard in another context that special relativity implies that the mass of an object depends on its velocity. This is an old-fashioned way of looking at mass that obscures some of the simplicity and beauty of relativity theory. See C. G. Adler, "Does Mass Really Depend on Velocity, Dad?" *Am. J. Phys.*, vol. 55, no. 8, pp. 739–743, August 1987, for a careful and entertaining look at the problems with the old way of thinking about mass in relativity theory. Most modern treatments of relativity treat an object's mass as being frame-*independent*.

The frame-independent four-magnitude of a four-vector

Now we know that although the coordinate differences Δt, Δx, Δy, Δz between two events are frame-*dependent* quantities, the spacetime interval Δs between the events given by

$$\Delta s^2 = \Delta t^2 - \Delta x^2 - \Delta y^2 - \Delta z^2 \tag{R8.15}$$

has a frame-*independent* value. Similarly, we can define the frame-independent **four-magnitude** of a four-vector as follows: if a four-vector $\boldsymbol{A}$ has components A_t, A_x, A_y, and A_z, then we define its squared four-magnitude to be

$$|\boldsymbol{A}|^2 \equiv A_t^2 - A_x^2 - A_y^2 - A_z^2 \tag{R8.16}$$

This is the relativistic analog to the definition (based on the Pythagorean theorem) of the magnitude of an ordinary vector.

You can easily show that the frame-independent four-magnitude of an object's four-momentum vector is simply its frame-independent mass:

$$m = |\boldsymbol{p}| = \sqrt{p_t^2 - p_x^2 - p_y^2 - p_z^2} \tag{R8.17}$$

Exercise R8X.3

Verify that equation R8.17 is correct, using the fact that for infinitesimally separated events, there is no distinction between proper time and the spacetime interval, so $d\tau = ds = (dt^2 - dx^2 - dy^2 - dz^2)^{1/2}$.

R8.5 Four-Momentum and Relativity

We now have a suitable candidate for a relativistic generalization of the concept of momentum. The final step is to verify that a law of conservation of four-momentum is in fact consistent with the principle of relativity.

Proof that four-momentum is conserved in *all* inertial frames if it is in any one frame

Consider an arbitrary collision of two objects moving along the x axis. The law of conservation of four-momentum says that

$$\boldsymbol{p}_1 + \boldsymbol{p}_2 = \boldsymbol{p}_3 + \boldsymbol{p}_4 \tag{R8.18}$$

where $\boldsymbol{p}_1$ and $\boldsymbol{p}_2$ are the objects' four-momenta *before* the collision and $\boldsymbol{p}_3$ and $\boldsymbol{p}_4$ are their four-momenta *after*. We can usefully rewrite this as follows:

$$\boldsymbol{p}_1 + \boldsymbol{p}_2 - \boldsymbol{p}_3 - \boldsymbol{p}_4 = 0 \tag{R8.19}$$

which essentially says that the *difference* between the system's initial and final total momenta is zero. In component form, this last equation tells us that

$$\begin{bmatrix} p_{1t} \\ p_{1x} \\ p_{1y} \\ p_{1z} \end{bmatrix} + \begin{bmatrix} p_{2t} \\ p_{2x} \\ p_{2y} \\ p_{2z} \end{bmatrix} - \begin{bmatrix} p_{3t} \\ p_{3x} \\ p_{3y} \\ p_{3z} \end{bmatrix} - \begin{bmatrix} p_{4t} \\ p_{4x} \\ p_{4y} \\ p_{4z} \end{bmatrix} = \begin{bmatrix} 0 \\ 0 \\ 0 \\ 0 \end{bmatrix} \quad \text{or} \quad \begin{aligned} p_{1t} + p_{2t} - p_{3t} - p_{4t} &= 0 \\ p_{1x} + p_{2x} - p_{3x} - p_{4x} &= 0 \\ p_{1y} + p_{2y} - p_{3y} - p_{4y} &= 0 \\ p_{1z} + p_{2z} - p_{3z} - p_{4z} &= 0 \end{aligned} \tag{R8.20}$$

When expressed in component form, we see that the single equation R8.19 is really a set of *four* equations, one for each of the four-momentum's four components. Each must be *independently* satisfied for four-momentum to be conserved.

Suppose we have observed a collision in the Home Frame and determined that it satisfies the law of conservation of four-momentum in that frame. The principle of relativity requires that the same law apply in *every other* inertial reference frame, that is,

$$\begin{aligned} &\text{If} \quad \boldsymbol{p}_1 + \boldsymbol{p}_2 - \boldsymbol{p}_3 - \boldsymbol{p}_4 = 0 \quad \text{in the Home Frame} \\ &\text{Then} \quad \boldsymbol{p}_1' + \boldsymbol{p}_2' - \boldsymbol{p}_3' - \boldsymbol{p}_4' = 0 \quad \text{in any Other Frame} \end{aligned} \tag{R8.21}$$

If this statement is *not* true, our proposed relativistic generalization of momentum is not any better than Newtonian momentum. If the statement *is* true, then the law of conservation of four-momentum represents at least a *possible* relativistic expression of the law of conservation of momentum.

We can in fact easily show that this is true for any collision as viewed in any inertial Other Frame. Consider the x component of the conservation law in the Other Frame. According to the transformation law for the components of the four-momentum given by equation R8.13*b*,

$$p'_{1x} + p'_{2x} - p'_{3x} - p'_{4x} = \gamma(-\beta p_{1t} + p_{1x}) + \gamma(-\beta p_{2t} + p_{2x}) - \gamma(-\beta p_{3t} + p_{3x}) - \gamma(-\beta p_{4t} + p_{4x}) \quad \text{(R8.22}a\text{)}$$

Collecting the terms on the right side of this equation that are multiplied by γ and those multiplied by $\gamma\beta$, we get

$$p'_{1x} + p'_{2x} - p'_{3x} - p'_{4x} = -\gamma\beta(p_{1t} + p_{2t} - p_{3t} - p_{4t}) + \gamma(p_{1t} + p_{2t} - p_{3t} - p_{4t}) \quad \text{(R8.22}b\text{)}$$

But if *both* the t and x components of the four-momentum are conserved in the Home Frame, then equations R8.20 tell us that the quantities in parentheses are equal to zero: $p'_{1x} + p'_{2x} - p'_{3x} - p'_{4x} = -\gamma\beta(0) + \gamma(0) = 0$. So *if* both the t and x components of the system's total four-momentum are conserved in the Home Frame, then the x component of the system's total four-momentum will also be conserved in the Other Frame, as hoped.

Exercise R8X.4

In the same way, verify that the t and y components of the four-momentum are conserved in the Other Frame if all components are conserved in the Home Frame.

Thus, conservation of four-momentum is consistent with the principle of relativity, but is it true?

What we have shown is that the law of conservation of four-momentum expressed by equation R8.19 is *consistent* with the principle of relativity in the sense that if it holds in one frame, it holds in all frames. That does not make the law *true*: it simply makes it *possible*. But now let me argue for the law's *truth*. (1) We know from a multitude of experiments at low velocities that *some* quantity that reduces to Newtonian momentum at such velocities is conserved. (2) But conservation of *Newtonian* momentum is inconsistent with the principle of relativity. (3) The hypothetical law of conservation of four-momentum *is* compatible with the principle of relativity. (4) The three spatial components of the four-momentum *do* reduce to the components of Newtonian momentum at low velocities. (5) Therefore, if the law of conservation of four-momentum were true, it would both explain the low-velocity experimental data and maintain compatibility with the principle of relativity. (6) Moreover, there must be *some* relativistically valid expression of the deep symmetry principle that gives rise to the law of conservation of momentum. In the absence of compelling alternatives, it makes *sense* to believe it is the total four-momentum of a system that is conserved.

Abundant experimental evidence suggests it is

Of course, no matter how suggestive a theoretical argument might be, experimental evidence is the final arbiter. Since the 1950s, physicists have been using particle accelerators to create beams of subatomic particles traveling at relativistic speeds that collide with stationary targets or other particle beams. At such speeds, the distinction between Newtonian momentum and four-momentum is very clear, and analysis of a typical experiment involves applying conservation of four-momentum to anywhere from thousands to even billions of particle collisions. Conservation of four-momentum is therefore implicitly tested thousands of times daily in the course of such research. In spite of this enormous wealth of data, no one has *ever* seen compelling evidence of a violation of the law of conservation of four-momentum.

R8.6 Relativistic Energy

How should we interpret the fourth conserved quantity p_t?

Equation R8.22*b* makes it clear that conservation of four-momentum is *only* consistent with the principle of relativity if all *four* components of the four-momentum (p_t as well as p_x, p_y, and p_z) are independently conserved. The three *spatial* components of an object's four-momentum correspond (at low velocities) to the three components of its Newtonian momentum. What is the physical interpretation of the additional conserved quantity p_t?

Examining the value of p_t at low velocities suggests that p_t is related to energy

According to equation R8.10*a*, in a frame where an object with mass m is moving with speed $|\vec{v}|$, the object's four-momentum has a t component

$$p_t = \frac{m}{\sqrt{1 - |\vec{v}|^2}} \tag{R8.23}$$

We know this reduces to the object's mass m at low velocities, but is not exactly equal to the mass. The binomial approximation says that

$$p_t = \frac{m}{\sqrt{1 - |\vec{v}|^2}} = m(1 - |\vec{v}|^2)^{-1/2} \approx m[1 - (-\tfrac{1}{2}v^2)] = m + \tfrac{1}{2}mv^2 \tag{R8.24}$$

The first term here is the particle's mass, and when $|\vec{v}| = 0$, that is what p_t becomes. But when $|\vec{v}|$ is nonzero but still very small, we have an additional term in p_t that corresponds to the particle's *kinetic energy*. If p_t is conserved in a collision at low velocities, what we are saying is that the sum of the particles' masses plus the sum of their kinetic energies is conserved.

So in a collision that preserves the particles' masses, conservation of p_t is (at low velocities) basically the same as conservation of Newtonian (kinetic) energy! As we have already generalized the concept of momentum, so now we generalize the concept of energy. We *define* the time component p_t of a particle's four-momentum to be that particle's **relativistic energy** E, and we assert that it is this relativistic energy that is the fourth quantity that is conserved in an isolated system. So the relativistic energy E of an object moving at a speed $|\vec{v}|$ as measured in a given inertial frame is

The definition of relativistic energy

$$E \equiv p_t = \frac{m}{\sqrt{1 - |\vec{v}|^2}} \tag{R8.25}$$

Note that if the object's speed $|\vec{v}|$ is an appreciable fraction of the speed of light, equation R8.24 does not hold. We define an object's **relativistic kinetic energy** K (for all $|\vec{v}|$) to be the difference between the object's total relativistic energy E and its mass energy m:

The definition of relativistic kinetic energy

$$K \equiv E - m = \frac{m}{\sqrt{1 - |\vec{v}|^2}} - m = m\left(\frac{1}{\sqrt{1 - |\vec{v}|^2}} - 1\right) \tag{R8.26}$$

The value of $K \approx \frac{1}{2}m|\vec{v}|^2$ *only* for $|\vec{v}| << 1$: in general, $K > \frac{1}{2}m|\vec{v}|^2$.

Conservation of four-momentum thus unifies Newtonian conservation of energy and momentum

Newtonian mechanics treats conservation of energy and momentum as separate concepts. But just as special relativity binds space and time into a single geometry, so here it binds the laws of conservation of momentum and energy into a single statement: *an isolated system's total four-momentum is conserved*. Conservation of energy is impossible without conservation of momentum, and vice versa: in the theory of relativity, energy and momentum are indissolubly bound together as parts of the same whole!

Processes that convert rest energy (mass) to kinetic energy are possible

Note, however, that an object's relativistic energy involves a particle's *mass* as well as its kinetic energy. The fact that $E = p_t$ is conserved does *not* imply that a system's total mass and its total kinetic energy are *separately* conserved, only that the *whole* (that is, the relativistic energy) is conserved. This implies that *processes that convert mass to kinetic energy, and vice versa, do not necessarily violate the law of conservation of four-momentum* and therefore might exist. We will explore this subject more fully in chapter R9.

Converting SR units for energy to SI units

Equation R8.24 is expressed in SR units, where velocity is unitless and both kinetic energy and mass are measured in kilograms. If we want to express a particle's relativistic energy in the SI unit of joules ($1\text{ J} = 1\text{ kg}\cdot\text{m}^2/\text{s}^2$), we must multiply the energy in kilograms by two powers of the spacetime conversion factor $c = 2.998 \times 10^8$ m/s to get the units to come out right. Therefore, equation R8.24 in SI units would read

$$E_{[SI]} \approx mc^2 + \tfrac{1}{2}m|\vec{v}|^2 \qquad \text{when } |\vec{v}|^2 << c^2 \tag{R8.27}$$

When the particle is at rest, its relativistic **rest energy** in SI units is

(The most famous equation of special relativity)

$$E_{\text{rest}[SI]} = mc^2 \tag{R8.28}$$

This is the famous equation that has served as an icon representing both the essence of special relativity and Einstein's achievement. This equation really is just a special case of the more general equation R8.25. Even so, it does focus our attention on the startling new idea implicit in equation R8.25: *an object at rest has relativistic energy simply by virtue of its mass*, and this mass energy is a part of the total energy conserved by interactions inside an isolated system. Conservation of four-momentum does not require that a system's total mass and its total kinetic energy be separately conserved, only that their *sum* be conserved. Therefore, conservation of four-momentum opens the possibility that there may be processes that convert mass energy to kinetic energy, and vice versa. We will see in chapter R9 that such processes do indeed exist!

Note that a 1.0-kg object has a rest energy of $(1.0\text{ kg})c^2 = 9.0 \times 10^{16}$ J, which an *enormous* amount of energy (larger than what most nuclear bombs release). Our safety therefore depends on the fact that everyday physical processes do *not* convert much mass energy into other forms of energy!

It is often helpful to split four-momentum into space and time components

So, just as special relativity teaches us that time and space are but different aspects of spacetime, we now see that energy and momentum are but different aspects of four-momentum. The fact remains, however, that just as we experience time and space very differently, so we experience energy and momentum differently. It is often convenient, therefore, to split a particle's four-momentum into time and space components.

We have seen (see equation R8.11) that at low velocities the spatial components of an object's four-momentum vector p_x, p_y, and p_z become approximately equal to the components of that object's Newtonian momentum vector $\vec{p}$. Thus, they represent as close a relativistic analog to $\vec{p}$ as we have. We therefore call the three-dimensional vector $\vec{p} \equiv [p_x, p_y, p_z]$ the object's **relativistic three-momentum** (to distinguish it from its four-momentum). More often, we are interested in the simple magnitude of these three components:

Definition of relativistic momentum

$$|\vec{p}| \equiv \sqrt{p_x^2 + p_y^2 + p_z^2} \tag{R8.29}$$

which we will call the "magnitude of the object's relativistic momentum" as opposed to m, which is the magnitude $|\boldsymbol{p}|$ of the object's four-momentum. The relativistic momentum magnitude $|\vec{p}|$ is thus the relativistic generalization of the magnitude $|\vec{p}|$ of an object's Newtonian momentum vector.

With the help of equations R8.10, we can express an object's relativistic momentum magnitude in terms of its mass m and speed $|\vec{v}|$ as follows:

$$|\vec{p}| = \sqrt{\left(\frac{mv_x}{\sqrt{1-|\vec{v}|^2}}\right)^2 + \left(\frac{mv_y}{\sqrt{1-|\vec{v}|^2}}\right)^2 + \left(\frac{mv_z}{\sqrt{1-|\vec{v}|^2}}\right)^2}$$

$$= \frac{m\sqrt{v_x^2+v_y^2+v_z^2}}{\sqrt{1-|\vec{v}|^2}} = \frac{m|\vec{v}|}{\sqrt{1-|\vec{v}|^2}} \tag{R8.30}$$

Note that $|\vec{p}|$ does indeed become approximately equal to the magnitude $m|\vec{v}|$ of the object's Newtonian momentum $\vec{p}$ when $|\vec{v}| << 1$.

Useful relationships between E, $|\vec{p}|$, and m

We can write equation R8.17, which shows how an object's frame-independent mass can be computed by using the frame-dependent components of its four-momentum, in terms of E and $|\vec{p}|$ as follows:

$$m^2 = p_t^2 - (p_x^2 + p_y^2 + p_z^2) = E^2 - |\vec{p}|^2 \tag{R8.31}$$

We can also use an object's relativistic energy and relativistic momentum magnitude in a given frame to determine the object's speed in that frame:

$$\frac{|\vec{p}|}{E} = \frac{m|\vec{v}|/\sqrt{1-|\vec{v}|^2}}{m/\sqrt{1-|\vec{v}|^2}} = |\vec{v}| \tag{R8.32}$$

This relationship applies to each individual spatial component as well:

$$\frac{p_x}{E} = \frac{mv_x}{\sqrt{1-|\vec{v}|^2}}\frac{\sqrt{1-|\vec{v}|^2}}{m} = v_x, \quad \text{similarly } \frac{p_y}{E} = v_y, \quad \frac{p_z}{E} = v_z \tag{R8.33}$$

Indeed, we can think of the spatial components of an object's four-momentum as expressing the rate at which relativistic energy is transported through space:

$$\begin{bmatrix} p_t \\ p_x \\ p_y \\ p_z \end{bmatrix} = \begin{bmatrix} m/\sqrt{1-|\vec{v}|^2} \\ mv_x/\sqrt{1-|\vec{v}|^2} \\ mv_y/\sqrt{1-|\vec{v}|^2} \\ mv_z/\sqrt{1-|\vec{v}|^2} \end{bmatrix} = \begin{bmatrix} E \\ Ev_x \\ Ev_y \\ Ev_z \end{bmatrix} \tag{R8.34}$$

Equations R8.25, R8.26, R8.30, R8.31,and R8.32 express relationships between the quantities E, $|\vec{p}|$, m, K, and $|\vec{v}|$ that are really helpful when one is working with the four-momentum. Figure R8.4 gathers them in one convenient and memorable diagram.

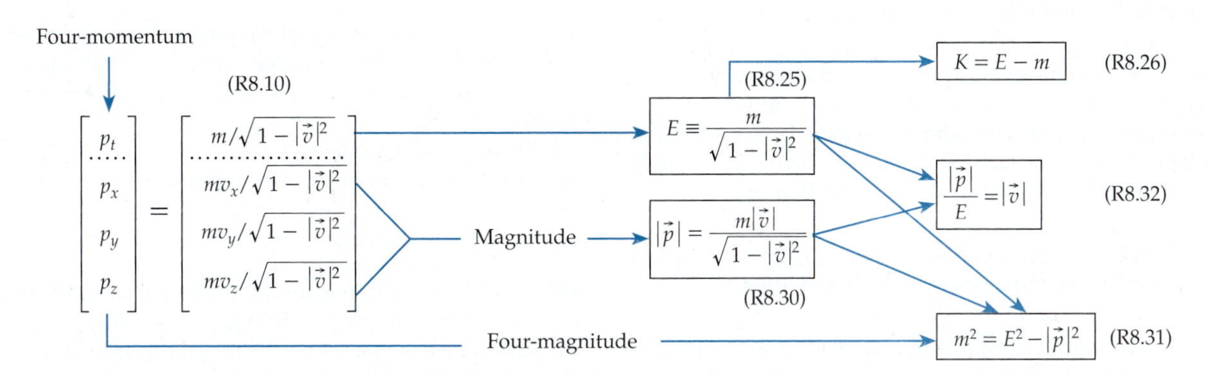

Figure R8.4
Virtually everything you need to know about four-momentum.

TWO-MINUTE PROBLEMS

R8T.1 A particle's Newtonian x-momentum $\not{p}_x$ is always either equal to or *smaller* than the x component of the particle's four-momentum p_x. T or F?

R8T.2 A particle's mass is always either equal to or smaller than the time-component of its four-momentum. T or F?

R8T.3 The absolute value of the x component of a particle's four-momentum vector is always either equal to or smaller than its t component. T or F?

R8T.4 **(a)** The squared magnitude of a four-vector could be negative. T or F? **(b)** The squared magnitude of a particle's four-momentum could be negative. T or F?

R8T.5 How does a particle's relativistic kinetic energy K compare to its Newtonian kinetic energy when the particle's speed $|\vec{v}| = \frac{4}{5}$?
A. $K < \frac{1}{2}m|\vec{v}|^2$
B. $K > \frac{1}{2}m|\vec{v}|^2$
C. $K = \frac{1}{2}m|\vec{v}|^2$

R8T.6 Consider a particle with mass m. As its speed approaches the speed of light, its relativistic momentum magnitude $|\vec{p}|$ approaches:
A. infinity
B. mc
C. m
D. zero

R8T.7 A particle's mass is the same in all inertial reference frames (even though its speed is not). T or F?

R8T.8 The components of a particle's four-momentum are the same in all inertial reference frames. T or F?

R8T.9 Suppose a particle's four-momentum vector has components $[p_t, p_x, p_y, p_z] = [5\text{ kg}, 3\text{ kg}, 0, 0]$. What is this particle's mass?
A. 5 kg
B. 8 kg
C. 4 kg
D. 3 kg
E. $\sqrt{34}$ kg
F. Other (specify)

R8T.10 A dust particle has a rest energy (mass) of 1.0 μg. In joules, this rest energy is closer to:
A. 10^{17} J
B. 10^{11} J
C. 10^{8} J
D. 10^{-3} J
E. 10^{-23} J
F. Other (specify)

R8T.11 As a particle's speed approaches that of light, the difference between its relativistic momentum magnitude $|\vec{p}|$ and its relativistic energy E becomes small (compared to E). T or F?

R8T.12 Particle A has mass m and speed $|\vec{v}|$. Particle B has mass $\frac{1}{3}m$ and speed $3|\vec{v}|$. How do the magnitudes of their relativistic momenta compare?
A. $|\vec{p}_A| > |\vec{p}_B|$
B. $|\vec{p}_A| < |\vec{p}_B|$
C. $|\vec{p}_A| = |\vec{p}_B|$

HOMEWORK PROBLEMS

Basic Skills

R8B.1 How fast must a particle move so that its total energy is 1% larger than its rest energy?

R8B.2 Two freight trains approach each other on the same track. Each has a mass of 10,000 metric tons (1 metric ton = 1000 kg) and is traveling at 30 m/s. How much total kinetic energy does each bring into the horrific train wreck that follows? Express your answer in milligrams.

R8B.3 A 2.0-kg object moves with $v_x = \frac{3}{5}$, $v_y = v_z = 0$. Find the components of the object's four-momentum.

R8B.4 An alien spaceship with a mass of 12,000 kg is traveling in the solar system frame at a speed of $\frac{4}{5}$ in the $-z$ direction. Find the components of the spaceship's four-momentum vector.

R8B.5 A 12-kg rock moves with a velocity whose components are $v_x = \frac{4}{15}$, $v_y = -\frac{3}{13}$, and $v_z = 0$ in a certain frame. Find the components of its four-momentum in that frame.

R8B.6 Suppose that in a certain reference frame we observe an object of mass 5.0 kg to have velocity components of $v_x = -0.866$ and $v_y = v_z = 0$. Evaluate the following in that reference frame:
(a) the object's total energy E,
(b) its relativistic momentum magnitude $|\vec{p}|$,
(c) the spatial components of its four-momentum, and
(d) its relativistic kinetic energy K.

R8B.7 In a certain frame, an object has a four-momentum whose components are $p_t = 5.0$ kg, $p_x = 4.0$ kg, $p_y = p_z = 0$.
(a) Calculate the object's x-velocity (in this frame).
(b) Calculate the object's mass.
(c) Calculate its kinetic energy.
(d) Calculate its relativistic momentum magnitude.

R8B.8 In a certain frame, an object has a four-momentum whose components are $p_t = 13$ kg, $p_x = -12$ kg, $p_y = p_z = 0$.
(a) Calculate the object's x-velocity in this frame.
(b) Calculate the object's mass.
(c) Calculate its kinetic energy.

R8B.9 In a certain inertial frame, an object has a four-momentum whose components are $p_t = 18$ kg, $p_x = 9.0$ kg, $p_y = 15$ kg, $p_z = 1.0$ kg.
(a) Find the object's velocity (vector) in this frame.
(b) Find the object's speed.
(c) Find the object's mass.
(d) Find the object's relativistic momentum magnitude.
(e) Find the object's kinetic energy.

R8B.10 In a given frame, an object's four-momentum has components $p_t = 5.0$ kg, $p_x = 3.0$ kg, $p_y = p_z = 0$. What is its four-momentum in an inertial frame that moves in the $+x$ direction at a speed of $\frac{3}{5}$ relative to the given frame?

R8B.11 In a given frame, an object's four-momentum has components $p_t = 5.0$ kg, $p_x = -3.0$ kg, $p_y = p_z = 0$. What is its four-momentum in an inertial frame that moves in the $+x$ direction at a speed of $\frac{3}{5}$ relative to the given frame?

Modeling

R8M.1 While solving a problem, a classmates claims that a particle's four-momentum vector has components $p_t = -0.10$ kg, $p_x = \frac{2}{5}$, $p_y = -\frac{1}{5}$ and $p_z = 0$. Even without knowing the problem's particulars, there are at least three things wrong with this four-vector as stated. What are they?

R8M.2 Suppose a dust particle of mass 2.0 μg is traveling at a speed of $\frac{4}{5}$ in the xy plane at an angle of 30° clockwise from the x axis in a certain inertial reference frame. Evaluate the following in that frame:
(a) the particle's relativistic energy E,
(b) its relativistic momentum magnitude $|\vec{p}|$,
(c) its three spatial four-momentum components, and
(d) its kinetic energy K (in joules).

R8M.3 A particle of mass m is moving in the $+z$ direction with a kinetic energy equal to $\frac{2}{3}m$. What are the components of its four-momentum, in terms of m?

R8M.4 A particle of mass m has a kinetic energy equal to $\frac{8}{5}m$. Its x-momentum is positive and is $-\frac{3}{4}$ times the value of its y-momentum, and its z-momentum is zero. What are the components of its four-momentum in terms of m?

R8M.5 If electrical energy can be sold at about \$0.06 per 10^6 J (the approximate current price in southern California), compute how much your rest energy is worth in dollars. That is, find the amount of money your survivors could put in your memorial fund if there was a way to convert your mass entirely to electrical energy. (It is probably a good thing that this is not easy to do.)

R8M.6 If electrical energy costs about \$0.06 per 10^6 J (the approximate current price in southern California) and you have \$1.5 million at your disposal to spend on energy to convert to kinetic energy, about how fast can you make a 1.0-g object travel?

R8M.7 In a number of problems in this text, we have blithely spoken about trains traveling at significant fractions of the speed of light. If electrical energy costs about \$0.06 per 10^6 J (the approximate current price in southern California), what would it cost to accelerate an electric train with a mass of about 100,000 kg to a speed of $\frac{3}{5}$?

R8M.8 Consider again the one-dimensional elastic collision discussed in section R8.2. In this collision, an object of mass m moving at $\frac{3}{5}$ in the $+x$ direction hits an object of mass $2m$ at rest. It turns out that conservation of *four-momentum* for this elastic collision implies that the x-velocities of the objects after the collision are $v_{3x} = -9/41$ and $v_{4x} = +39/89$ (instead of the Newtonian predictions of $-\frac{1}{5}$ and $+\frac{2}{5}$, respectively) in the laboratory frame.
(a) Show that if the objects have these final x-velocities, both the total t-momentum and the total x-momentum are indeed conserved in the laboratory frame.
(b) Use the Einstein velocity transformation equation to find the velocities in the frame in which the lighter object is initially at rest, and show that the t' and x' components of the system's four-momentum are conserved in this frame also.

R8M.9 Suppose that in a certain particle physics experiment, a particle of mass m_0 moving with speed $|\vec{v}_0| = \frac{3}{5}$ in the $+x$ direction in the laboratory frame is observed to decay into two particles, one with mass m_1 moving with speed $|\vec{v}_1| = \frac{4}{5}$ in the $+x$ direction and another with mass m_2 that is essentially at rest.
(a) Show that if total particle mass and Newtonian momentum are conserved in the laboratory frame, then we must have $m_1 = \frac{3}{4}m_0$ and $m_2 = \frac{1}{4}m_0$.
(b) Show that if four-momentum is to be conserved in the laboratory frame, we must have $m_1 = \frac{9}{16}m_0$ and $m_2 = \frac{5}{16}m_0$ (note that this decay converts some mass energy to kinetic energy if conservation of four-momentum is true).
(c) Consider now an inertial reference frame (the Other Frame) that moves with the same speed in the $+x$ direction as the initial particle. Use the Einstein velocity transformation equations to show that the particles' x-velocities in this frame must be $v'_{0x} = 0$, $v'_{1x} = \frac{5}{13}$, and $v'_{2x} = -\frac{3}{5}$.
(d) Show that if we use the outgoing particle masses that ensure that Newtonian momentum and particle mass are conserved in the laboratory frame [see part (a)], then Newtonian momentum is *not* conserved in the Other Frame.
(e) Show that if we use the masses that ensure that four-momentum is conserved in the laboratory frame [see part (b)], we find that four-momentum *is* conserved in the Other Frame.

R8M.10 Note that in SR units, energy, momentum, and mass all have the *same* units. When working with subatomic particles, physicists commonly use units of MeV (where 1 MeV = 1.6×10^{-13} J) instead of kilograms for all of these quantities. Suppose that a pion (a subatomic particle with mass $m_\pi = 140$ MeV) and an initial momentum magnitude of $|\vec{p}_0| = 900$ MeV hits a proton (with mass $m_p = 938$ MeV) at rest. The colored arrows below show the proton's and pion's momentum vectors that (my source implied) were actually measured after such an interaction in a certain experiment. All momentum magnitudes are accurate to within about ±40 MeV and the angles to within about ±1°.

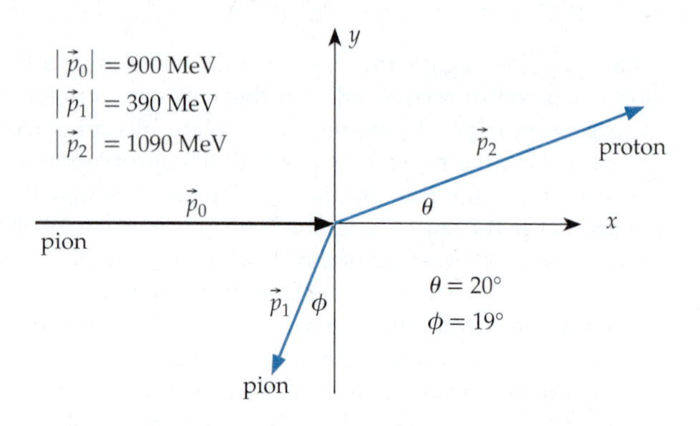

(a) Show that in this situation, we have $p_{1x} \approx -127$ MeV, $p_{1y} \approx -369$ MeV, $p_{2x} = 1024$ MeV, and $p_{2y} = 373$ MeV.

(b) Is spatial momentum conserved (considering the uncertainties in the numbers)?

(c) Note that according to quantum mechanics, a subatomic particle can have no internal energy, so in this collision, kinetic energy should be conserved. If we assume that the momenta measured here are Newtonian momenta, then is Newtonian kinetic energy $K = |\vec{p}|^2/2m$ conserved within our uncertainties?

(d) Assume that the momenta being measured here are actually relativistic momenta. If this is so, is relativistic energy $E = \sqrt{|\vec{p}|^2 + m^2}$ conserved in this collision (within our uncertainties)?

Derivations

R8D.1 Verify equation R8.17 by squaring the definitions of the four-momentum components given in equations R8.10 and combining as required.

R8D.2 Argue that as $|\vec{v}| \to 1$, $|\vec{p}| \to E$. What speed is required for these quantities to be equal to within 1% of E?

R8D.3 One can use the four-momentum transformation law to do a correct relativistic transformation of velocities. Consider an object of mass m moving at a speed of $|\vec{v}|$ in the $+x$ direction in the Home Frame. Using the Lorentz transformation equations, find the components of this object's four-momentum in an Other Frame moving in the $+x$ direction with an x-velocity of β relative to the Home Frame. Then, use equation R8.32 to find the object's speed in that Other Frame. Compare your result with the Einstein velocity transformation equations.

R8D.4 At nonrelativistic speeds ($|\vec{v}| \ll 1$), a particle's kinetic energy is $K \approx \frac{1}{2}m|\vec{v}|^2 = (m|\vec{v}|)^2/2m \approx |\vec{p}|^2/2m$, so we have $|\vec{p}| \approx (2mK)^{1/2}$. Show that the exact relativistic expression for the particle's relativistic momentum magnitude $|\vec{p}|$ in terms of its relativistic kinetic energy K is

$$|\vec{p}| = \sqrt{K(K + 2m)} \qquad \text{(R8.35)}$$

and argue that at low speeds, this reduces to $|\vec{p}| \approx (2mK)^{1/2}$.

R8D.5 Prove that the four-magnitude of any four-vector is a frame-independent number. (*Hint:* What is the formal definition of a four-vector?)

Rich-Context

R8R.1 In this text, we have blithely described people running at substantial fractions of the speed of light. Consider a runner who has somehow managed to accelerate to a speed of $\frac{3}{5}$. Even assuming that this highly trained athlete uses food energy very efficiently, describe the meal that the runner must have eaten before the race. (*Hint:* Remember that one food calorie = 1 Cal = 1000 cal = 4186 J.)

R8R.2 A typical household might use about 2×10^{10} J of electrical energy per year. About 1 in every 5000 hydrogen atoms in a quantity of water is actually deuterium, and the fusion of two deuterium nuclei converts about 0.5% of the rest energy of the deuterium nuclei to other forms of energy. Avogadro's number of hydrogen atoms has a mass of 1 g, the same number of deuterium atoms has a mass of 2 g, and the same number of oxygen atoms has a mass of 16 g. About how long could you run a typical household on the energy that would be produced by the fusion of the deuterium in 1 gal of water?

R8R.3 Cosmic rays are subatomic particles that have been accelerated (by obscure astrophysical processes) to large energies. Physicists have detected individual cosmic-ray protons with energies as high as several joules. Suppose a certain cosmic-ray proton has an energy of 1.5 J.

(a) How long would it take such a proton to travel across the diameter of our galaxy (about 100,000 ly) according to a clock traveling with the proton?

(b) Suppose a photon begins the journey at one end of the galaxy alongside such a proton. How far is it ahead of the proton at the end (in the galaxy's frame)?

[*Hint:* For $|\vec{v}| \approx 1$, $1 - |\vec{v}|^2 = (1 + |\vec{v}|)(1 - |\vec{v}|) \approx 2(1 - |\vec{v}|)$.]

Advanced

R8A.1 Prove conclusively that a particle's relativistic kinetic energy $K \equiv E - m$ is always larger than $\frac{1}{2}m|v|^2$ (although the values are close for $|\vec{v}| \ll 1$). (*Hint:* Both go to zero as $|\vec{v}|$ goes to zero, but you can show that the derivative of K with respect to $|\vec{v}|$ is always greater than the derivative of $\frac{1}{2}m|\vec{v}|^2$ with respect to $|\vec{v}|$. How does this help?)

ANSWERS TO EXERCISES

R8X.1 We have

$$v'_{4x} = \frac{v_{4x} - \beta}{1 - \beta_{4x}} = \frac{+\frac{2}{5} - \frac{3}{5}}{1 - (\frac{2}{5})(\frac{3}{5})} = -\frac{\frac{1}{5}}{\frac{19}{25}} = -\frac{5}{19} \quad \text{(R8.36)}$$

R8X.2 Using the same basic approach outlined in equation R8.13*a*, we get

$$p'_x = m\frac{dx'}{d\tau} = m\frac{\gamma(-\beta dt + dx)}{d\tau} = -\gamma\beta m\frac{dt}{d\tau} + \gamma m\frac{dx}{d\tau}$$
$$= -\gamma\beta p_t + \gamma p_x = \gamma(-\beta p_t + p_x) \quad \text{(R8.37)}$$

R8X.3 Substituting the definition of the four-momentum components into the definition of the four-magnitude of the four-momentum, we see that

$$|\boldsymbol{p}| \equiv \sqrt{p_t^2 - p_x^2 - p_y^2 - p_z^2}$$
$$= \sqrt{\left(m\frac{dt}{d\tau}\right)^2 - \left(m\frac{dx}{d\tau}\right)^2 - \left(m\frac{dy}{d\tau}\right)^2 - \left(m\frac{dz}{d\tau}\right)^2}$$
$$= m\frac{\sqrt{dt^2 - dx^2 - dy^2 - dz^2}}{d\tau} = m\frac{d\tau}{d\tau} = m \quad \text{(R8.38)}$$

(where in the next-to-last step, I used the provided hint).

R8X.4 We can prove conservation of p'_t as follows:

$$p'_{1t} + p'_{2t} - p'_{3t} - p'_{4t}$$
$$= \gamma(p_{1t} - \beta p_{1x}) + \gamma(p_{2t} - \beta p_{2x})$$
$$-\gamma(p_{3t} - \beta p_{3x}) - \gamma(p_{4t} + \beta p_{4x})$$
$$= \gamma(p_{1t} + p_{2t} - p_{3t} - p_{4t}) - \gamma\beta(p_{1x} + p_{2x} - p_{3x} - p_{4x})$$
$$= \gamma(0) - \gamma\beta(0) = 0 \quad \text{(R8.39)}$$

since $p_{1t} + p_{2t} - p_{3t} - p_{4t} = 0$ and $p_{1x} + p_{2x} - p_{3x} - p_{4x} = 0$ if four-momentum is conserved in the Home Frame. Proving conservation of p'_y is even easier:

$$p'_{1y} + p'_{2y} - p'_{3y} - p'_{4y} = p_{1y} + p_{2y} - p_{3y} - p_{4y} = 0 \quad \text{(R8.40)}$$

R9 Conservation of Four-Momentum

Chapter Overview

Introduction

Chapter R8 introduced the concept of conservation of four-momentum and showed that such a law could be fully consistent with the principle of relativity while reducing to the Newtonian form of the law at low speeds. In this chapter, we explore some of the surprising consequences and experimental tests of this law.

Section R9.1: Energy–Momentum Diagrams

An **energy–momentum diagram** represents the four-momenta of particles or systems as arrows on a diagram whose vertical and horizontal axes are E and p_x respectively. Such a diagram displays these four-momenta in the same way an ordinary spacetime diagram displays the coordinates of events.

A given object's four-momentum arrow on such a diagram has a slope of $1/v_x$ where v_x is the object's x-velocity. We will use a flag attached to the shaft of the arrow to indicate the arrow's four-magnitude (that is, its mass). We can qualitatively estimate the magnitude corresponding to a given four-momentum vector by following a hyperbola from the vector's tip back to the vertical axis. See figure R9.5 for a summary of how to read such a diagram.

Section R9.2: Solving Conservation Problems

One can qualitatively solve a conservation of four-momentum problem by using an energy–momentum diagram, or quantitatively by using four-dimensional column vectors. As an example of the latter, consider a collision between incoming particles

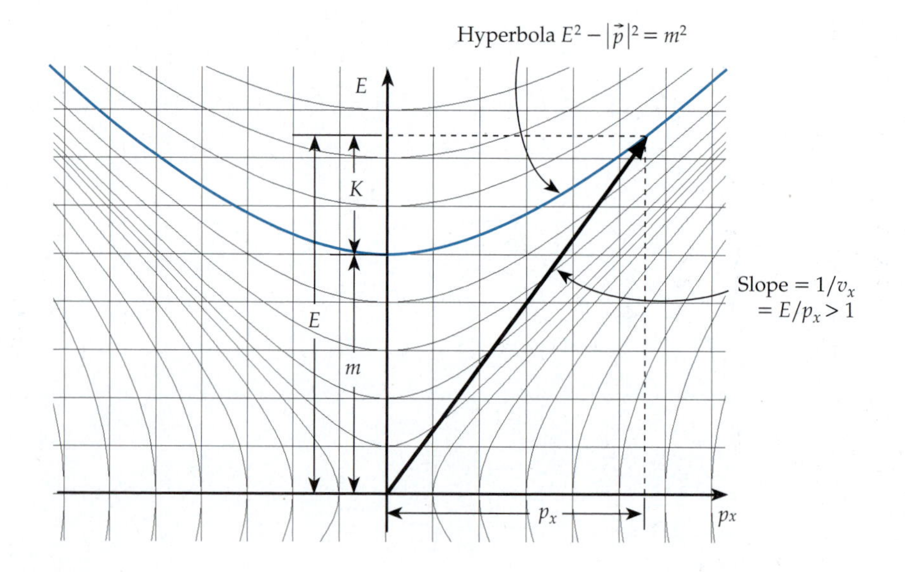

Figure R9.5
Virtually everything you need to know about four-momentum diagrams. No matter what the x-velocity of an object of mass m might be, the tip of the arrow representing its four-momentum lies on the hyperbola $E^2 - |\vec{p}|^2 = m^2$. The inverse slope of the arrow representing the four-momentum is equal to v_x, which always has a magnitude less than 1.

with four-momenta $\boldsymbol{p}_1$ and $\boldsymbol{p}_2$ that produces final particles with four-momenta $\boldsymbol{p}_3$ and $\boldsymbol{p}_4$. Conservation of four-momentum requires that

$$\begin{bmatrix} E_1 \\ p_{1x} \\ p_{1y} \\ p_{1z} \end{bmatrix} + \begin{bmatrix} E_2 \\ p_{2x} \\ p_{2y} \\ p_{2z} \end{bmatrix} = \begin{bmatrix} E_3 \\ p_{3x} \\ p_{3y} \\ p_{3z} \end{bmatrix} + \begin{bmatrix} E_4 \\ p_{4x} \\ p_{4y} \\ p_{4z} \end{bmatrix} \tag{R9.3}$$

As usual, each row of this column-vector equation must be satisfied independently.

This section discusses an example that illustrates both approaches. The diagrams can be easier than doing the algebra (especially if you use hyperbola graph paper), and often provide a useful visual check on algebraic solutions.

Section R9.3: The Mass of a System of Particles

In chapter R8, we defined an object's mass to be the frame-independent magnitude of its four-momentum. This definition implies, however, that *mass is not additive*: the mass of a system is *not* equal to the sum of the masses of its particles. The total mass of a system at rest is in fact equal to the sum of the *energies* of all its particles, which (for noninteracting particles at least) is usually larger than the sum of the particle masses. The extra mass does not really reside anywhere that we can locate: it is rather a property of the system as a whole.

In a collision between two balls of putty of mass m that stick together afterward, for example, the mass of the final putty glob is $M > 2m$. However, the *system's* mass is M both before and after the putty balls collide!

Section R9.4: The Four-Momentum of Light

A sufficiently short and spatially confined flash of light is the approximate equivalent of a particle. Such a light flash carries energy and so must have a four-momentum vector. Its four-momentum arrow must be parallel to the light flash's worldline, so (1) the slope of the four-momentum arrow must be ± 1 on an energy–momentum diagram, which means that (2) the relativistic momentum $|\vec{p}|$ of a flash must be $|\vec{p}| = E$, which means that (3) the flash's mass m is $m = (E^2 - |\vec{p}|^2)^{1/2} = 0$! To put it another way, a flash of light has all its energy in the form of kinetic energy and none in the form of mass energy.

An antimatter–matter rocket would convert the mass of its fuel entirely to light that travels out the engine's nozzle with speed c. Such an engine would be the best possible rocket engine. Example R9.3 discusses such a rocket as an illustration of doing a conservation of four-momentum problem involving light.

Section R9.5: Applications to Particle Physics

Most real applications and tests of relativity involve subatomic particles, because such particles are about the only things that are light enough to be accelerated to near-light speeds without a prohibitively high energy budget. Particle physicists typically use the electronvolt (eV) as the unit of mass, energy, and momentum, where $1\text{ eV} = 1.602 \times 10^{-19}\text{ J} = 1.782 \times 10^{-36}\text{ kg}$. This section discusses the decay of a particle called a *kaon* as an example of a realistic application of relativity.

Section R9.6: Parting Comments

This section discusses a few places where interested readers can go to learn more about relativity.

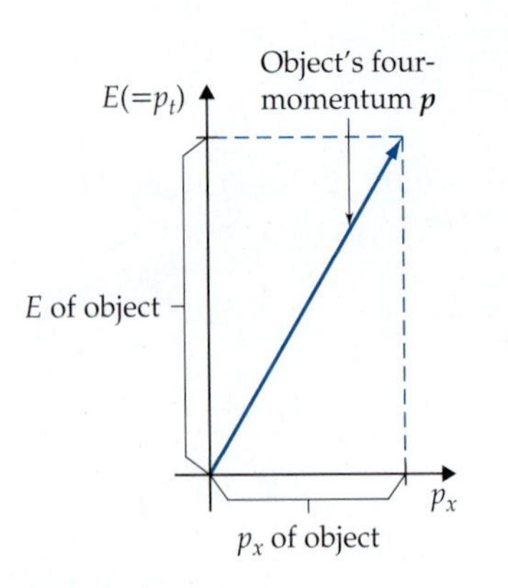

Figure R9.1
An energy–momentum diagram displaying the four-momentum of an object moving in the $+x$ direction. The object's four-momentum is shown on the diagram as an arrow whose projections on the vertical and horizontal axes represent the values of the object's relativistic energy $E = p_t$ and its relativistic x-momentum p_x, respectively.

R9.1 Energy–Momentum Diagrams

We can visually represent an object's four-momentum as an arrow on a special kind of spacetime diagram called an **energy–momentum diagram** (see figure R9.1). We can do this conveniently only if the object is moving in the spatial x direction (so that $p_y = p_z = 0$ and $|\vec{p}| = |p_x|$). In the rest of the chapter, we will assume that this is true (unless otherwise specified).

Just as the direction of the arrow representing an object's ordinary momentum is tangent to its path through space, the direction of the arrow representing an object's four-momentum is tangent to its worldline in spacetime (because the object's four-momentum vector $\boldsymbol{p} = m\,d\boldsymbol{s}/d\tau$ at any given point along its worldline is proportional to the object's differential displacement in spacetime $d\boldsymbol{s}$ along that worldline around that point; see figure R9.2).

Since the inverse slope of an object's worldline at any instant is equal to its x-velocity at that instant, the inverse slope of the object's four-momentum arrow at a given time (i.e., run/rise $= p_x/E$) should also be equal to its x-velocity at that time if the two vectors are to be parallel. Equation R8.33 says essentially the same thing:

$$\frac{p_x}{E} = \frac{mv_x}{\sqrt{1-|\vec{v}|^2}}\frac{\sqrt{1-|\vec{v}|^2}}{m} = v_x \tag{R9.1}$$

The four-magnitude of an object's four-momentum is its mass m (see equation R8.17): this value is frame-independent and independent of the object's motion. But the *length* of the arrow that represents the object's four-momentum on an energy–momentum diagram *does* depend on the object's

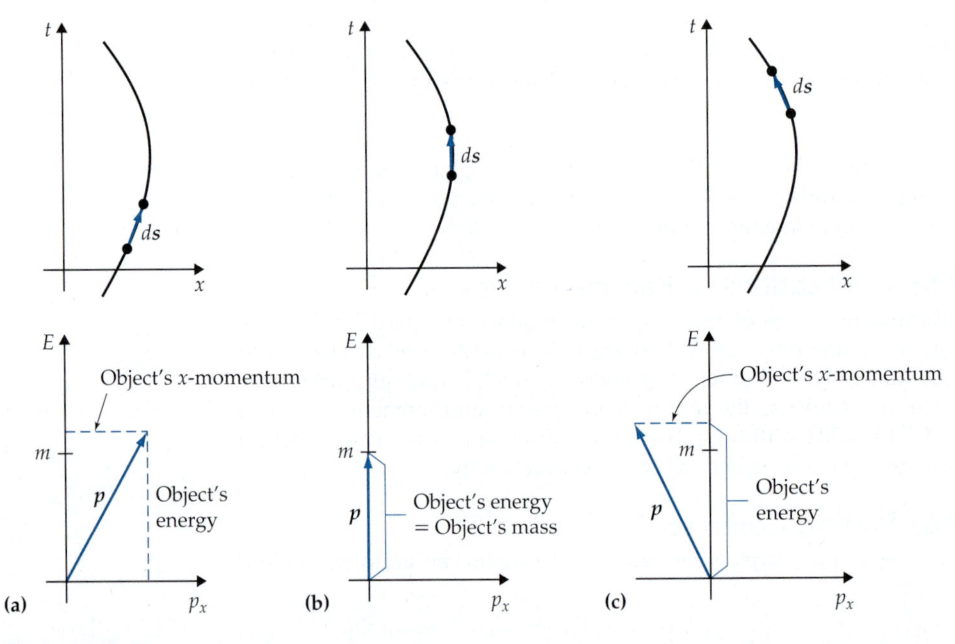

Figure R9.2
(a) At any given time, the arrow that represents an object's four-momentum on an energy–momentum diagram points in a direction tangent to the object's worldline, because $\boldsymbol{p}$ is proportional to $d\boldsymbol{s}$. (b) When an object is at rest (even at just an instant), its four-momentum arrow is vertical and its energy is equal to its mass (see equation R8.10*a*). (c) When the object moves in the $-x$ direction, its x-momentum is negative (see equation R8.10*b*), but its energy remains positive.

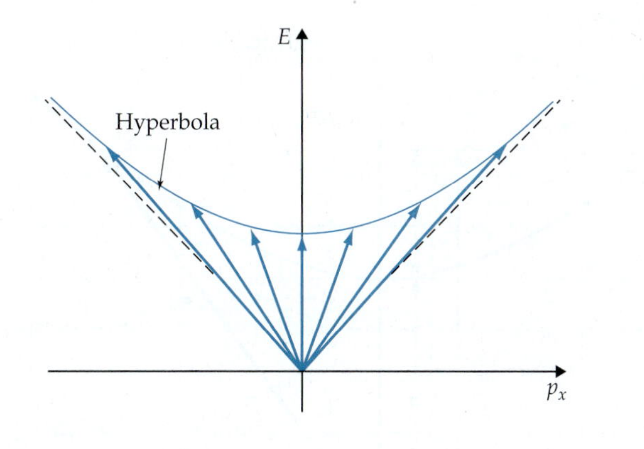

Figure R9.3
An energy-momentum diagram showing four-momentum arrows for a set of identical objects of mass m moving at different x-velocities in the Home Frame. The tips of all these arrows lie on a hyperbola whose equation is $m^2 = E^2 - |\vec{p}|^2$. Note that as the object's x-velocity approaches ± 1 (and thus $|\vec{p}|/E$ approaches 1), both $|\vec{p}|$ and E must become very large if the difference of their squares is to remain fixed.

velocity: it is *not* proportional to the four-magnitude of the corresponding four-momentum. (This is similar to the problem with the spacetime interval discussed in section R3.6.) According to equation R8.31, we in fact have

$$m^2 = E^2 - |\vec{p}|^2 = E^2 - (p_x)^2 \tag{R9.2}$$

This means that the tips of the four-momentum arrows for objects of identical mass m traveling at different x-velocities (or the four-momentum arrows for a single accelerating object observed at different times) lie along the hyperbola on the diagram defined by the equation $m^2 = E^2 - |\vec{p}|^2$ as shown in figure R9.3.

If you know an object's x-velocity and its mass m, you can easily draw an energy–momentum diagram showing its four-momentum vector as follows:

How to draw the four-momentum for an object with known mass and speed

1. Set up E and p_x axes (with the E axis vertical).
2. Draw a line from the origin of those axes having the slope $1/v_x$.
3. Compute the value of $E = m/(1-|\vec{v}|^2)^{1/2} = m/(1-v_x^2)^{1/2}$ for the object.
4. Draw a horizontal line from this value on the E axis until it intersects the line you drew in step 2.
5. The arrow representing the object's four-momentum lies along the line drawn in step 2, with its tip at the intersection found in step 4.

Alternatively, if you have access to hyperbola graph paper (which you can copy from the end of chapter R5 or download from the *Six Ideas* website), then you can replace the middle steps in the list above with the following:

3. Find the hyperbola corresponding to the object's mass.
4. The object's four-momentum arrow along the line drawn in step 2, with its tip where that line intersects the hyperbola identified in step 3.

We can easily read an object's relativistic kinetic energy $K = E - m$ directly from an energy–momentum diagram. For example, K for an object of mass m moving at a speed $|\vec{v}| = \frac{3}{5}$ is $\frac{1}{4}m$, as shown in figure R9.4. (Note that $K \neq \frac{1}{2}m(\frac{3}{5})^2 = \frac{18}{100}m$, which is substantially smaller!) Figures R9.3 and R9.4 together clearly show that as an object's speed $|\vec{v}|$ approaches 1, both the object's total relativistic energy E and its kinetic energy K go to infinity. This means *you would have to supply an infinite amount of energy to accelerate an object of nonzero mass to the speed of light*. (This is the most practical reason that no object can go faster than the speed of light: all the energy in the universe could not accelerate even a mote of dust to that speed!)

Virtually all you need to know to construct and interpret an energy–momentum diagram is summarized in figure R9.5 on the next page.

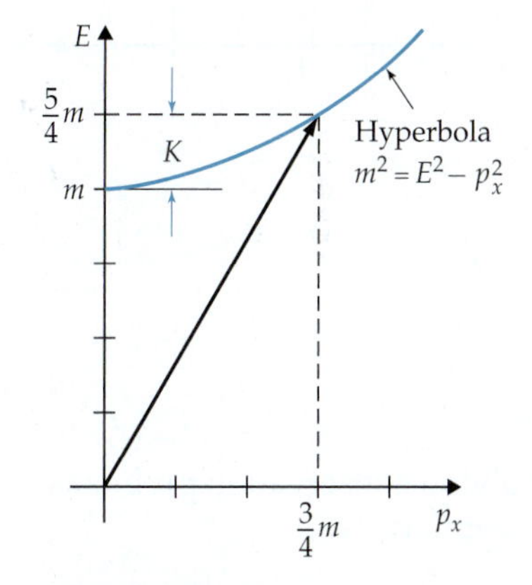

Figure R9.4
An energy–momentum diagram of an object of mass m traveling with an x-velocity $v_x = \frac{3}{5}$. The arrow representing the object's four-momentum has a slope of 5/3 and [since $(1-|\vec{v}|^2)^{1/2} = \frac{4}{5}$ in this case] an energy of $E = m(1-|\vec{v}|^2)^{-1/2} = \frac{5}{4}m$. For the arrow to have the correct slope, we must have $p_x = \frac{3}{4}m$. We can see that $K = E - m = \frac{1}{4}m$.

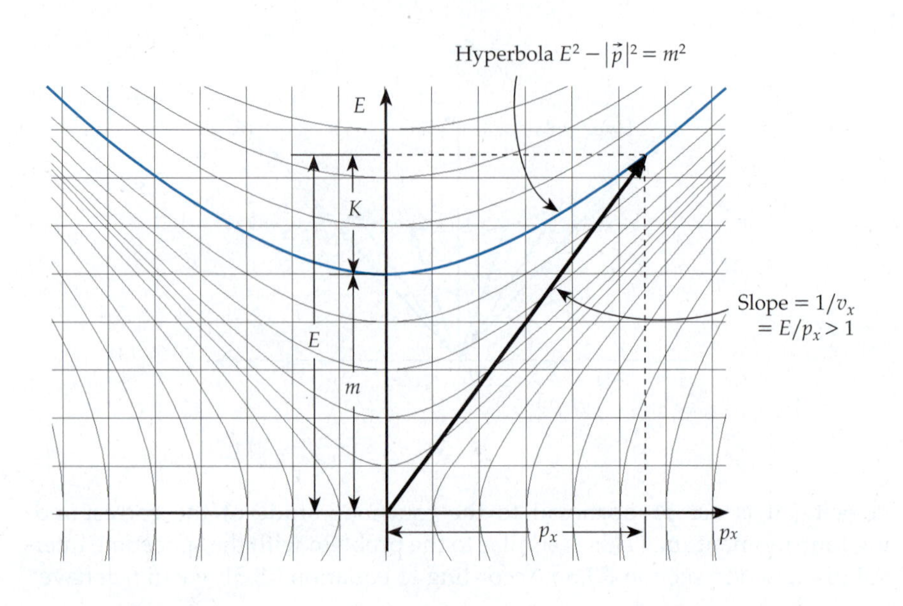

Figure R9.5
Virtually everything you need to know about four-momentum diagrams. No matter what the x-velocity of an object of mass m might be, the tip of the arrow representing its four-momentum lies on the hyperbola $E^2 - |\vec{p}|^2 = m^2$. The inverse slope of the arrow representing the four-momentum is equal to v_x, which always has a magnitude less than 1.

Exercise R9X.1

On the hyperbola graph paper to the left, draw an energy–momentum diagram of an object whose mass is 1.0 kg and which moves in the $-x$ direction at a speed of $\frac{3}{5}$. Read its relativistic energy E, its relativistic kinetic energy K, and the x component of its four-momentum p_x from the diagram.

R9.2 Solving Conservation Problems

The law of conservation of four-momentum (like the law of conservation of ordinary momentum) is most useful when applied to an isolated system of objects undergoing some type of *collision* process (that is, a kind of sudden interaction between the objects in the system that may be strong and complicated but limited in time). In such a case, the system has a clearly defined state "before" and "after" the collision, making it easy to compute the system's total four-momentum both before and after the collision. The law of conservation of four-momentum states that the system should have the *same* total four-momentum after the collision process as it had before.

Solving conservation problems algebraically

What does this really mean mathematically? Since four-momentum is a (four-dimensional) vector quantity, conservation of four-momentum means that *each component of the system's total four-momentum is separately conserved.* For example, consider a system consisting of two objects, and let the objects' four-momenta before the collision be $\boldsymbol{p}_1$ and $\boldsymbol{p}_2$, and after the collision be $\boldsymbol{p}_3$ and $\boldsymbol{p}_4$. Conservation of four-momentum then requires that

$$\begin{bmatrix} E_1 \\ p_{1x} \\ p_{1y} \\ p_{1z} \end{bmatrix} + \begin{bmatrix} E_2 \\ p_{2x} \\ p_{2y} \\ p_{2z} \end{bmatrix} + \begin{bmatrix} E_3 \\ p_{3x} \\ p_{3y} \\ p_{3z} \end{bmatrix} + \begin{bmatrix} E_4 \\ p_{4x} \\ p_{4y} \\ p_{4z} \end{bmatrix} \tag{R9.3}$$

remembering that the time component of a four-momentum vector (that is, the relativistic energy) is usually given the more evocative symbol E instead of p_t. As usual, each row of this equation must be *separately* true for four-momentum to be conserved.

Example R9.1

Problem: Suppose that somewhere in deep space a rock with mass $m_1 = 12$ kg is moving in the $+x$ direction with $v_{1x} = +\frac{4}{5}$ in some inertial frame. This rock then strikes another rock of mass $m_2 = 28$ kg at rest ($v_{2x} = 0$). Pretend that the first rock, instead of instantly vaporizing into a cloud of gas (as any *real* rocks colliding at this speed would), simply bounces off the more massive rock and is subsequently observed to have an x-velocity $v_{3x} = -\frac{5}{13}$. What is the larger rock's x-velocity v_{4x} after the collision?

Solution The first step in solving this problem is to calculate the energy E_1 and the x-momentum p_{1x} of the smaller rock before the collision. Using the definitions of these four-momentum components, we find that

$$E_1 \equiv \frac{m_1}{\sqrt{1-v_{1x}^2}} = \frac{m_1}{\sqrt{1-\frac{16}{25}}} = \frac{m_1}{\sqrt{\frac{9}{25}}} = \frac{5}{3}(12\text{ kg}) = 20\text{ kg} \quad \text{(R9.4}a\text{)}$$

$$p_{1x} \equiv \frac{m_1 v_{1x}}{\sqrt{1-v_{1x}^2}} = \frac{m_1(+\frac{4}{5})}{\frac{3}{5}} = +\frac{4}{3}(12\text{ kg}) = +16\text{ kg} \quad \text{(R9.4}b\text{)}$$

Similarly, the larger rock's energy and x-momentum before the collision are

$$E_2 \equiv \frac{m_2}{\sqrt{1-v_{2x}^2}} = \frac{m_2}{\sqrt{1-0^2}} = m_2 = 28\text{ kg} \quad \text{(R9.5}a\text{)}$$

$$p_{2x} \equiv \frac{m_2 v_{2x}}{\sqrt{1-v_{2x}^2}} = \frac{m_2(0)}{\sqrt{1-0^2}} = 0\text{ kg} \quad \text{(R9.5}b\text{)}$$

The smaller rock's energy and x-momentum *after* the collision are

$$E_3 \equiv \frac{m_1}{\sqrt{1-v_{3x}^2}} = \frac{m_1}{\sqrt{1-(-\frac{5}{13})^2}} = \frac{m_1}{\sqrt{\frac{144}{169}}} = \frac{13}{12}(12\text{ kg}) = 13\text{ kg} \quad \text{(R9.6}a\text{)}$$

$$p_{3x} \equiv \frac{m_1 v_{3x}}{\sqrt{1-v_{3x}^2}} = \frac{m_1(-\frac{5}{13})}{\frac{12}{13}} = -\frac{5}{12}(12\text{ kg}) = -5\text{ kg} \quad \text{(R9.6}b\text{)}$$

Conservation of four-momentum requires that the four-momentum vectors before the collision add up to the same value after the collision, meaning that $\boldsymbol{p}_1 + \boldsymbol{p}_2 = \boldsymbol{p}_3 + \boldsymbol{p}_4$ or

$$\boldsymbol{p}_4 = \boldsymbol{p}_1 + \boldsymbol{p}_2 - \boldsymbol{p}_3 = \begin{bmatrix} 20\text{ kg} \\ 16\text{ kg} \\ 0 \\ 0 \end{bmatrix} + \begin{bmatrix} 28\text{ kg} \\ 0 \\ 0 \\ 0 \end{bmatrix} - \begin{bmatrix} 13\text{ kg} \\ -5\text{ kg} \\ 0 \\ 0 \end{bmatrix} = \begin{bmatrix} 35\text{ kg} \\ 21\text{ kg} \\ 0 \\ 0 \end{bmatrix} \quad \text{(R9.7)}$$

Knowing the energy and x-momentum of an object is sufficient to determine both its energy and x-velocity. Using equation R8.31, we see that the larger rock's mass is still

$$m = \sqrt{E_4^2 - p_{4x}^2} = \sqrt{(35\text{ kg})^2 - (21\text{ kg})^2} = (7\text{ kg})\sqrt{5^2 - 3^2} = 28\text{ kg} \quad \text{(R9.8}a\text{)}$$

after the collision. (Since energy is not being transformed to other forms, this is an elastic collision.) According to equation R8.33, its final x-velocity is

$$v_{4x} = \frac{p_{4x}}{E_4} = \frac{+21\text{ kg}}{35\text{ kg}} = +\frac{3}{5} \quad \text{(R9.8}b\text{)}$$

This completes the solution.

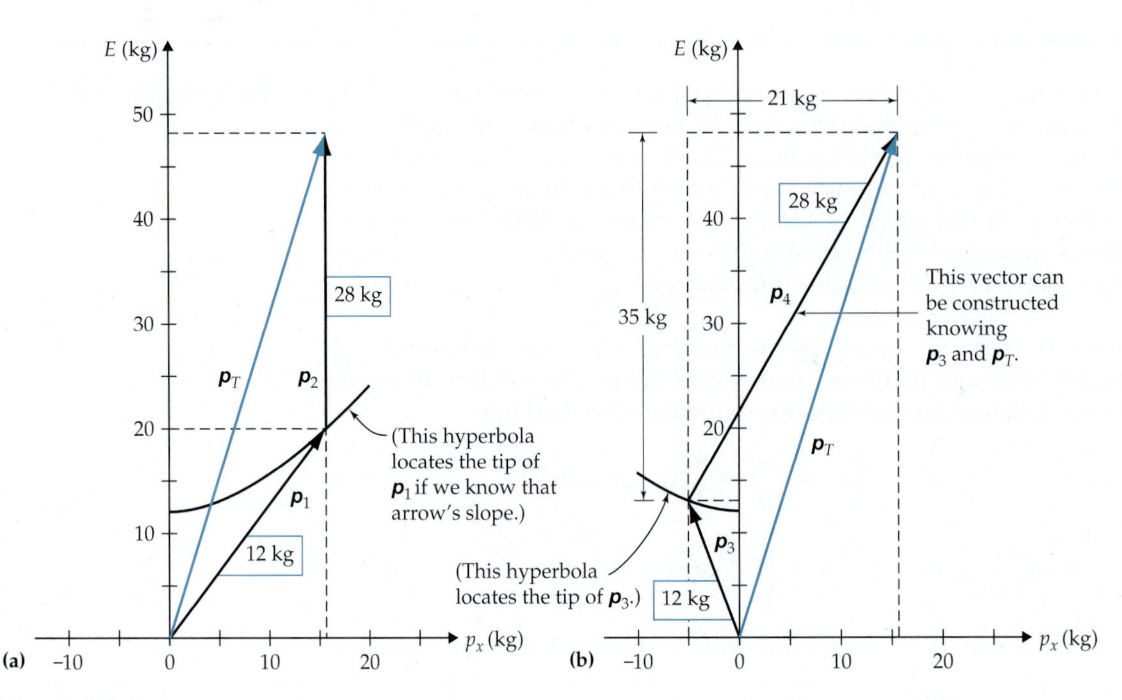

Figure R9.6
(a) The four-momenta of the rocks *before* the collision. The vector sum of these four-momenta is represented by the arrow $\boldsymbol{p}_T$. Since the four-magnitudes of the individual four-momentum arrows (which equal the masses of the corresponding rocks) cannot be read directly from the diagram, I have adopted the expedient of attaching a "flag" to each four-momentum arrow that states its magnitude. (b) The four-momenta of the rocks *after* the collision. The vector sum of these four-momenta is still $\boldsymbol{p}_T$ by conservation of four-momentum. Since we know $\boldsymbol{p}_3$, we can construct the unknown four-momentum arrow for $\boldsymbol{p}_4$, read its components from the diagram as shown, and use these to compute the corresponding rock's mass and x-velocity.

Example R9.2

Problem: Solve the rock collision problem discussed in example R9.1, using an energy–momentum diagram. (Remember that a rock with mass $m_1 = 12$ kg moving with $v_{1x} = \frac{3}{5}$ hit a rock with mass 28 kg at rest. After the collision, the first rock, whose mass is unchanged, moves with x-velocity $v_{3x} = -\frac{5}{13}$.)

Solution Since the sum of four-momentum arrows is defined as the sum of ordinary vectors (we simply add the components), we can add four-momenta arrows on an energy–momentum diagram just as we would ordinary vector arrows (by putting the tail of one vector on the tip of the other while preserving their directions). Using this technique, we see in figure R9.6a that in the rock example, the system's total four-momentum *before* the collision has components $E_T = 48$ kg and $P_{Tx} = 16$ kg. The two rocks' four-momentum arrows after the collision have to add up to the *same* total four-momentum arrow; and since we know the smaller rock's four-momentum after the collision, we can *construct* the larger rock's final four-momentum arrow $\boldsymbol{p}_4$ (figure R9.6b). We can then read the components of this arrow right off the diagram, getting the same results as in equations R9.7. One can then use these results (as we did before) to compute that rock's mass and x-velocity.

Advantages and disadvantages of the graphical method

This graphical approach to the problem can be somewhat easier than the algebraic approach if one has access to hyperbola graph paper (which makes calculating the object's energies unnecessary). However, even if one lacks hyperbola graph paper, drawing such a diagram does have some advantages: (1) It provides a more visual and concrete way of dealing with the problem, and may be helpful to you if you find the algebraic approach rather abstract. (2) When used in conjunction with the algebraic method, it serves as a useful check on the algebraic results: it is harder to make errors while using the graphical method. (3) In some cases, as we will see, simply *looking* at the diagram can yield qualitative information that is very difficult to get from the algebraic equations alone.

In short, the graphical method represents an alternative method for solving problems involving conservation of four-momentum that often complements the algebraic approach. Armed with both of these techniques, we are now ready to explore some of the strange and interesting consequences of the law of conservation of four-momentum.

R9.3 The Mass of a System of Particles

Conversion of rest energy (mass) into other kinds of energy (or vice versa) is possible

As we have seen, an object's relativistic energy is not simply equal to its kinetic energy (even at low velocities) but involves the object's mass as well. The fact that $E = p_t$ is conserved by an isolated system's internal interactions does not imply that the object's mass and its kinetic energy are separately conserved—only that their *sum* is conserved. In special relativity, mass and kinetic energy are simply two aspects of the same whole (the relativistic energy). We have no reason to presuppose a barrier between these two aspects of relativistic energy that would preclude the conversion of one to the other.

In much of the remainder of this chapter, we consider examples of processes that do just that. We will begin with a simple example that illustrates something we must understand about "mass" before we can go on: *a system's mass is not the same as the sum of the masses of its parts.*

An example showing conversion of kinetic energy to mass

Consider the collision of two identical balls of putty with mass $m = 4$ kg which in some inertial frame are observed to have x-velocities of $v_{1x} = +\frac{3}{5}$ and $v_{2x} = -\frac{3}{5}$; that is, these putty balls are approaching each other with equal speeds. Imagine that when these putty balls collide, they stick together, as shown in figure R9.7.

Note that before the collision, the x component of the system's total four-momentum is zero:

$$p_{1x} + p_{2x} = \frac{m(+\frac{3}{5})}{\sqrt{1-(\frac{3}{5})^2}} + \frac{m(-\frac{3}{5})}{\sqrt{1-(\frac{3}{5})^2}} = \frac{m(\frac{3}{5}-\frac{3}{5})}{\sqrt{1-(\frac{3}{5})^2}} = 0 \quad \text{(R9.9)}$$

so conservation of four-momentum implies that the x component of the final mass's four-momentum is zero as well, meaning that it must be at rest.

What of relativistic energy conservation in this case? A Newtonian analysis of this collision would speak of the kinetic energy being converted to thermal energy in this inelastic collision. Such an analysis would also assert that the mass of the coalesced particle is $M = m + m = 2m$. But we have more constraints to consider in a relativistic solution to this problem. If the spatial components of the four-momentum are conserved in this collision, the time

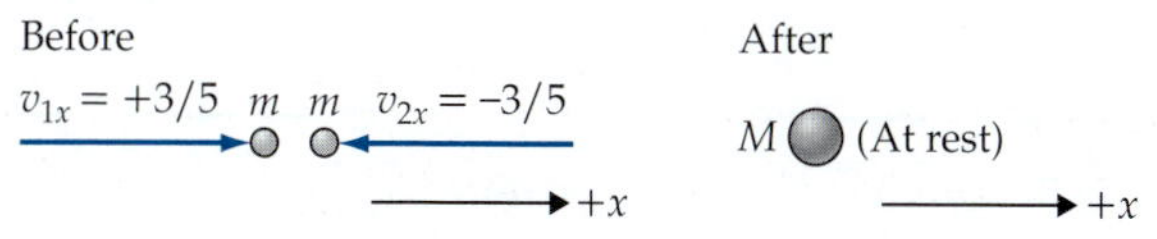

Before

$v_{1x} = +3/5$ m m $v_{2x} = -3/5$

$+x$

After

M (At rest)

$+x$

Figure R9.7
The inelastic collision of two balls of putty (each having mass $m = 4$ kg) as seen in the frame where they initially have equal speeds but opposite directions.

component must also be conserved, whether the collision is elastic or not. But how can we think of the relativistic energy being conserved in this case, since no mention has been made of thermal energy in the definition of the relativistic energy given in chapter R8?

The answer is direct and surprising. Since the final object is motionless, its relativistic energy is simply equal to its mass M. But by conservation of the time component of four-momentum, we have

$$M = E_1 + E_2 = \frac{m}{\sqrt{1-(\frac{3}{5})^2}} + \frac{m}{\sqrt{1-(\frac{3}{5})^2}} = \frac{2m}{\sqrt{\frac{16}{25}}} = \frac{10}{4}m = 10 \text{ kg} \qquad \text{(R9.10)}$$

which is *not* equal to $2m$ = 8 kg! Conservation of four-momentum thus requires that the final object have a *greater* mass than the sum of the masses that collided to form it!

We know from experience with collisions at low speeds that when two objects collide and stick together, their energy of motion gets converted to thermal energy, making the final object a little warmer than the original objects. (In this case, actually, the final object will be a *lot* warmer than the original objects, so much so that any *real* putty balls colliding at such speeds would vaporize instantly!) What equation R9.10 is telling us is that the final object *must* be more massive than the original objects, and that the increased thermal energy is somehow correlated with this.

The mass of a system is not the same as the sum of the masses of its parts

But where does this extra mass actually reside? The final object has the same number of atoms as the original objects did. Does each atom somehow gain some mass? This seems absurd. The final object's larger thermal energy means that its atoms will jostle around more vigorously. Can the motion of these atoms "have mass" in some sense? This seems crazy. *Individual* particles have the same mass no matter how they move. *So where is this extra mass?*

There is only one fully self-consistent answer to this question: *the extra mass belongs to the system as a whole;* it does not reside in any of its parts.

We can see this more vividly as follows. Consider the system consisting of the two balls of putty *before* they collide. If we consider them as separate objects, the putty balls *each* have a mass m of 4 kg and a relativistic energy of 5 kg, and one has an x-momentum of −3 kg and the other +3 kg. But if we consider the balls as a single system, the *system* has a total x-momentum of zero and a total energy of 10 kg, so the system's mass $M \equiv (E_T^2 - |\vec{p}_T|^2)^{1/2}$ is equal to 10 kg (see figure R9.8). So we see that the thermal energy produced by the collision is *not* the source of this extra mass: the extra mass was present in the "system" *before* the collision and remains the same after the collision.

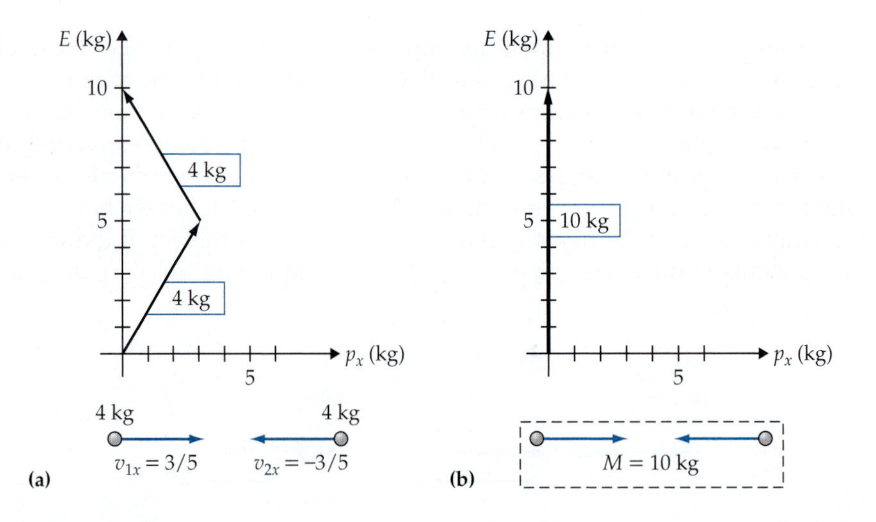

Figure R9.8
(a) Putty balls before the collision, considered as two individual objects. Each has its own mass, energy, and x-momentum.
(b) Putty balls before the collision, considered as a single system. The system's x-momentum is zero, meaning that its total energy of 10 kg is the same as the system's mass.

So mass is not "created" by the collision process at all: the collision simply *manifests* the system's mass in the mass of a *single* final object. If we focus on the masses of the individual objects in the system before and after the collision, we think of mass being created. But if we focus on the *system* before and after the collision, we see that its mass remains the same.

One can get unnecessarily hung up on the difference between the mass of a system and the masses of its parts. The reason this seems screwy is that we are *used* to treating mass as if it were additive: the mass of a jar of beans is the sum of the individual beans' masses plus the jar's mass, right? This is close enough if the beans travel at low velocities. But if we had common experience with a jar full of beans that bang around inside the jar with speeds close to that of light, then we would be *used* to the idea that such a jar would have a different mass than the sum of the individual beans' masses. Mass is simply *not* additive in the way that energy and x-momentum are.

There are actually many examples of things in the world where the whole is greater than the sum of its parts. The meaning of a poem is not the same as the sum of the meaning of the individual letters in its words. The "life" of an organism cannot be localized in any of its parts. We simply need to start thinking about mass in the same way that we think about such things.

A system's mass can't be localized: it is a property of the system as a whole

The best way to look at this is to think of the mass of a system of particles as *a property of the system as a whole* (that is, the magnitude of the system's total four-momentum vector) and something that simply doesn't have very much to do with the masses of its parts. The only self-consistent way to define the mass of a *system* is as the magnitude of the system's total four-momentum; and if this definition leads to the mass of a system being greater or less than the masses of its parts, well, that's the way it is!

Exercise R9X.2

Let's look at the system described in figure R9.8 in a frame moving with the left-hand ball. Use the Einstein velocity transformation to show that the other ball's x-velocity is $v'_{2x} = -\frac{15}{17}$. Find the *system's* four-momentum components E'_T and p'_{Tx} in this frame, and show that the system's mass [found using $M' = (E'^2_T - |\vec{p}'_T|^2)^{1/2}$] is still 10 kg. (A system's mass may not be equal to the sum of the masses of its parts, but it *is* frame-independent.)

R9.4 The Four-Momentum of Light

We all have experienced the fact that light carries energy: we have felt sunlight warm our skin, or seen an electric motor powered by solar cells. Since we have seen that energy is the time component of four-momentum, it follows that light should have an associated four-momentum vector. What does the four-momentum of light look like?

A light flash as a "particle" of light

Previously, we have explored the four-momenta of objects (rocks or putty balls or the like) that could be considered to be *particles* that have a well-defined position in space and thus a well-defined worldline through spacetime. The analogous thing in the case of light would be a "flash" or "burst" of light energy that is similarly localized in space. We can consider a continuous beam of light to be composed of a sequence of closely spaced flashes, much as we might imagine a stream of water to be a sequence of closely spaced drops. Quantum physics indeed teaches us that light actually is comprised of particles called photons, which we can consider to be tiny

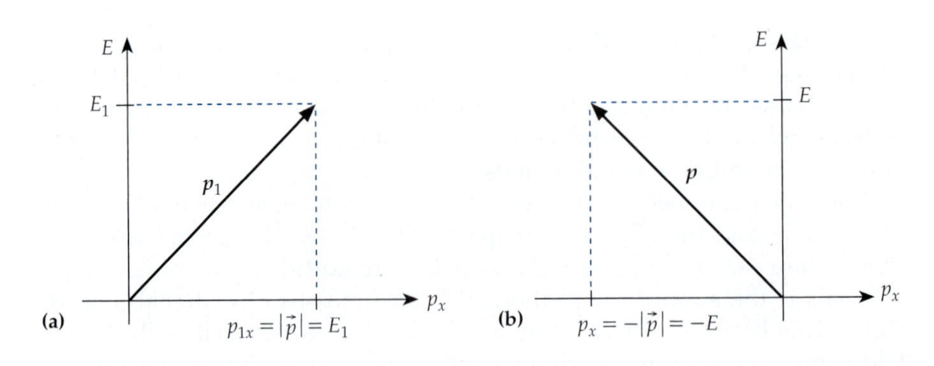

Figure R9.9
Energy–momentum diagrams showing the four-momentum arrows for light flashes moving in the (a) $+x$ and (b) $-x$ directions, respectively. In each case, for the four-momentum to be parallel to that flash's worldline, it must have slope ± 1 on the diagram. This implies that a light flash's relativistic momentum magnitude $|\vec{p}|$ must have the same value (in SR units) as its energy: $|\vec{p}| = E$.

flashes of light. Alternatively, we might model a flash of light as being one gigantic photon.

A light-flash's four-momentum vector has slope ± 1

So what might the four-momentum of a photon or a flash of light look like? Arguably, the most basic feature of any object's four-momentum is that it is parallel to that object's worldline. If this is true for a light flash, then it follows that the flash's four-momentum vector must have a slope of ± 1 on an energy–momentum diagram. The four-momentum of a flash with a given energy E moving in the $+x$ direction will thus look as shown in figure R9.9a. If a light flash is moving in the $-x$ direction instead of the $+x$ direction, the slope of its four-momentum arrow on an energy–momentum diagram is -1 instead of $+1$, and its x-momentum is negative, as shown in figure R9.9b.

You can see from figure R9.9 that if either flash's four-momentum vector is to have such a slope, it must have a relativistic momentum magnitude $|\vec{p}|$ that is equal to its relativistic energy E. This is in fact consistent with equation R8.32, which in the case of light tells us that

A light flash's relativistic momentum is equal to its energy

$$\frac{|\vec{p}|}{E} = |\vec{v}| = 1 \quad \Rightarrow \quad |\vec{p}| = E \tag{R9.11}$$

One immediate implication of this important formula is that *light must carry momentum* (as well as energy). Light bouncing off a mirror will thus transfer momentum to the mirror (causing it to recoil) in much the same way as a ball bouncing off an object transfers momentum to the object and causes it to recoil. This has been experimentally verified,* and it is now known that the pressure exerted by light due to its momentum plays an important part in the evolution of stars, the physics of the early universe, and a number of other astrophysical processes.

Another immediate consequence is that a flash of light has *zero* mass. We have defined the mass of an object in special relativity to be the magnitude of its four-momentum. According to equation R9.11, the flash's mass is

A light flash's mass is zero

$$m^2 = E^2 - |\vec{p}|^2 = 0 \tag{R9.12}$$

This is actually good. If a flash of light were to have some *nonzero* mass m, then its relativistic energy $E = m/(1 - |\vec{v}|^2)^{1/2}$ and relativistic momentum $p = m|\vec{v}|/(1 - |\vec{v}|^2)^{1/2}$ would both be infinite, since $|\vec{v}| = 1$ for light, which makes the denominator zero in each expression. But since $m = 0$, these equations actually read $E = 0/0$ and $|\vec{p}| = 0/0$, and as $0/0$ is technically *undefined* instead of being infinite, these equations simply don't say anything useful about the four-momentum of light instead of yielding absurdities.

*Maxwell's theory of electromagnetic waves also predicted (before relativity did) that light should carry momentum of this magnitude. Experiments performed in 1903 by Nichols and Hull in the United States and Lebedev in Russia confirmed this prediction. See G. E. Henry, "Radiation Pressure," *Sci. Am.*, June 1957, p. 99.

Example R9.3

Problem: When a particle collides with its corresponding antiparticle, they annihilate each other, converting their rest energy entirely to light (photons). A perfect rocket engine might mix antimatter with an equal amount of matter of the same type, and direct the resulting light in a tight beam out of the engine nozzle. No other kind of exhaust could possibly carry more momentum out the rear of the rocket per unit energy expended than light can. Imagine a rocket of original mass $M = 90{,}000$ kg sitting at rest in some frame in deep space (M includes the mass of the matter–antimatter fuel). Imagine that it fires its engines, emitting a burst of light having a total (unknown) energy E_L. If after this the ship's final speed is $|\vec{v}| = \frac{4}{5}$, what is its final mass m?

Solution Here are initial and final drawings for the situation:

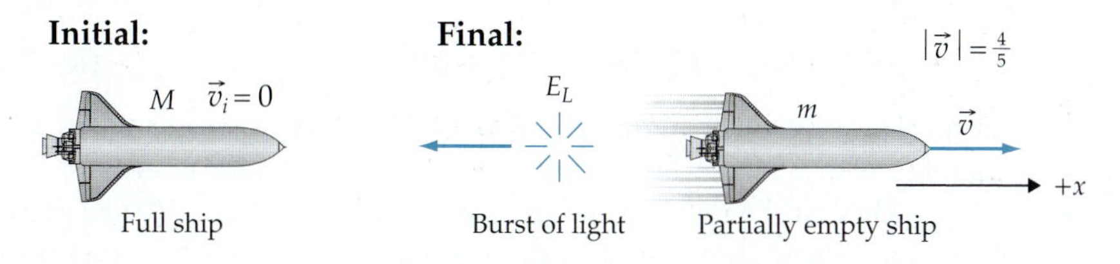

The system here is the ship and the flash. This system's initial four-momentum vector is that of a mass M at rest, which has components $[M, 0, 0, 0]$. After the engines fire, the system consists of a light flash and the somewhat lighter ship. According to equation R9.11, the flash's relativistic momentum magnitude $|\vec{p}_L|$ is equal to its energy E_L, and since the flash is moving in the $-x$ direction, the light's spatial four-momentum components are $p_{Lx} = -|\vec{p}_L| = -E_L$ and $p_{Ly} = p_{Lz} = 0$. We don't know the final ship mass m, but we do know that it is moving with $|\vec{v}| = \frac{4}{5}$ in the $+x$ direction, so by equations R8.10, we know that the ship's final four-momentum vector has a time component of $E = m/(1 - |\vec{v}|^2)^{1/2} = m/(1 - \frac{16}{25})^{1/2} = m/(\frac{9}{25})^{1/2} = \frac{5}{3}m$ and an x component of $Ev_x = (\frac{5}{3}m)\frac{4}{5} = \frac{4}{3}m$. Conservation of four-momentum therefore implies that

$$\begin{bmatrix} M \\ 0 \\ 0 \\ 0 \end{bmatrix} = \begin{bmatrix} E_L \\ -E_L \\ 0 \\ 0 \end{bmatrix} + \begin{bmatrix} \frac{5}{3}m \\ +\frac{4}{3}m \\ 0 \\ 0 \end{bmatrix} \quad \Rightarrow \quad \begin{aligned} M &= E_L + \tfrac{5}{3}m \\ E_L &= \tfrac{4}{3}m \\ 0 &= 0 \\ 0 &= 0 \end{aligned} \tag{R9.13}$$

The top two component equations represent two equations in the unknowns m and E_L. Substituting the second into the first yields

$$M = \frac{4}{3}m + \frac{5}{3}m = \frac{9}{3}m = 3m \quad \Rightarrow \quad m = \frac{1}{3}M = 30{,}000 \text{ kg} \tag{R9.14}$$

So even this ideal rocket must use 60,000 kg of matter–antimatter fuel to boost the remaining 30,000 kg to a speed of 4/5.

Exercise R9X.3

Solve example R9.3 by using an energy–momentum diagram. (*Hint:* We know that the slope of the light flash's four-momentum is -1, and that the slope of the rocket's final four-momentum vector has slope $+\frac{4}{5}$.) Where do these lines intersect on the diagram? Choose your scale so that one mark $= \frac{1}{9}M$.

R9.5 Applications to Particle Physics

Particle physics as the most important practical application of relativity

In spite of our imaginative talk in this book about relativistic trains, spaceships, runners, and so on, special relativity has few genuinely practical applications other than in the realm of subatomic particle physics (the GPS system and some astrophysical processes being the main exceptions). Subatomic particles (such as electrons, protons, and neutrons) have small enough masses that common processes can give them relativistic speeds. Most of the experimental tests of relativity theory involve such particles.

Appropriate units for doing particle physics

The kilogram is an inappropriately large unit of mass or energy when one deals with elementary particles. Elementary particle physicists more commonly express the mass, energy, and momentum of such particles in terms of the energy unit of **electronvolts**, where 1 eV is the energy an electron gains by going through a 1-V battery. In terms of more common units,

$$1 \text{ eV} = 1.602 \times 10^{-19} \text{ J} = 1.782 \times 10^{-36} \text{ kg} \tag{R9.15}$$

So, for example, an electron (whose mass is 0.511 MeV) moving at $|\vec{v}| = \frac{4}{5}$ has an energy $E = m/(1 - |\vec{v}|^2)^{1/2} = \frac{5}{3}m = 0.852$ MeV, a kinetic energy of $K = E - m = 0.341$ MeV, and a momentum magnitude of $|\vec{p}| = E|\vec{v}| = 0.682$ MeV.

Example R9.4 shows how we can apply conservation of four-momentum to a very real problem in subatomic physics.

Example R9.4

Problem: The most stable version of the subatomic particle called the K^0 meson or *kaon* (which has a mass $M = 498$ MeV) decays with a half-life of about 36 ns to two identical π^0 *pions* (which have a mass $m = 135$ MeV). If the original kaon is at rest, what are the speeds of the pions after the decay?

Solution Let's number the pions 1 and 2, and choose the orientation of our reference frame so that the first pion moves in the $+x$ direction. Conservation of four-momentum then implies that

$$\begin{bmatrix} M \\ 0 \\ 0 \\ 0 \end{bmatrix} = \begin{bmatrix} E_1 \\ +|\vec{p}_1| \\ 0 \\ 0 \end{bmatrix} + \begin{bmatrix} E_2 \\ p_{2x} \\ p_{2y} \\ p_{2z} \end{bmatrix} \tag{R9.16}$$

(kaon) (pion 1) (pion 2)

(Note that since the kaon is at rest, its spatial four-momentum components are zero and its energy is just its rest energy, which is its mass.) This equation's bottom two components tell us that $p_{2y} = p_{2z} = 0$, so the second pion also moves along the x axis. The x component line tells us that $p_{2x} = -|\vec{p}_1|$, which says that the second pion moves in the $-x$ direction with the same relativistic momentum magnitude as the first. Since the pions have the same mass m, this means that the pions' relativistic *energies* are the same:

$$E_2 = \sqrt{m^2 + |\vec{p}_2|^2} = \sqrt{m^2 + |\vec{p}_1|^2} = E_1 \tag{R9.17}$$

where the first step here follows from $m^2 = E^2 - |\vec{p}|^2$ (equation R8.31). Equation R9.16's first component then tells us that $M = 2E_1$, or $E_1 = \frac{1}{2}M = 249$ MeV. If we plug this into $m^2 = E_1^2 - |\vec{p}_1|^2$, we can solve for $|\vec{p}_1|$:

$$|\vec{p}_1| = \sqrt{E_1^2 - m^2} = \sqrt{(249 \text{ MeV})^2 - (135 \text{ MeV})^2} = 209 \text{ MeV} \tag{R9.18}$$

Finally, we can find the pions' speeds by using equation R8.32:

$$|\vec{v}_1| = \frac{|\vec{p}_1|}{E_1} = \frac{209\,\text{MeV}}{249\,\text{MeV}} = 0.839 \quad |\vec{v}_2| = \frac{|\vec{p}_2|}{E_2} = \frac{|\vec{p}_1|}{E_1} = |\vec{v}_1| \qquad \text{(R9.19)}$$

These speeds end up being unitless and smaller than that of light, which are both good signs. (Note that this process "converts" kaon mass energy to pion kinetic energy.)

Figure R9.10 shows the aftermath of a collision between two protons recorded by the CMS detector associated with the Large Hadron Collider at CERN in Switzerland. The particle tracks are curved because the detector is placed in a magnetic field which causes each charged particle to follow a circular path whose radius of curvature is equal to the particle's relativistic momentum magnitude. These particle tracks are registered electronically and computers do the measuring and calculating. This allows experimenters to sift through billions of collision events to look for evidence of rare particles. The computer analysis of trillions of events like this led to the discovery of the Higgs boson, the last missing piece of the Standard Model of particle physics, in July of 2012.

Figure R9.10
The aftermath of a collision between two protons as recorded by the CMS detector associated with the Large Hadron Collider. (Credit: © CERN/Science Source)

The particles involved in such collisions are extremely relativistic, so their analysis involves conservation of four-momentum and would make no sense without it. E. F. Taylor and J. A. Wheeler estimated in 1963 that the annual work of physicists analyzing particle collisions at that time was already testing the law of conservation of four-momentum more often than the annual work of surveyors in the United States was testing the laws of Euclidean geometry.* Since then, the number of particle collisions per year that have been analyzed by computers has *enormously* increased. Any one of these tests could have spotted a violation of this law, but none have. As a result, the law of conservation of four-momentum is one of the best-tested laws in all physics.

R9.6 Parting Comments

The principle of relativity is rich with fascinating implications, and this book has touched on just the most basic of these implications. I close with some suggestions as to where an interested reader can go from here.

An excellent and somewhat more advanced and detailed exploration of special relativity is found in E. F. Taylor and J. A. Wheeler's *Spacetime Physics,* 2nd ed., New York: Freeman, 1992. The *American Journal of Physics* (which is found in many college libraries and electronic journal collections and is written primarily for college physics professors) is a great place to look for articles on current issues in relativity theory. Search for *special relativity* in a physical or online index. (Many of these articles are accessible to students.)

The next step beyond special relativity is *general* relativity. A book that one can use when beginning a study of general relativity is Taylor, Wheeler, and Bertschiger, *Exploring Black Holes* (Addison Wesley Longman). Look for the second edition, which (at this writing) is due to be published shortly.

Higher-level textbooks that are still accessible to undergraduates are Hartle's *Gravity* (Addison-Wesley, 2003), and my own book *A General Relativity Workbook* (University Science Books, 2012).

More delights await: I encourage everyone to continue the exploration!

*E. F. Taylor and J. A. Wheeler, *Spacetime Physics,* New York: Freeman, 1963, p. 123.

TWO-MINUTE PROBLEMS

R9T.1 A particle moves along the x axis. The slope of the arrow representing an object's four-momentum on an energy–momentum diagram is equal to
A. The particle's speed.
B. The particle's x-velocity.
C. The inverse of the particle's speed.
D. The inverse of the particle's x-velocity.
E. The particle's energy.

R9T.2 The length of the arrow representing an object's four-momentum on an energy–momentum diagram is directly proportional to
A. The four-momentum's four-magnitude $|\boldsymbol{p}|$.
B. The relativistic momentum's magnitude $|\vec{p}|$.
C. The object's mass.
D. The object's speed.
E. The object's inverse speed.
F. None of the above.

R9T.3 Suppose we know a particle's four-momentum p_t and p_x components. If we draw the particle's four-momentum arrow on an energy–momentum diagram, we can use a hyperbola to determine
A. The particle's energy.
B. The particle's speed.
C. The particle's x-velocity.
D. The particle's momentum magnitude $|\vec{p}|$.
E. The particle's mass.
F. The scaling on the t axis.
T. Something else (specify).

R9T.4 A particle with a mass of 3.0 kg is accelerated to a speed of 0.80. The mass of this particle is now:
A. Greater than 3.0 kg
B. Less than 3.0 kg
C. Still 3.0 kg

R9T.5 A sealed cup of water is placed in a microwave oven. The water absorbs microwave energy, which causes its atoms to vibrate more vigorously, making the water warmer. In this process, the mass of the water in the cup:
A. Increases
B. Decreases
C. Does not change

R9T.6 In example R9.3, the spaceship engines convert 60,000 kg of matter–antimatter fuel to massless light. The mass of the total system (empty spaceship plus flash of light, considered as a unit) thus decreases. T or F?

R9T.7 In example R9.4, a kaon (whose mass is 498 MeV) decays to two pions (each with a mass of 135 MeV).
(a) Mass is converted to energy in this process. T or F?
(b) The total system's mass decreases in this process. T or F?

R9T.8 Special relativity teaches us that energy is the same as mass. T or F?

R9T.9 Consider three different systems. Each consists of two particles that have equal mass m and equal energy $3m$. In system A, both particles move in the $+x$ direction. In system B, the particles have opposite velocities. In system C, the particle's velocities are at right angles to each other.
(a) Which system has the largest total mass?
(b) Which system has the smallest total mass?
A. System A
B. System B
C. System C
D. Systems A and B (which have equal total masses)
E. Systems A and C (which have equal total masses)
F. Systems B and C (which have equal total masses)
T. All three systems have the same total mass.

R9T.10 A system consists of two photons moving in opposite directions. One photon has energy E and the other has energy $4E$. The total system's mass M is
A. Zero, of course.
B. $3E$.
C. $4E$.
D. $5E$.
E. Something else (specify).

R9T.11 A photon hitting an electron (mass m) at rest can create an electron–antielectron pair in addition to the original electron. Assume that the two electrons and the positron move away from the collision at rest with respect to each other afterward. What was the photon's initial energy?
A. Zero
B. m
C. $2m$
D. $3m$
E. $4m$

R9T.12 Which of the following processes is consistent with the law of conservation of four-momentum? (Answer C if it is consistent with the law; F if it violates it.)
(a) A particle moving eastward collides with and sticks to an identical particle at rest. The resulting single particle remains at rest but is more massive than the sum of the original particles' masses.
(b) Two identical particles with equal speeds, one moving east and one moving west, collide. After the collision, the same particles move north and south, respectively, with the same speeds that they had originally.
(c) Two electrons with equal speeds, one moving east and one moving west, collide. After the collision, the electrons move north and south with smaller speeds than they had originally. Nothing else is emitted.
(d) A spaceship of mass $2m$ originally at rest burns some matter–antimatter fuel until the ship's mass is m. The light energy ejected from the ship in this process has total energy m as well.
(e) A particle of mass $2m$ at rest decays to two identical particles each of mass m that move in opposite directions away from the decay position at a speed of 0.5.

HOMEWORK PROBLEMS

Basic Skills

R9B.1 Draw an energy–momentum diagram for an object with a mass of 12 kg moving in the $+x$ direction at a speed of $\frac{4}{5}$. Estimate its relativistic kinetic energy from the graph.

R9B.2 Draw an energy–momentum diagram for an object with a mass of 5 kg moving in the $-x$ direction at a speed of $\frac{3}{5}$. Estimate its relativistic kinetic energy from the graph.

R9B.3 A particle of mass m at rest decays into two identical particles, each with mass $\frac{1}{3}m$. Conservation of spatial momentum means that the product particles must move off in opposite directions with the same speed. What is the relativistic kinetic energy of each particle?

R9B.4 A flash of light moves in the $+x$ direction. The flash has a total energy of 2.0×10^{-18} kg.
(a) What is this in joules?
(b) What is the x component of the flash's four-momentum vector (in kilograms)?
(c) What is this in kg·m/s?

R9B.5 Two balls of putty, each of mass m, move in opposite directions toward each other with speeds of 0.95. The balls stick together, forming a single motionless ball of putty at rest. What is the mass of this final ball of putty? (Express your answer as a multiple of m.)

R9B.6 A ball of putty with mass m moves in the $+x$ direction with speed $|\vec{v}| = \frac{4}{5}$. It hits another ball of putty of mass m at rest. The balls stick together after the collision, forming a single ball.
(a) What is this final ball's final x-momentum (as a multiple of m)?
(b) What is its relativistic energy (as a multiple of m)?
(c) What is its x-velocity?

R9B.7 A spaceship with a mass m originally at rest burns matter–antimatter fuel, radiating light with a total energy of $E_L = \frac{1}{3}m$ in the $-x$ direction.
(a) What is the total relativistic energy of the partially empty spaceship now (as a multiple of m)?
(b) What is its x-momentum (as a multiple of m)?
(c) What is the ship's final speed?
(d) What is its final mass (as a fraction of m)?

R9B.8 Calculate the total mass of each of the systems described in problem R9T.9.

R9B.9 Answer problem R9T.10 and show your work.

R9B.10 Do problem R9T.12. Explain your reasoning for your answer for each part.

R9B.11 The sun radiates energy at a rate of 3.9×10^{26} W. At about what rate is the sun's mass decreasing per year?

Modeling

R9M.1 A particle of mass m decays into two identical particles that move in opposite directions, each with a speed of $\frac{12}{13}$. What is the mass of each of the product particles (expressed as a fraction of m)?

R9M.2 Consider problem R9T.11.
(a) Solve it graphically using hyperbola graph paper.
(b) Solve it algebraically.

R9M.3 An object with a mass m sits at rest. A light flash moving in the $+x$ direction with a total energy of $4m$ hits this object and is completely absorbed. Find the final object's mass and x-velocity.
(a) Do this graphically using hyperbola graph paper.
(b) Do this algebraically.

R9M.4 An object with mass $m_1 = 8$ kg traveling with an x-velocity of $v_{1x} = \frac{15}{17}$ collides with an object with mass $m_2 = 12$ kg traveling with an x-velocity of $v_{2x} = -\frac{5}{13}$. After this elastic collision, the 8-kg object is measured to have an x-velocity of $v_{3x} = -\frac{3}{5}$. Find the other object's x-velocity, and show that it has the same mass as it started with.
(a) Do this graphically using hyperbola graph paper.
(b) Do this algebraically.

R9M.5 A ball of putty with mass $m = 8$ kg traveling with an x-velocity of $v_{1x} = \frac{15}{17}$ collides with an identical ball of putty at rest and sticks to it, forming one glob. Find the final glob's mass and final x-velocity.
(a) Do this graphically using hyperbola graph paper.
(b) Do this algebraically.

R9M.6 A photon with energy $2m$ hits a particle of mass m at rest. The photon "back-scatters" from this interaction (that is, it moves in the opposite direction) while the particle moves forward to conserve momentum. Find the back-scattered photon's energy E and the particle's speed $|\vec{v}|$.
(a) Do this graphically using hyperbola graph paper. (*Hint:* Mark your E and p_x axes in units of $0.2m$.)
(b) Do this algebraically.

R9M.7 A spaceship of mass M is traveling through an uncharted region of deep space. Suddenly its sensors detect a black hole dead ahead. In a desperate attempt to stop the spaceship, the pilot fires the forward matter–antimatter engines. These engines convert the mass energy of matter–antimatter fuel entirely to light, which is emitted in a tight beam in the direction of the ship's motion. The spaceship's initial speed toward the black hole is $|\vec{v}| = \frac{4}{5}$. Find the fraction of its mass M that must be converted to energy to bring the spaceship to rest with respect to the black hole. (*Hint:* Treat the emitted light as one big flash.)
(a) Do this graphically by using hyperbola graph paper. (*Hint:* Mark your E and p_x axes in units of $\frac{1}{3}M$.)
(b) Do this algebraically.

R9M.8 Starfleet Academy cadets are taking practice shots at some beat-up old freighters. One cadet hits a freighter of mass m (initially at rest) with a phasor blast that delivers a total energy $E = \frac{9}{32}m$, which the freighter's shields successfully absorb. (Model a phasor blast as a burst of light.)

(a) What is the freighter's final velocity?

(b) What is its final mass?

R9M.9 Suppose a subatomic particle with mass M decays into a photon with energy $\frac{3}{8}M$ and a second particle. What is this second particle's velocity (magnitude and direction) and its mass (as a fraction of m)?

R9M.10 A spaceship with rest mass m_0 is traveling with an x-velocity $v_{0x} = +\frac{4}{5}$ in the frame of the earth. It collides with a photon torpedo (an intense burst of light) moving in the $-x$ direction relative to the earth. Assume that the ship's shields totally absorb the photon torpedo.

(a) The oncoming torpedo is measured by terrified observers on the ship to have an energy of $0.75m_0$. What is the photon torpedo's energy in the frame of the earth? (*Hint:* How do the components of four-momentum transform when we go from one inertial frame to another?)

(b) Use hyperbola graph paper to find the damaged ship's final x-velocity (in the earth frame) and mass (in terms of m_0) after it absorbs the torpedo.

(c) Find the ship's final mass and x-velocity algebraically using four-dimensional column vectors.

R9M.11 *Starship Design I.* Suppose you want to design a starship using the best possible rocket engine (the matter–antimatter engine discussed in example R9.4) that can boost a payload of mass m = 25 metric tons (25,000 kg) to a final cruising speed of 0.95. Show that the ship's initial mass must be $M = 6.24m$. (*Hint:* The ship can essentially be considered to be a particle of mass M at rest that decays into a big flash of light and a smaller particle (the payload) of known mass m traveling at the known speed $|\vec{v}|$. Use conservation of four-momentum to determine M.)

R9M.12 A π^- pion (mass 140 MeV) normally decays to a π^- muon (mass 106 MeV) and a neutrino. The neutrino is a particle that is so light you can treat its mass as being essentially zero (like a flash of light). If the pion is at rest, find the speed of the emitted muon.

R9M.13 Suppose a photon with energy E_0 is traveling in the $+x$ direction and hits an electron of mass m at rest. The photon scatters from the electron and travels in the $-x$ direction after the collision. Find a formula for the final energy of the photon E in terms of E_0 and m. (Physicists call this process *Compton scattering*. The fact that the formula correctly describes the behavior of light scattering from electrons is one of the most important pieces of evidence supporting the photon model of light.)

R9M.14 A sufficiently energetic electron colliding with an electron at rest can create an additional electron/antielectron pair out of energy. All these particles have the same mass m. How much energy must the incoming electron have to make this possible? [*Hints*: It turns out that the process requires the least energy if the four final particles move off together, so treat them as a single particle with mass $4m$. On hyperbola graph paper, one can solve this problem fairly easily by finding where a certain pair of hyperbolas have a certain vertical separation: think about it. To solve mathematically, I recommend writing the final "particle's" energy in the form $\sqrt{|\vec{p}|^2 + (4m)^2}$, where $|\vec{p}|$ is that particle's momentum magnitude.]

Rich-Context

R9R.1 *Starship Design II.* Consider the starship discussed in problem R9M.11 (which I recommend doing first).

(a) Suppose you can find astronauts who are willing to travel for up to 50 years (as measured by their watches) on a round trip to the stars. About how many light-years could the ship go out and return (assuming that the ship spends a negligible time accelerating)? Is this very far compared to the galaxy as a whole?

(b) The take-off mass calculated in problem R9M.11 only included enough fuel to boost the payload to the cruising speed. For a complete round trip, one must boost the payload (mass m) to the cruising speed, decelerate it to rest at the destination, boost it to cruising speed again for the return trip, and decelerate it upon reaching earth. How much fuel (as a multiple of m) do we need for a complete round trip? (*Hint:* The answer is *not* 4 times the fuel calculated in problem R9M.11.)

(c) Comment on the practicality of visiting distant stars by using a rocket that must carry its own fuel.

R9R.2 One way to get around the difficulties discussed in problems R9M.11 and R9R.1 is to use light pressure to accelerate a payload. Suppose you attach a perfect mirror to the back of your payload, and then you accelerate the payload by bouncing a powerful laser beam off the mirror. (This has the big advantage that you don't have to carry the mass of the fuel or the rocket engine!) The lasers producing the beam could be massive things powered by solar energy, so neither the size nor mass nor power of these driving lasers is a limitation (at least in principle). Suppose we wish to accelerate a 2000-kg scientific payload outward from the earth's orbit around the sun at a rate of 1 m/s^2 (at this rate, it would take about a year to reach 10% of the speed of light). Assume the payload has already been delivered at rest to a point far enough from the earth that the earth's gravity is negligible (but *don't* ignore the sun's gravity). How many watts of light power must the driving laser produce?

(The next two problems are adapted from ones in Taylor and Wheeler, *Spacetime Physics*, 2/*e*, Freeman, 1992.)

R9R.3 Suppose you mount an electric motor on one end of a platform. The motor is connected via a belt to a paddlewheel on the other end of the platform that stirs an enclosed container of water, causing the water's temperature to increase. The motor is powered by a battery placed directly on top of the motor. The center-to-center distance between the battery and the water is 1.0 m. The battery contains about 9000 J of electrical energy. The platform is mounted on completely frictionless wheels and is initially at rest. The entire system has a mass of 10 kg.

(a) Explain why the platform must move in a certain direction when the motor is turned on.

(b) Calculate how far the system's center of mass will have moved by the time the battery dies.

R9R.4 An electron (mass m) moving with a kinetic energy of $K = m$ hits an antielectron (of mass m) at rest. The particles annihilate, producing two photons. One photon travels perpendicular to the electron's original direction of motion, while the other travels at an angle of θ with respect to that direction. Find the photons' energies and the angle θ.

Advanced

R9A.1 Rework the Compton scattering problem (problem R9M.13) to find the energy of the scattered photon if its trajectory after scattering from the electron makes an angle of θ with its original direction. (This is not easy!)

ANSWERS TO EXERCISES

R9X.1 The graph appears below.

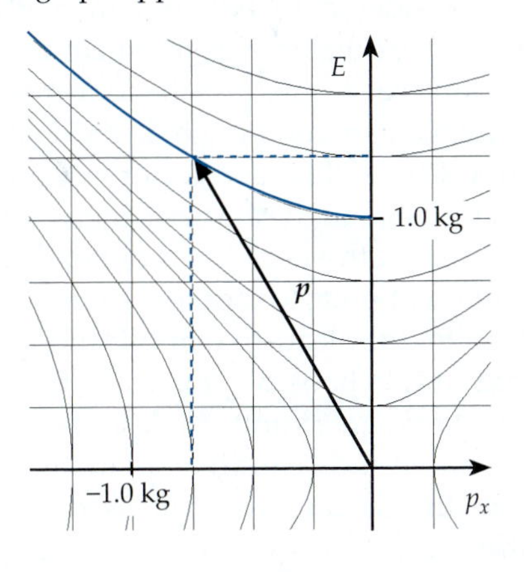

The trick is to draw a line with slope $1/v_x = -5/3$ (colored dashed lines) and see where that line intersects the hyperbola corresponding to 1.0 kg (colored solid curve). We can see from the diagram that the object's energy is $E = \frac{5}{4}$ kg, its kinetic energy is $K = \frac{1}{4}$ kg, and $p_x = -\frac{3}{4}$ kg. One can easily check this with a calculation.

R9X.2 Let the Home Frame be the frame where the collision looks as shown in figure R9.8. The Other Frame in question then moves relative to the Home Frame in the $+x$ direction at $\beta = \frac{3}{5}$. According to equation R7.14*a*, then,

$$v'_{2x} = \frac{v_{2x} - \beta}{1 - \beta v_{2x}} = \frac{-3/5 - 3/5}{1 - (3/5)(-3/5)} = \frac{-6/5}{34/25} = -\frac{15}{17} \quad \text{(R9.20)}$$

Because $\sqrt{1 - v_{2x}^2} = \sqrt{1 - (-\frac{15}{17})^2} = \sqrt{1 - \frac{225}{269}} = \sqrt{\frac{64}{269}} = \frac{8}{15}$,

$$E'_2 = \frac{m}{\sqrt{1 - v_{2x}^2}} = \frac{17}{8}m \quad \text{(R9.21)}$$

$$p'_{2x} = E'_2 v'_{2x} = \left(\frac{17}{8}m\right)\left(-\frac{15}{17}\right) = -\frac{15}{8}m \quad \text{(R9.22)}$$

The first ball is at rest in this frame, so $E'_1 = m$ and $p'_{1x} = 0$. So the system's total four-momentum components are

$$E'_T = E'_1 + E'_2 = m + \frac{17}{8}m = \frac{25}{8}m \quad \text{(R9.23a)}$$

$$p'_{Tx} = p'_{1x} + p'_{2x} = 0 - \frac{15}{8}m = -\frac{15}{8}m \quad \text{(R9.23b)}$$

and its total mass is

$$M' = \sqrt{E'^2_T - p'^2_{Tx}} = \frac{m}{8}\sqrt{25^2 - 15^2} = \frac{m}{8}\sqrt{625 - 225}$$
$$= \frac{m}{8}\sqrt{400} = \frac{20}{8}m = \frac{10}{4}m \quad \text{(R9.24)}$$

Since $m = 4$ kg, $M' = 10$ kg $= M$, as claimed.

R9X.3 Since we don't know either the light flash's or the ship's final relativistic energy, we cannot immediately draw their four-momentum vectors on the diagram. But we do know that the slopes of these vectors are -1 and $\frac{4}{5}$, respectively, and that the two vectors have to add up to the ship's original four-momentum vector, which (since the ship is initially at rest) is vertical. So sketch a line with slope $\frac{4}{5}$ from the origin and another with slope -1 down from the tip of the ship's original four-momentum vector, (see figure R9.11a). Since the flash's and ship's final vectors have to add to the ship's original four-momentum, the vectors have to lie on these lines and stretch to their intersection (see figure R9.11b below). From the diagram, then, we can see that E and $|\vec{p}|$ for the ship after the engines fire are about $\frac{5}{9}M$ and $\frac{4}{9}M$, respectively, and using $m = (E^2 - |\vec{p}|^2)^{1/2}$ (or a hyperbola if one uses hyperbola graph paper), one can show that $m = \frac{1}{3}M$.

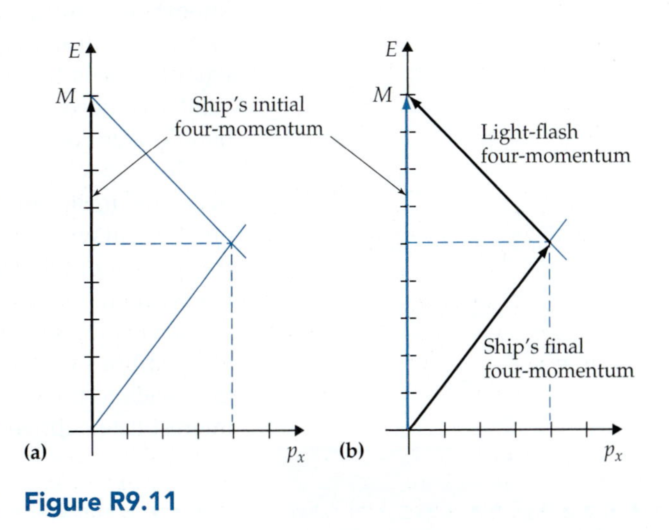

Figure R9.11

RA Converting Equations to SI Units

RA.1 Why Use SR Units?

SR units make relativistic relationships clearer

The equations of special relativity are greatly simplified when one uses SR units to measure distance, as we have seen. But the purpose of using SR units is not merely to simplify a few equations: using such units also vividly draws one's attention to the connections that special relativity makes between quantities that were previously considered to be fundamentally distinct. For example, special relativity teaches us that energy, momentum, and mass are in fact different aspects of the same basic quantity: the four-momentum. It is not merely convenient to measure the time component E, the spatial components p_x, p_y, and p_z, and the magnitude m of this four-vector in the same units, it is fundamentally appropriate as well. Similarly, the basic metaphor of spacetime geometry that lies at the root of both special and general relativity is obscured if one insists on using different units to measure time and distance.

But physicists use SI units in most practical applications

Nonetheless, this choice of units does lead to complications when one tries to apply the ideas presented in this book to practical situations, since practicing physicists in their daily work use SI units to describe quantities. It is important to be able to use the simple and beautiful equations in this book in situations where the quantities in question are expressed in traditional units. Fortunately, it is straightforward to convert equations appearing in this text to equivalent equations involving quantities measured in SI units. The purpose of this appendix is to describe an easy method for doing this.

RA.2 Conversion of Basic Quantities

How to convert basic quantities between SR and SI units

SR units, as defined in chapter R2, differ from SI units only in the substitution of the *second* for the *meter* as the basic unit of distance. Mass and time thus have the same units in both unit systems and need no conversion. The most important quantities that are affected by the shift in units as one changes systems are *distance*, *velocity*, *energy*, and *momentum*. We also should consider what happens to values of universal constants such as c and Planck's constant h. (Indeed, we will treat h in what follows as an example of how we can handle such constants in general.)

To help keep things straight in what follows, let me denote quantities measured in SR units with an "(SR)" subscript; for example, (in this appendix only!) I will write an object's speed in SR units as $|\vec{v}_{(SR)}|$, an object's energy in the SR unit of kilograms as $E_{(SR)}$, and so on. You should assume that quantities without this subscript are expressed in SI units.

In chapter R2, we saw that the general rule for converting SI quantities to SR quantities was to multiply the SI quantity by the appropriate power of c that leads to the correct SR units. Let us apply this rule to the quantities of interest listed above.

Distance in SR units is measured in seconds. Distance in SI units is measured in meters. To convert an SI distance x to an SR distance $x_{(SR)}$, we must divide x by one factor of c (in meters per second). The SR unit of energy is the kilogram, but the SI unit is the joule, where $1\ \text{J} \equiv 1\ \text{kg}\cdot\text{m}^2/\text{s}^2$. To convert from E (in joules) to $E_{(SR)}$ (in kilograms), we must divide E by two powers of c: $E_{(SR)} = E/c^2$. Planck's constant h has units of energy multiplied by time. In SR units, energy is measured in kilograms instead of joules, so again we have to divide the SI version of Planck's constant by two powers of c to get the correct SR units: that is, $h_{(SR)} = h/c^2$. Conversion equations involving velocity and momentum can be derived in a similar manner. The results are summarized in table RA.1.

RA.3 Converting SR Unit Equations to SI Unit Equations

To convert equations, substitute SI quantities from table RA.1 into the SR equation

The trick for converting equations from SR to SI units is now very simple: you simply replace the SR quantities in an equation by the SI equivalents given in table RA.1. For example, consider the metric equation

$$\Delta s^2_{(SR)} = \Delta t^2_{(SR)} - \Delta x^2_{(SR)} - \Delta y^2_{(SR)} - \Delta z^2_{(SR)} \tag{RA.1}$$

Both $\Delta s_{(SR)}$ and $\Delta t_{(SR)}$ have units of seconds and so have the same value in both systems. But the SI units of Δx, Δy, and Δz are meters; therefore, $\Delta x_{(SR)} = \Delta x/c$, $\Delta y_{(SR)} = \Delta y/c$, and $\Delta z_{(SR)} = \Delta z/c$. The metric equation in SI units is thus

$$\Delta s^2 = \Delta t^2 - \left(\frac{\Delta x}{c}\right)^2 - \left(\frac{\Delta y}{c}\right)^2 - \left(\frac{\Delta z}{c}\right)^2 \tag{RA.2}$$

As another example, in unit Q we will study equations that give the energy of a light photon in terms of the frequency f or the wavelength λ of the light involved. In SR units, these equations are $E_{(SR)} = h_{(SR)}f_{(SR)} = h_{(SR)}/\lambda_{(SR)}$. Both energy and Planck's constant gain a factor of $1/c^2$ when we switch from SR to SI units, so this factor divides out in these equations above. The wavelength, on the other hand, has SI units of meters but SR units of seconds, so $\lambda_{(SR)} = \lambda/c$. The second equation thus becomes $E = hc/\lambda$ in SI units, while the first equation becomes simply $E = hf$.

Table RA.1 SI equivalents for SR quantities

Quantity	SR Symbol	SI Equivalent
Time coordinate	$t_{(SR)}$	t
Spatial coordinate	$x_{(SR)}$	x/c
Speed (of an object)	$\lvert\vec{v}_{(SR)}\rvert$	$\lvert\vec{v}/c\rvert$
Frame x-velocity	$\beta_{(SR)}$	β/c
Mass	$m_{(SR)}$	m
Momentum	$\lvert\vec{p}_{(SR)}\rvert$	$\lvert\vec{p}/c\rvert$
Energy	$E_{(SR)}$	E/c^2
Speed of light	1	c
Planck's constant	$h_{(SR)}$	h/c^2

In some cases, we can make equations prettier by multiplying through by powers of c

Table RA.2 lists some of the important equations in this unit and their SI equivalents. In many of the cases described there, the SI equations are simply found by substituting the SI unit equivalents from table RA.1 for the SR unit quantities in the equation from the text. However, in many cases, the SI unit equations have been further simplified by dividing out common factors of c. For example, the SR unit version of the equation giving the magnitude of a particle's relativistic momentum in terms of its speed reads

$$|\vec{p}_{(SR)}| = \frac{m_{(SR)}|\vec{v}_{(SR)}|}{\sqrt{1-|\vec{v}_{(SR)}|^2}} \tag{RA.3}$$

If we simply perform the substitutions called for in table RA.1, we get

$$\frac{|\vec{p}|}{c} = \frac{m|\vec{v}/c|}{\sqrt{1-|\vec{v}/c|^2}} \tag{RA.4}$$

The equation can be made prettier, however, by multiplying through by c:

$$|\vec{p}| = \frac{m|\vec{v}|}{\sqrt{1-|\vec{v}/c|^2}} \tag{RA.5}$$

This is the simplified equation given in table RA.2. I have simplified many of the equations in the table's right-hand column in this manner.

RA.4 Energy-Based SR Units

In energy-based SR units, we express $E, \vec{p}$, and m in energy units instead of mass units

Most of the practical applications of special relativity are in nuclear and particle physics. Physicists in these fields typically focus on *energy* as the most important dynamic quantity, so they usually modify SR units so that the preferred unit for energy, momentum, and mass is an energy unit (typically the electronvolt) instead of the kilogram (see section R9.5). Let us call a unit system where four-momentum quantities are measured in units of energy *energy-based SR units* (ESR). Note that in the SR equations dealing with four-momentum quantities (the last five equations in table RA.2), it doesn't really matter what units one uses to express the quantities $m, |\vec{p}|$, and E as long as one uses the same units for these quantities. Note also that $mc^2, |\vec{p}c|$, and E all have SI units of energy, and thus are the SI quantities that most directly correspond to the quantities $m_{(ESR)}, |\vec{p}_{(ESR)}|$, and $E_{(ESR)}$.

Planck's constant is the same in SI units and ESR units

In regard to the last equation in the table, if you use energy units instead of mass units to express quantities related to four-momentum, you should note that Planck's constant h (which has SI units of J·s or eV·s) has the same value in ESR units: $h_{(ESR)} = h$, whereas $h_{(SR)} = h/c^2$ in ordinary SR units. Other constants involving mass or energy will also (of course) be different in energy-based and ordinary SR units.

RA.5 Exercises for Practice

Here are some exercises that will help you practice equation conversions. Answers appear upside-down at the bottom of the next page.

Exercise RAX.1

Check that $|\vec{p}_{(SR)}| = |\vec{p}/c|$, as claimed in table RA.1.

Table RA.1 Some important equations and their SI equivalents

Equation	SR Version	SI Equivalent
Metric	$\Delta s^2 = \Delta t^2 - \Delta x^2 - \Delta y^2 - \Delta z^2$	$\Delta s^2 = \Delta t^2 - \dfrac{\Delta x^2 + \Delta y^2 + \Delta z^2}{c^2}$
Proper time	$d\tau = dt\sqrt{1-\lvert\vec{v}\rvert^2}$	$d\tau = dt\sqrt{1-\lvert\vec{v}/c\rvert^2}$
Lorentz transformations (t and x)	$\gamma = 1/\sqrt{1-\beta^2}$ $t' = \gamma(t - \beta x)$ $x' = \gamma(-\beta t + x)$	$\gamma = 1/\sqrt{1-(\beta/c)^2}$ $t' = \gamma(t - \beta x/c^2)$ $x' = \gamma(-\beta t + x)$
Lorentz contraction	$L = L_R\sqrt{1-\lvert\vec{v}\rvert^2}$	$L = L_R\sqrt{1-\lvert\vec{v}/c\rvert^2}$
Transformation for x-velocity	$v'_x = \dfrac{v_x - \beta}{1 - \beta v_x}$	$v'_x = \dfrac{v_x - \beta}{1 - \beta v_x/c^2}$
Energy in terms of speed	$E = \dfrac{m}{\sqrt{1-\lvert\vec{v}\rvert^2}}$	$E = \dfrac{m}{\sqrt{1-\lvert\vec{v}/c\rvert^2}}$
Relativistic momentum magnitude	$\lvert\vec{p}\rvert = \dfrac{m\lvert\vec{v}\rvert}{\sqrt{1-\lvert\vec{v}\rvert^2}}$	$\lvert\vec{p}\rvert = \dfrac{m\lvert\vec{v}\rvert}{\sqrt{1-\lvert\vec{v}/c\rvert^2}}$
Mass in terms of E and $\lvert\vec{p}\rvert$	$m^2 = E^2 - \lvert\vec{p}\rvert^2$	$(mc^2)^2 = E^2 - \lvert\vec{p}c\rvert^2$
Speed in terms of E and $\lvert\vec{p}\rvert$	$\lvert\vec{v}\rvert = \dfrac{\lvert\vec{p}\rvert}{E}$	$\dfrac{\lvert\vec{v}\rvert}{c} = \dfrac{\lvert\vec{p}c\rvert}{E}$
Photon energy in terms of f and λ	$E = hf = \dfrac{h}{\lambda}$	$E = hf = \dfrac{hc}{\lambda}$

Exercise RAX.2

Convert the equation $K_{(ESR)} = E_{(ESR)} - m_{(ESR)}$ (where K is an object's kinetic energy) into its equivalent in SI units.

Exercise RAX.3

The spacetime separation $\Delta\sigma$ between two events is given by the equation $\Delta\sigma^2_{(SR)} = \lvert\Delta d_{(SR)}\rvert^2 - \Delta t^2_{(SR)}$. Because $\Delta\sigma$ is actually directly measured by a ruler as opposed to a clock, it makes more sense to express its value in meters rather than in seconds. With this in mind, what would be the equivalent of this equation in SI units?

Exercise RAX.4

What are the SR units of the universal gravitational constant $G_{(SR)}$? Derive an equation expressing $G_{(SR)}$ in terms of G in SI units.

Short Answers for the Exercises:

RAX.1 Note that $\vec{p}$ has units of kg·m/s, so $\vec{p}_{(SR)} = \vec{p}/c$.

RAX.3 $\Delta\sigma^2 = \lvert\Delta\vec{d}\rvert^2 - (c\Delta t)^2$

RAX.2 $K = E - mc^2$

RAX.4 The SR units of $G_{(SR)}$ are s/kg, $G_{(SR)} = G/c^3$.

RB The Relativistic Doppler Effect

RB.1 Introduction to the Doppler Effect

A basic description of the Doppler effect

An important practical application of the metric equation is computing the shift in the wavelength of light emitted by a relativistic moving source. You may already know that the wavelength of light emitted by something moving with respect to an observer is measured by that observer to be red-shifted or blue-shifted if the emitting object is moving away from or toward the observer, respectively. This shift in wavelength is called a **Doppler shift** and the general effect the **Doppler effect** (after Christian Johann Doppler, the 19th-century physicist who first described the effect for light). This effect has many important applications in all areas of science, and it forms the basis of such technologies as Doppler weather radars (that can detect severe weather) and the radar guns that police use to detect speeders.

Overview of the appendix

This appendix examines this effect in some detail, using the tools and ideas we have developed through chapter R4. Section RB.2 uses a basic model of the emission/detection process and a spacetime diagram to find a formula for the shift in wavelength. Section RB.3 looks at an example application, and section RB.4 explores the nonrelativistic limit of our formula. Finally, section RB.5 looks at the slightly different situation involved in Doppler-shift radar.

RB.2 Deriving the Doppler Shift Formula

Some basic assumptions

To simplify matters, let's consider a source that moves directly toward or away from the observer in question, and let's take the x axis of the observer's frame (the Home Frame) to be the line connecting the source and observer. Let us also assume that the observer is located at $x = 0$ in that frame.

A simplified light-flash model

First consider a clock that emits brief flashes of light (we will generalize this to continuous light waves shortly). Let the event of the emission of any one flash be event A, and the emission of the next flash be event B. Since the clock is present at both events, it measures a proper time between these flashes. If the time between flashes happens also to be so short that the clock follows an essentially straight worldline between the events, then the relationship between the proper time $d\tau_{AB}$ measured in the emitting clock's frame and the coordinate time dt_{AB} measured in the observer's frame is

$$d\tau_{AB} = \sqrt{1-|\vec{v}|^2}\, dt_{AB} = \sqrt{1-v_x^2}\, dt_{AB} \qquad \text{(RB.1)}$$

(see equation R4.5), where $|\vec{v}|$ is the emitting clock's speed in the observer's frame. (The last step follows because a clock moving directly toward or away from the observer is moving along the x axis, so $|\vec{v}| = v_x$.) Therefore,

Implication of the proper time formula

$$dt_{AB} = \frac{d\tau_{AB}}{\sqrt{1-v_x^2}} \qquad \text{(RB.2)}$$

Now, dt_{AB} is the time our observer would measure between the *emission* of the flashes. But how much time dt_R passes between the observer *receiving* those flashes? These times are not necessarily the same! As both flashes travel at a speed of 1 back to the observer, the time (in the observer's frame) that it takes a flash to get from its emission event to the observer is equal to the emission event's *distance* from the origin (in that frame). If the source is moving relative to the observer, then the two events will not happen at the same place in the observer's frame, meaning that the light travel times for the two flashes will not be the same, implying that $dt_R \neq dt_{AB}$.

Exactly how should we correct for this effect? We can answer this question fairly easily with the help of a spacetime diagram. Figure RB.1 shows the worldlines of the light flashes in question as they travel back to the observer from a clock that happens to be moving in the $+x$ direction. In the time dt_{AB} between the emission events (as measured in the observer's frame), the emitting clock moves a distance $v_x dt_{AB}$ away from the observer. This means (since the speed of light is 1 in the observer's frame) that it takes $v_x dt_{AB}$ *more* time for the second flash to make it back to the observer than it took for the first flash. The time between the observer's reception of the flashes is simply the time between their emission (as measured in the observer's frame) plus the extra light travel time required for the second flash to reach the observer:

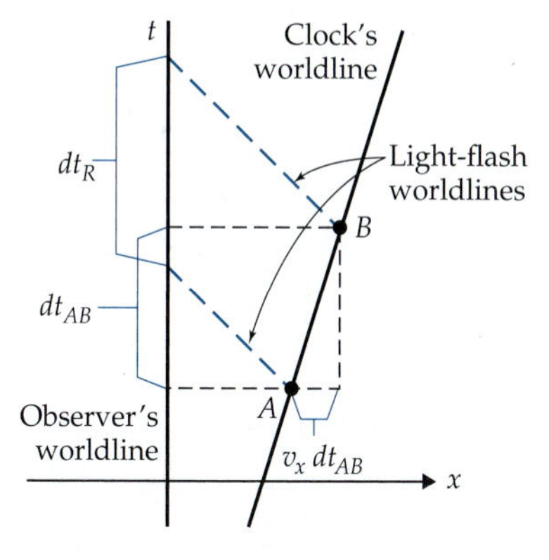

Figure RB.1
How to calculate the time dt_R between the flash receptions.

$$dt_R = dt_{AB} + v_x dt_{AB} = (1 + v_x)dt_{AB} \tag{RB.3}$$

Combining this with the result given in equation RB.2, we get

$$dt_R = \frac{dt_{AB}}{\sqrt{1 - v_x^2}}(1 + v_x) = d\tau_{AB}\frac{\cancel{\sqrt{1+v_x}}\,\sqrt{1+v_x}}{\cancel{\sqrt{1+v_x}}\,\sqrt{1-v_x}} = \sqrt{\frac{1+v_x}{1-v_x}}\,d\tau_{AB} \tag{RB.4}$$

Connecting the light-flash model to continuous waves

Now what has all this to do with the wavelength of a continuous beam of light waves? Consider! Each crest of a light wave moves at the speed of light, just as a flash would. Therefore, for the purposes of this calculation, a wave crest is analogous to a light flash, and there is a direct analogy between a sequence of light flashes and a continuous series of light wave crests. Therefore, equation RB.4 should also apply to light waves if we interpret $d\tau_{AB}$ as the time between the emission of light wave crests in the emitter's frame and dt_R as the time between the reception of light wave crests in the observer's frame.

Moreover, we define a light wave's wavelength λ to be the distance between successive crests in the wave. Since the crests move at a speed of 1, this means the distance λ between adjacent crests is equal to the time dt it takes two successive crests to pass a given point in space (or emerge from the emitter): $\lambda = dt$. Substituting this into equation RB.4 yields

The relativistic Doppler shift formula

$$\frac{\lambda_R}{\lambda_E} = \sqrt{\frac{1+v_x}{1-v_x}} \tag{RB.5}$$

where λ_R is the wavelength of the light measured in the observer's frame and λ_E is its wavelength as measured in the emitter's frame.

Equation RB.5 is the **relativistic Doppler shift formula**. Although figure RB.1 (and certain phrases in the argument above) assume that the emitter is moving away from the observer ($v_x > 0$), the formula also applies if the emitter is moving toward the observer ($v_x < 0$).

Exercise RBX.1

Review the argument and argue that equation RB.5 is valid even if $v_x < 0$.

Note also that if $v_x > 0$, then $\lambda_R = \lambda_E\sqrt{1 + v_x}/\sqrt{1 - v_x} > \lambda_R$, meaning that the received light has a *longer* (red-shifted) wavelength than its wavelength observed in the emitter frame. On the other hand, if $v_x < 0$, then $\lambda_R < \lambda_E$: the received light has a shorter (blue-shifted) wavelength than that observed in the emitter frame. This coincides with what you have probably heard before.

Exercise RBX.2

We have assumed in this derivation that the time between emission events (or successive crests of the light wave) is small compared with the time required for significant changes in the emitter's motion (so that it can be assumed to follow an approximately straight worldline during the interval in question). Is this approximation likely to be valid in the case of light waves (which have a wavelength $\lambda \approx 600$ nm)? Justify your answer.

RB.3 Astrophysical Applications

Applications in astrophysics

Since excited atoms emit light having a characteristic set of wavelengths (in their own frame), equation RB.5 is commonly used by astrophysicists to compute the radial velocity of astronomical objects relative to the earth. This equation technically only applies to an emitter that we know is moving directly toward or away from an observer (a somewhat more complicated formula applies when the emitter has a tangential velocity as well), but it is a useful approximation in most cases (since tangential velocities are rarely large enough to matter much). Here is an example application.

Example RB.1

Problem: Light from excited atoms in a certain quasar is received by observers on earth. The wavelength of a certain spectral line of this light is measured by those observers to be 1.12 times longer than it would be if the atoms were at rest in the laboratory (that is, the light has been red-shifted by about 12%). What is the quasar's speed relative to earth (assuming it is moving directly away from earth)?

Solution We are told that $\lambda_R/\lambda_E = 1.12$. Equation RB.5 then implies that

$$\frac{\lambda_R}{\lambda_E} = \sqrt{\frac{1 + v_x}{1 - v_x}} = 1.12 \tag{RB.6}$$

To solve this for v_x, let us define $u \equiv \lambda_R/\lambda_E = 1.12$ and square both sides of equation RB.6. Doing this yields

$$u^2 = \frac{1 + v_x}{1 - v_x} \quad \Rightarrow \quad u^2(1 - v_x) = 1 + v_x$$

$$\Rightarrow \quad u^2 - u^2 v_x = 1 + v_x \quad \Rightarrow \quad u^2 - 1 = v_x + u^2 v_x = v_x(1 + u^2)$$

$$\Rightarrow \quad v_x = \frac{u^2 - 1}{u^2 + 1} = \frac{(1.12)^2 - 1}{(1.12)^2 + 1} = 0.11 \tag{RB.7}$$

The quasar's speed relative to the earth is thus 11% of the speed of light.

RB.4 The Nonrelativistic Limit

The derivation of the relativistic Doppler shift formula given by equation RB.5 actually combines two effects: (1) the relativistic distinction between the emitter's proper time and the observer's coordinate time between successive pulses (see equation RB.2) and (2) the delay of successive received pulses due to the changing separation between emitter and observer (see equation RB.3). The latter effect would apply even if time were absolute (as long as light moves with a speed of roughly 1 in the observer's frame), but the first effect arises only in the theory of relativity. Indeed, when $v_x << 1$, $\sqrt{1 - v_x^2} \approx 1 + \frac{1}{2}v_x^2$ by the binomial approximation, and equation RB.5 becomes

$$dt_R = (1 + v_x)dt_{AB} \approx (1 + v_x)(1 - \tfrac{1}{2}v_x^2)d\tau_{AB} = (1 + v_x)d\tau_{AB} \qquad \text{(RB.8)}$$

where I have dropped terms of order v_x^2 and smaller in favor of the much larger terms 1 and v_x. So the nonrelativistic effect dominates, and

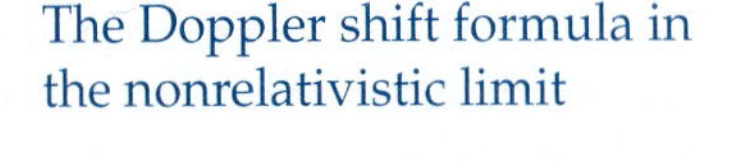

$$\frac{\lambda_R}{\lambda_E} \approx 1 + v_x \quad \text{when } v_x << 1 \qquad \text{(RB.9)}$$

This is the nonrelativistic Doppler formula you will see in many textbooks.

RB.5 Doppler Radar

Doppler radar (used by weather observers and police) involves a somewhat different situation. In this case, the emitter and observer are typically in the same frame, and the emitted waves are reflected by an object moving relative to both. You can use the spacetime diagram shown in figure RB.2 (and an argument analogous to that given in section RB.2) to show that in this situation

$$\frac{\lambda_R}{\lambda_E} = \frac{1 + v_x}{1 - v_x} \qquad \text{(RB.10)}$$

Exercise RBX.3

Use figure RB.2 to verify equation RB.10. (*Hint:* Note that all the times shown on the diagram are coordinate times measured in the observer's frame. The time measured by a clock traveling with the reflector is irrelevant in this case.)

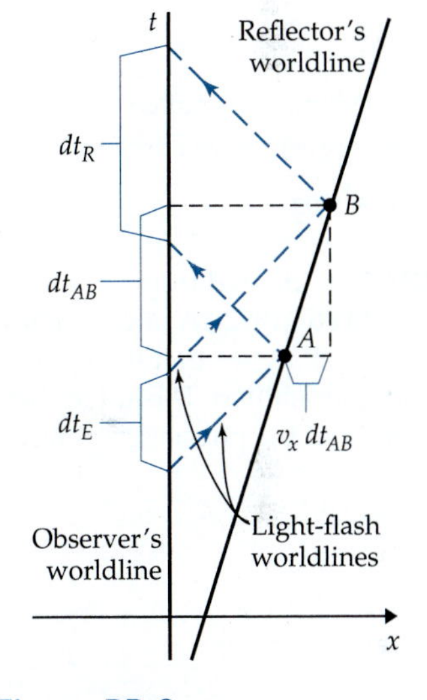

Figure RB.2
A spacetime diagram showing the worldlines of successive wave crests emitted by a Doppler radar system.

Note that because the time measured by a clock traveling with the reflector is not involved, this formula does not employ any relativity except for the fact that the speed of light is 1 in the observer's frame (whatever frame that might be).

HOMEWORK PROBLEMS

Basic Skills

RBB.1 A spaceship moves directly away from the earth at a speed of 0.5. By what factor is light from the spaceship's taillight red-shifted?

RBB.2 Earth observers detect light reflected from the hull of an alien starship and observe that the light is blue-shifted by 35% compared to normal. About how rapidly is the alien ship approaching?

RBB.3 A spectrogram of light from a distant galaxy shows a spectral line that can confidently be identified as a spectral line of hydrogen. In the laboratory, this line would have a wavelength of 486 nm, but in the light from the galaxy, it is observed to have a wavelength of 729 nm.

(a) Is the galaxy moving toward or away from the earth?

(b) Assuming the x axis points toward the galaxy from the earth, what is the galaxy's x-velocity v_x?

(c) Astronomers define a galaxy's redshift z to be

$$z \equiv \frac{\lambda_R}{\lambda_E} - 1 \tag{RB.11}$$

What is this galaxy's redshift?

Modeling

RBM.1 (Adapted from Taylor and Wheeler, *Spacetime Physics*, 2d ed., 1992.) A physicist brought to court for running a red light argues that the light looked green because it was Doppler-shifted. The judge changes the charge to speeding and fines the physicist a penny for every mile per hour the physicist was traveling over the speed limit of 45 mi/h. What was the fine (roughly)? (*Hints*: Red light has a wavelength of 650 nm; green light has a wavelength of 530 nm. Also, 2.2 mi/h = 1.0 m/s.)

RBM.2 Top scientists observe light reflected from an alien starship and find it to be blue-shifted by 6% compared to normal. Observations of the shift in the blue-shift suggest the aliens' speed relative to us is decreasing at a rate of $3|\vec{g}|$. Assuming the aliens are moving directly toward us, they continue decelerating at a constant rate, and intend that their ship comes to rest exactly as they arrive at the earth, how far are they away now and when will they arrive?

Derivation

RBD.1 Check that equation RB.9 is correct by using the binomial approximation to rewrite both the numerator and the denominator of equation RB.5 in the limit that $v_x \ll 1$.

RBD.2 How should we correct equation RB.5 if the emitter does not move directly toward or away from the observer? Assume that we still choose the observer's x axis to go through the emitter's position at the time it emits the flashes observed. (*Hints*: Where did we assume in section RB.2 that the motion was along the x axis? Argue that the emitter's motion in the y or z direction during a small time interval will not significantly affect the light travel time.)

RBD.3 Quantum mechanics tells us that a photon's energy E is related to its wavelength λ by $E = h/\lambda$ (in SR units). Use this result and the Lorentz transformation for four-momentum to derive the Doppler shift formula.

ANSWERS TO EXERCISES

RBX.1 If the emitter is approaching, the time between reception events is $dt_R = dt_{AB} - |v_x| dt_{AB} = dt_{AB} + v_x dt_{AB}$ still, because $v_x = -|v_x|$ in this case. The rest of the derivation follows as before.

RBX.2 If the distance between visible light crests is about 600 nm, the time (in SI units) between crest emission events will be the time it takes the first crest to travel 600 nm away from the emitter at the speed of light, which is about $(6.0 \times 10^{-7}\ \text{m})/(3.0 \times 10^{8}\ \text{m/s}) = 2.0 \times 10^{-15}$ s. It is hard to imagine something accelerating so violently that its motion would change very much during such a time interval!

RBX.3 Note that $dt_{AB} = dt_E + v_x dt_{AB}$. Solve this for dt_E and combine with equation RB.3 to get equation RB.10.

Index

NOTE: Entries followed by *t* and *f* refer to tables and figures, respectively.

Periodic Table of the Elements

Key: 1 — Atomic number; H — Symbol; 1.008 — Atomic mass

1 1A	2 2A	3 3B	4 4B	5 5B	6 6B	7 7B	8 8B	9 8B	10 8B	11 1B	12 2B	13 3A	14 4A	15 5A	16 6A	17 7A	18 8A
1 **H** 1.008																	2 **He** 4.003
3 **Li** 6.941	4 **Be** 9.012											5 **B** 10.81	6 **C** 12.01	7 **N** 14.01	8 **O** 16.00	9 **F** 19.00	10 **Ne** 20.18
11 **Na** 22.99	12 **Mg** 24.31											13 **Al** 26.98	14 **Si** 28.09	15 **P** 30.97	16 **S** 32.07	17 **Cl** 35.45	18 **Ar** 39.95
19 **K** 39.10	20 **Ca** 40.08	21 **Sc** 44.96	22 **Ti** 47.88	23 **V** 50.94	24 **Cr** 52.00	25 **Mn** 54.94	26 **Fe** 55.85	27 **Co** 58.93	28 **Ni** 58.69	29 **Cu** 63.55	30 **Zn** 65.39	31 **Ga** 69.72	32 **Ge** 72.59	33 **As** 74.92	34 **Se** 78.96	35 **Br** 79.90	36 **Kr** 83.80
37 **Rb** 85.47	38 **Sr** 87.62	39 **Y** 88.91	40 **Zr** 91 .22	41 **Nb** 92.91	42 **Mo** 95.94	43 **Tc** (98)	44 **Ru** 101.1	45 **Rh** 102.9	46 **Pd** 106.4	47 **Ag** 107.9	48 **Cd** 112.4	49 **In** 114.8	50 **Sn** 118.7	51 **Sb** 121.8	52 **Te** 127.6	53 **I** 126.9	54 **Xe** 131.3
55 **Cs** 132.9	56 **Ba** 137.3	57 **La** 138.9	72 **Hf** 178.5	73 **Ta** 180.9	74 **W** 183.9	75 **Re** 186.2	76 **Os** 190.2	77 **Ir** 192.2	78 **Pt** 195.1	79 **Au** 197.0	80 **Hg** 200.6	81 **Tl** 204.4	82 **Pb** 207.2	83 **Bi** 209.0	84 **Po** (210)	85 **At** (210)	86 **Rn** (222)
87 **Fr** (223)	88 **Ra** (226)	89 **Ac** (227)	104 **Rf** (257)	105 **Db** (260)	106 **Sg** (263)	107 **Bh** (262)	108 **Hs** (265)	109 **Mt** (266)	110	111	112	(113)	114	(115)	116	(117)	

58 **Ce** 140.1	59 **Pr** 140.9	60 **Nd** 144.2	61 **Pm** (147)	62 **Sm** 150.4	63 **Eu** 152.0	64 **Gd** 157.3	65 **Tb** 158.9	66 **Dy** 162.5	67 **Ho** 164.9	68 **Er** 167.3	69 **Tm** 168.9	70 **Yb** 173.0	71 **Lu** 175.0
90 **Th** 232.0	91 **Pa** (231)	92 **U** 238.0	93 **Np** (237)	94 **Pu** (242)	95 **Am** (243)	96 **Cm** (247)	97 **Bk** (247)	98 **Cf** (249)	99 **Es** (254)	100 **Fm** (253)	101 **Md** (256)	102 **No** (254)	103 **Lr** (257)

Short Answers to Selected Problems

[Note that most of the derivation (D) problems as well as a number of other problems have answers given in the problem statement. These problems are also useful for practice, but their answers are not reiterated here.]

Chapter R1

B1 $t' = 15$ s, $x' = 175$ m. **B3** $v_x = +18$ m/s. **B5** (a) 7 m/s, (b) 43 m/s. **M3** (a) Yes. (b) 125 m, (c) 200 m. **M5b** $1.5c$. **M7** (b) $-2|\vec{v}_1|$, (c) $-|\vec{v}_1|$. **M9** (c) -30 m/s, $+70$ m/s, (e) 400 m.

Chapter R2

B1 (a) 43 ms, (b) 40 mi/h, (c) $30{,}600{,}000|\vec{g}|$. **B3b** 2.2 g. **B7** (a) Yes. (b) -5 h. **B9b** 12:00:15. **B12** Rear firecracker. **M1c** $t = 400$ s. **M3** Photo received at $t = 40$ min. **M5c** $1420|\vec{g}|$ **M7** $t = 140$ μs, $x = -115$ μs, $y = 80$ μs, $z = 13$ μs. **M9** Mars. **R2** Ship arrives at midnight. **R3** Rear light blinks first.

Chapter R3

B1 (a) P, (b) both, (c) P. **B3** (a) Alice and Brian, (b) Brian and Cara/Dave, (c) Brian. **B5** 3 h. **B7** 8 h. **B9** No separation. **M1** (a) 2:17 pm, (c) 0.80. **M2a** 0.90. **M5b** 9500 m. **M7b** 0.815. **M9** 4.41 y **M11b** 1/4 of the particles. **R1b** 0.2 kg.

Chapter R4

B1 4.6 h. **B3b** 56.7 ns. **B5** 5.2 ns. **B7a** 0.23 ps. **M1a** 230 ns. **M3** (b) 2.0 ps, (c) 0.5 ps longer. (d) No. **M5b** 403 μs. **M7b** 1.4 min. **R1** (a) 269 ns, (b) -143 ns, (c) 107 ns, -304 ns. **R3** Yes.

Chapter R5

B3 E before W. **B5** $t' = 4.6$ s, $x' = 1.15$ s. **B7** $t' = 4.5$ s, $x' = 3.5$ s. **M1c** 300 y. **M3c** $t'_C = 10$ min, $t'_D = 8$ min. **M5a** Alan. **R1d** 78 m.

Chapter R6

B1 0.866. **B3** 18 ns. **B5** 8.66 ns. **B7** 6.3 cm. **M1a** 0.99999950. **M3** 460 m. **M5** for paper 86 μm thick, speed $= 1 - 4 \times 10^{-8}$. **R3d** $\frac{4}{3}L_R$. **R5** (b) 250 ns, (d) 160 ns.

Chapter R7

B1b 3/5. **B3** No. **B5** True. **B7** 30/33. **B9** 35/37. **B11** 0.65 **M1** (a) 477 km, (b) no violation. **M5** No. **M7** (a) $v'_{1x} = 0$, $v'_{2x} = -0.882$, $v'_{3x} = -0.60$. (c) Not conserved, yes. **M9a** 0.95, 0.59. **M11b** 24° **R1** Yes.

Chapter R8

B1 0.14. **B3** $p_t = 2.5$ kg, $p_x = 1.5$ kg, $p_y = p_z = 0$. **B5** $p_t = 13$ kg, $p_x = 4$ kg, $p_y = -3$ kg, $p_z = 0$. **B7** (a) 4/5, (b) 3 kg, (c) 2 kg, (d) 4 kg. **B9c** 4.1 kg. **B11** $p_t = 37/3$ kg, $p_x = -35/3$ kg, $p_y = p_z = 0$. **M2d** 120 MJ. **M3** $p_z = \frac{4}{3}m$. **M5** \$(300–500) billion. **M7** \$135 trillion. **R2** 3.8 y. **R3b** 47 μm.

Chapter R9

B1 20 kg. **B3** $\frac{1}{6}m$. **B5** $6.4m$. **B7** (a) $\frac{2}{3}m$, (b) $\frac{1}{3}m$, (c) $\frac{1}{2}$, (d) $m/\sqrt{3}$. **M1** $\frac{5}{26}m$. **M3** $3m$, 4/5. **M5** 20 kg, 3/5. **M7** $\frac{2}{3}M$ converted. **M9** 3/5, $M/2$. **M12** 0.27. **R1b** $1520m$. **R3** 10 fm.

Appendix RB

B1 1.73. **B3** (a) away, (b) 0.385, (c) 0.5. **M1** \$130,000. **M2** 5.4 h, 7.3 d.